AF780513

Synthesis and Applications of Nanomaterials for Photocatalysis and Electrocatalysis

Synthesis and Applications of Nanomaterials for Photocatalysis and Electrocatalysis

Special Issue Editors

Giuseppe Cappelletti

Claudia Letizia Bianchi

MDPI • Basel • Beijing • Wuhan • Barcelona • Belgrade • Manchester • Tokyo • Cluj • Tianjin

MDPI

Special Issue Editors
Giuseppe Cappelletti
Universita' degli Studi Di Milano
Italy

Claudia Letizia Bianchi
University of Milan
Italy

Editorial Office
MDPI
St. Alban-Anlage 66
4052 Basel, Switzerland

This is a reprint of articles from the Special Issue published online in the open access journal *Nanomaterials* (ISSN 2079-4991) (available at: https://www.mdpi.com/journal/nanomaterials/special_issues/nano_photocata_electrocata).

For citation purposes, cite each article independently as indicated on the article page online and as indicated below:

LastName, A.A.; LastName, B.B.; LastName, C.C. Article Title. *Journal Name* **Year**, *Article Number*, Page Range.

ISBN 978-3-03928-831-1 (Pbk)
ISBN 978-3-03928-832-8 (PDF)

Cover image courtesy of Claudia L. Bianchi.

© 2020 by the authors. Articles in this book are Open Access and distributed under the Creative Commons Attribution (CC BY) license, which allows users to download, copy and build upon published articles, as long as the author and publisher are properly credited, which ensures maximum dissemination and a wider impact of our publications.
The book as a whole is distributed by MDPI under the terms and conditions of the Creative Commons license CC BY-NC-ND.

Contents

About the Special Issue Editors . . . . **vii**

Preface to "Synthesis and Applications of Nanomaterials for Photocatalysis and Electrocatalysis" . . . . **ix**

Prabhakarn Arunachalam, Maged N. Shaddad, Mohamed A. Ghanem, Abdullah M. Al-Mayouf and Mark T. Weller
Zinc Tantalum Oxynitride ($ZnTaO_2N$) Photoanode Modified with Cobalt Phosphate Layers for the Photoelectrochemical Oxidation of Alkali Water
Reprinted from: *Nanomaterials* **2018**, *8*, 48, doi:10.3390/nano8010048 . . . . **1**

Juan Yang, Jun You, Jun Dai, Yumei Chen and Yao Li
In-Situ Synthesis of Hydrogen Titanate Nanotube/Graphene Composites with a Chemically Bonded Interface and Enhanced Visible Photocatalytic Activity
Reprinted from: *Nanomaterials* **2018**, *8*, 229, doi:10.3390/nano8040229 . . . . **17**

Shihao Chen, Yang Xiao, Yinhai Wang, Zhengfa Hu, Hui Zhao and Wei Xie
A Facile Approach to Prepare Black TiO_2 with Oxygen Vacancy for Enhancing Photocatalytic Activity
Reprinted from: *Nanomaterials* **2018**, *8*, 245, doi:10.3390/nano8040245 . . . . **37**

Yizeng Zhang, Zhiwu Chen and Zhenya Lu
A Facile Method for the Preparation of Colored $Bi_4Ti_3O_{12-x}$ Nanosheets with Enhanced Visible-Light Photocatalytic Hydrogen Evolution Activity
Reprinted from: *Nanomaterials* **2018**, *8*, 261, doi:10.3390/nano8040261 . . . . **53**

Jin Liu, Huihui Gan, Hongzhang Wu, Xinlei Zhang, Jun Zhang, Lili Li and Zhenling Wang
Effect of Organic Substrates on the Photocatalytic Reduction of Cr(VI) by Porous Hollow Ga_2O_3 Nanoparticles
Reprinted from: *Nanomaterials* **2018**, *8*, 263, doi:10.3390/nano8040263 . . . . **75**

Tran Thi Thuong Huyen, Tran Thi Kim Chi, Nguyen Duc Dung, Hendrik Kosslick and Nguyen Quang Liem
Enhanced Photocatalytic Activity of {110}-Faceted TiO_2 Rutile Nanorods in the Photodegradation of Hazardous Pharmaceuticals
Reprinted from: *Nanomaterials* **2018**, *8*, 276, doi:10.3390/nano8050276 . . . . **87**

Tiekun Jia, Ming Liu, Dongsheng Yu, Fei Long, Shuyi Mo, Zhao Deng and Weimin Wang
A Facile Approach for the Synthesis of Zn_2SnO_4/BiOBr Hybrid Nanocomposites with Improved Visible-Light Photocatalytic Performance
Reprinted from: *Nanomaterials* **2018**, *8*, 313, doi:10.3390/nano8050313 . . . . **99**

Meiyin Wang, Hui Wang, Yuanhang Ren, Cheng Wang, Zhewei Weng, Bin Yue and Heyong He
Construction of g-C_3N_4-mNb_2O_5 Composites with Enhanced Visible Light Photocatalytic Activity
Reprinted from: *Nanomaterials* **2018**, *8*, 427, doi:10.3390/nano8060427 . . . . **113**

Siyu Xu, Jun Dai, Juan Yang, Jun You and Jingyi Hao
Facile Synthesis of Novel $CaIn_2S_4$/$ZnIn_2S_4$ Composites with Efficient Performance for Photocatalytic Reduction of Cr(VI) under Simulated Sunlight Irradiation
Reprinted from: *Nanomaterials* **2018**, *8*, 472, doi:10.3390/nano8070472 . . . . **127**

Alexander Müller, Sandra Peglow, Michael Karnahl, Angela Kruth, Henrik Junge, Volker Brüser and Christina Scheu
Morphology, Optical Properties and Photocatalytic Activity of Photo- and Plasma-Deposited Au and Au/Ag Core/Shell Nanoparticles on Titania Layers
Reprinted from: *Nanomaterials* **2018**, *8*, 502, doi:10.3390/nano8070502 . . . . **145**

Chiara Anna D'Amato, Rita Giovannetti, Marco Zannotti, Elena Rommozzi, Marco Minicucci, Roberto Gunnella and Andrea Di Cicco
Band Gap Implications on Nano-TiO_2 Surface Modification with Ascorbic Acid for Visible Light-Active Polypropylene Coated Photocatalyst
Reprinted from: *Nanomaterials* **2018**, *8*, 599, doi:10.3390/nano8080599 . . . . **161**

Mariangela Longhi, Camilla Cova, Eleonora Pargoletti, Mauro Coduri, Saveria Santangelo, Salvatore Patanè, Nicoletta Ditaranto, Nicola Cioffi, Anna Facibeni and Marco Scavini
Synergistic Effects of Active Sites' Nature and Hydrophilicity on the Oxygen Reduction Reaction Activity of Pt-Free Catalysts
Reprinted from: *Nanomaterials* **2018**, *8*, 643, doi:10.3390/nano8090643 . . . . **175**

Marta Stucchi, Daria C. Boffito, Eleonora Pargoletti, Giuseppina Cerrato, Claudia L. Bianchi and Giuseppe Cappelletti
Nano-MnO_2 Decoration of TiO_2 Microparticles to Promote Gaseous Ethanol Visible Photoremoval
Reprinted from: *Nanomaterials* **2018**, *8*, 686, doi:10.3390/nano8090686 . . . . **189**

About the Special Issue Editors

Giuseppe Cappelletti is Associate Professor of Chemistry and Industrial Chemistry at the University of Milan in the field of Physical Chemistry. He is an expert in Colloids, Interfaces, and Formulations Science. His expertise also extends to electrocatalytic studies; composite materials for sensors in gas and liquid media; thin layers for medical applications; synthesis of ceramic materials; synthesis of nanomaterials for energetic applications; synthesis of nanocrystalline photocatalysts with tailored properties for water and air remediation; superhydrophobic and superhydrophilic transparent layers for the protection and conservation of natural and artificial stone materials; colloid stability and study of interfaces, including in addition of biosurfactant. He has authored more than 100 articles published in various international high-impact journals. His full profile and a complete list of publications can be accessed on Scopus: https://www-scopus-com/authid/detail.uri?authorId=56274806100.

Claudia Letizia Bianchi is Full Professor of Industrial Chemistry, LCA Expert and Head of the Process and Plants Research Group for Industrial Chemistry at the University of Milan. She is Coordinator of the master's degree in Industrial Chemistry, a Board Member of the European Society of Sonochemistry and Ultrasonics Sonochemistry, and an Italian member of the ISO table (TC206) on Advanced Ceramics. Her expertise is in materials chemistry (preparation, characterisation, and testing of inorganic materials) for environmental applications. Her research group at the University of Milan includes researchers with interests covering an array of different fields, such as inorganic materials, surface science, and analytical and physical chemistry, mainly in the industrial chemistry sector. The kind of research being developed in these laboratories has spillover effects in both environmental and energy field. The competences of the group deal with the design and preparation of solid catalysts and microstructured materials, structure modulation during nanopowder syntheses, catalyst deposition techniques on different supports, catalytic and photocatalytic tests, transformation and reactivity of wastes in liquid and gas phase, and assessment of the environmental impact of products and processes through LCA calculations. She has authored more than 230 original research papers published in various international high-impact journals mainly in the field of applied materials science and environmental catalysis. Her full profile and complete list of publications can be accessed on Scopus: https://www.scopus.com/authid/detail.uri?authorId=7202398527.

Preface to "Synthesis and Applications of Nanomaterials for Photocatalysis and Electrocatalysis"

This book supplies fundamental aspects regarding the use of different nanostructures as heterogeneous catalysts for energy and environmental applications. In recent decades, the attention of both scientific and industrial communities has become increasingly focused on the implementation of groundbreaking nanomaterials in all fields of human activity, especially toward improving energy efficiency and fulfilling environmental demands. Energy and environment represent a perfect blend: energy-saving environmental remediations and promising energetic devices meeting environmental concerns represent potential future challenges that humankind will face. In this context, the fine control of the nanosized species is the real tool to overcome the current issues and to improve the final performances. Herein, from an energetic point of view, oxygen evolution and reduction reactions (OER and ORR) are key to deeply understanding the behaviour of water splitting devices and fuel cells as well as zinc/air batteries, respectively. Zinc tantalum oxynitride-based photoanodes and nitrogen-modified carbon doped with different metals will be presented and fully characterised. Concurrently, bismuth titanate nanosheets and noble metal core–shell nanoparticles can be adopted to enhance hydrogen evolution through photocatalytic water splitting, exploiting solar energy. Instead, for what concerns the environmental remediation, the use of pure (black, modified, and faceted TiO2, Ga2O3) and composite (graphene/titanate, Zn2–SnO4/BiOBr, g-C3N4/Nb2O5, MnO2/TiO2 and CaIn2S4/ZnIn2S4) nanomaterials allow for air and water purification, especially under solar irradiation. Particularly, the complete photodegradation of noxious species (benzylic acid), organic dyes (rhodamine B, methylene blue and alizarin red), heavy metals (chromium), recalcitrant pharmaceutical active principles (cinnamic acid, ibuprofen and tetracycline), and VOCs (ethanol) will be thoroughly discussed. Finally, we would like to acknowledge all the authors who have contributed to this book with their scientific expertise, and we hope that the readers will find the arguments both useful and interesting.

Giuseppe Cappelletti, Claudia Letizia Bianchi
Special Issue Editors

nanomaterials

MDPI

Article

Zinc Tantalum Oxynitride ($ZnTaO_2N$) Photoanode Modified with Cobalt Phosphate Layers for the Photoelectrochemical Oxidation of Alkali Water

Prabhakarn Arunachalam [1,*], Maged N. Shaddad [1], Mohamed A. Ghanem [1], Abdullah M. Al-Mayouf [1] and Mark T. Weller [2]

[1] Electrochemistry Research Group, Chemistry Department, College of Science, King Saud University, Riyadh 11451, Saudi Arabia; mshadad@ksu.edu.sa (M.N.S.); mghanem@ksu.edu.sa (M.A.G.); amayouf@ksu.edu.sa (A.M.A.-M.)

[2] Department of Chemistry, University of Bath, Bath BA2 7AY, UK; m.t.weller@bath.ac.uk

* Correspondence: parunachalam@ksu.edu.sa or prabhuchemist@hotmail.com; Tel.: +966-114696026

Received: 18 November 2017; Accepted: 28 December 2017; Published: 18 January 2018

Abstract: Photoanodes fabricated by the electrophoretic deposition of a thermally prepared zinc tantalum oxynitride ($ZnTaO_2N$) catalyst onto indium tin oxide (ITO) substrates show photoactivation for the oxygen evolution reaction (OER) in alkaline solutions. The photoactivity of the OER is further boosted by the photodeposition of cobalt phosphate (CoPi) layers onto the surface of the $ZnTaO_2N$ photoanodes. Structural, morphological, and photoelectrochemical (PEC) properties of the modified $ZnTaO_2N$ photoanodes are studied using X-ray diffraction (XRD), scanning electron microscopy (SEM), ultraviolet visible (UV−Vis) diffuse reflectance spectroscopy, and electrochemical techniques. The presence of the CoPi layer significantly improved the PEC performance of water oxidation in an alkaline sulphate solution. The photocurrent-voltage behavior of the CoPi-modified $ZnTaO_2N$ anodes was improved, with the influence being more prominent at lower oxidation potentials. A stable photocurrent density of about 2.3 $mA \cdot cm^{-2}$ at 1.23 V vs. RHE was attained upon visible light illumination. Relative to the $ZnTaO_2N$ photoanodes, an almost three-fold photocurrent increase was achieved at the CoPi/$ZnTaO_2N$ photoelectrode. Perovskite-based oxynitrides are modified using an oxygen-evolution co-catalyst of CoPi, and provide a new dimension for enhancing the photoactivity of oxygen evolution in solar-assisted water-splitting reactions.

Keywords: cobalt phosphate; tantalum oxynitride; water oxidation; photoelectrochemistry; solid-state synthesis

---

1. Introduction

The development of active and proficient semiconductor photocatalysts for the direct conversion of solar energy to chemical energy has gained attention in the desire to fulfil future energy and fuel demand [1,2]. The use of solar photocatalytic water splitting to produce H_2 and O_2 over a heterogeneous photocatalyst has been studied extensively for several years, and has been found to be a favorable process for clean and renewable hydrogen generation. The heterogeneous photocatalytic process function at a solar-energy conversion efficiency of 10% is theoretically able to generate hydrogen at a rate of \$1.63/kg H_2 [3]. However, the development of a photocatalyst for the oxidation of water to yield oxygen gas remains the main hurdle that needs to be overcome in order to establish a technology that is based on water-splitting. The water oxidation reaction is the most difficult half reaction process as it includes four-electron transfer to generate an oxygen molecule, O_2; therefore there are significant thermodynamic and kinetic limitations to this process [3]. Several polycrystalline photocatalyst materials have been established for solar-assisted water splitting [1,4]; on the other hand, the materials available for the fabrication of efficient photoelectrodes are still limited. The widely investigated photocatalyst is the TiO_2 photoelectrode, with a band gap (BG) of 3.0 eV; however, this large band is not suitable for the absorption of visible light, and an external bias is generally required to achieve surface redox reactions [5–8]. Therefore, for improved photoelectrodes, it is indispensable to narrow the BG of the semiconductors used to expand light absorption into the visible region of the solar spectrum.

Significant attempts have been carried out to develop new types of oxide-based semiconductors [9–12]. In particular, n-type semiconductor photoanodes, such as Fe_2O_3, WO_3, and $BiVO_4$ [13–15], have been employed to accumulate the solar visible-light photons in stable and proficient water-splitting reactions. However, Scaife reported that it is inherently harder to progress an oxide-based photocatalyst that has an appropriately negative conduction band (CB) and a narrow BG (i.e., <3 eV) for visible-light absorption due to their higher positive valence band (VB) (at ca. +3.0 V vs. NHE) created by the O 2p orbital [16,17]. Other oxide-based photoelectrodes, such as La-doped $NaTaO_3$ [18] and Rh-doped $SrTiO_3$ [19], show reasonable quantum efficiencies (QEs) in the presence of UV light, and have been utilized as H_2 evolution photocatalysts in visible-light irradiation.

Oxynitride-based photoanodes offer an alternative to oxide-based anodes for visible-light absorption to produce H_2 and O_2 from water at a stoichiometric ratio [17,20–25]. In these materials, atomic nitrogen is introduced into the oxygen sites, and it shifts the VB edge potentials to the more negative region via the hybridization of N 2p and O 2p orbitals. Domen et al. demonstrated TaON/co-catalyst-based photoanodes with an incident photon-to-electron conversion efficiency (IPCE) of 76% at 400 nm for water oxidation at a minimal or no external applied bias [20–26]. Since then, oxynitrides, such as $LaTiO_2N$ [22], $SrNbO_2N$ [27,28], $BaNbO_2N$ [29], among other materials, have been similarly developed as photoanodes [30–35]. Few of these have attained superior IPCEs of the order of magnitude of several tens of percent with an externally applied bias that is less than the thermodynamically mandatory oxidation potential (1.23 V) for water electrolysis. Currently, significant research is underway to develop the smaller band-gap photoanodes with a BG of <2 eV (equivalent to 600 nm in the visible absorption spectrum) to oxidize water with the application of minimal bias [33]. Several such photoanodes have been developed, e.g., $BaNbO_2N$ [21], $CoO_x/LaTiO_2N$ [22], and $BaZrO_3/BaTaO_2N$ [33], which have been found to utilize the absorption of visible-light in addition to appropriate sacrificial reagents. $SrNbO_2N$-based photoanodes (BG < 2 eV), function as water-oxidation catalysts with the support of a co-catalyst in visible-light illuminations [27,28,34]. However, the recombination of the photoexcited carriers and poor photon absorption efficiencies are considered as major drawbacks with these systems, and these attributes need to be improved for application in water-splitting reactions. The incorporation of oxygen-evolution electro-catalysts (OECs) with a photon-absorbing substrate shows it to be a suitable plan to develop the field of photocatalysis chemistry, thus increasing the overall solar conversion efficiency [35]. Recently, the use of cobalt-based materials has also been promising for electrochemical applications [36–38]. To this end, the amorphous cobalt-phosphate (Co-Pi) appears to be a promising candidate as a co-catalyst material that possesses the advantages of relatively high elemental earth abundances, a self-healing feature, and functionality under benign conditions [39,40]. Recently, we reported the PEC activity of ZrO_2- and Fe_2O_3- modified $BiVO_4$ and $CoPi/La(Ta,Nb)O_2N$ photoelectrodes for water oxidation in alkaline media [32,41], which achieved up to a 5-fold enhancement in the water-oxidation photocurrent at a lower oxidation potential. In contribution to this research, the present work conveys a significance of engaging a cobalt-based electrocatalyst over tantalum oxynitrides via the electrophoretic deposition route to fabricate the photoanodes. Among the various oxynitrides, $ZnTaO_2N$ belongs to a $LiNbO_3$-type structure, which crystallizes in a high temperature paraelectric phase (HTLN-type) structure. $ZnTaO_2N$ is a perovskite-type oxynitride containing comparatively earth-abundant metals and it offers a remarkable unique property compared to other available oxynitrides. Our aim in this work is to activate the $ZnTaO_2N$ catalyst for the water oxidation process by incorporating the oxygen evolution reaction co-catalysts. The loading of the CoPi OER co-catalyst could amplify the charge separation and collection of holes generated at the surface of the photoanodes, subsequently improving photocatalytic performance.

Herein, we describe about the cocatalytic influence as well as the PEC performance of the Co-Pi incorporated $ZnTaO_2N$ photoanodes. The PEC properties of the $ZnTaO_2N$ electrodes are enhanced predominantly when modified with the Co-Pi catalyst, and its effect is greatest at a lower potential, while the photocurrent is enriched by a factor of 2.5 times relative to that of a bare electrode. The enhancement of the performance can be credited to an improvement in hole collection and charge-separation efficiency at the surface of the oxynitride photoanodes. The present work also demonstrates a viable way for an improvement in the energy-conversion efficiency by coupling OECs with oxynitride photoanodes.

2. Results and Discussion

2.1. XRD and DRS Analysis

The $ZnTaO_2N$ catalyst was prepared using the conventional ammonolysis method, and for comparison, TaON was prepared using a similar procedure. The as-synthesized $ZnTaO_2N$ powders show an orange-yellow color, while the TaON powder had a yellow color. The electrophoretic deposition (EPD) method was employed to prepare photoanodes of these materials. The average thickness of the $ZnTaO_2N$-deposited film on the indium tin oxide (ITO) substrate was measured using a profilometer, and was about 2.5 µm. Figure 1 demonstrates the XRD patterns for the TaON, $ZnTaO_2N$, and CoPi/$ZnTaO_2N$ photoanodes that were fabricated by EPD followed by loading CoPi using photodeposition methods on ITO substrates, and annealing in a nitrogen atmosphere at 450 °C. The XRD pattern of as-synthesized $ZnTaO_2N$ reveals that all the major peaks can be assigned to a perovskite-type phase that is structurally similar to that of $ZnTaO_2N$ [28]. The TaON XRD pattern obtained was consistent with the formation of the baddeleyite (monoclinic ZrO_2) structure. The XRD pattern of TaON photoanodes (Figure 1a) can be well indexed with the phase of TaON (JCPDS # 70-1193). In $ZnTaO_2N$ films, a small level of peak broadening was observed, which implies a reduction in particle size compared to the TaON photoanodes. The XRD pattern of the electrodeposited $ZnTaO_2N$ photoanodes (Figure 1b) displays the mixed phase of $ZnTa_2O6$ (JCPDS # 39-1484) and TaON.

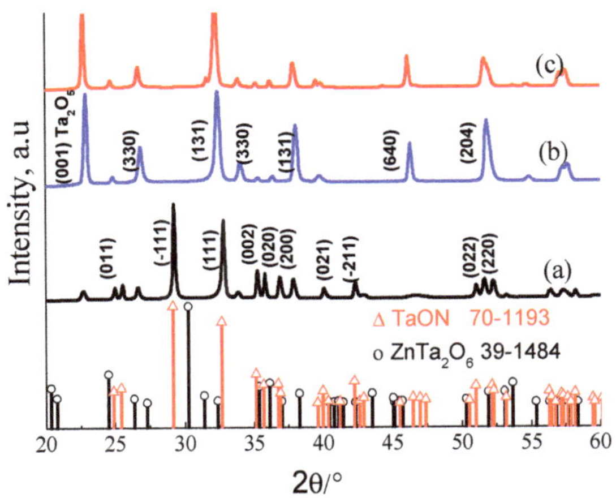

Figure 1. X-ray diffraction (XRD) patterns of (**a**) reference TaON; (**b**) $ZnTaO_2N$ and (**c**) CoPi/$ZnTaO_2N$ photo-anodes prepared via electrophoretic and photodeposition.

The XRD patterns of the CoPi-loaded $ZnTaO_2N$ photoanodes show that the CoPi deposition did not alter the crystalline phase of $ZnTaO_2N$ photoanodes, and no XRD peaks corresponding to metal impurities or simple metal oxides were observed. This may be credited to the good dispersion of nanocrystalline CoPi particles over the $ZnTaO_2N$. A small shift towards the lower 2θ values was observed in the presence of CoPi-loaded photoanodes, compared with the parent photoanodes, and this was due to a small difference in the sample height in the powder diffractometer. Figure 2 displays diffuse reflectance spectroscopy (DRS) spectra for the various TaON-based photoanodes. All fabricated TaON-based photoanodes have visible light, and the onset of light-absorption features of the band-gap excitation of TaON, $ZnTaO_2N$, and CoPi/$ZnTaO_2N$ was observed at 420, 447, and 452 nm, respectively.

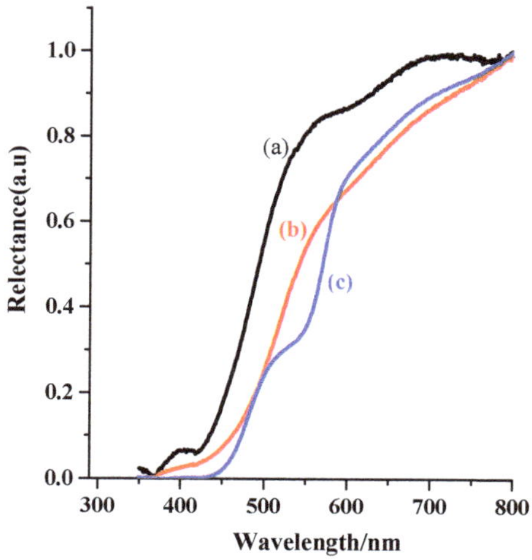

Figure 2. Ultraviolet visible (UV-Vis) reflectance spectra of perovskite-type tantalum oxynitride modified photoanodes of (**a**) TaON; (**b**) $ZnTaO_2N$ and (**c**) CoPi/$ZnTaO_2N$.

However, CoPi/$ZnTaO_2N$ photoanodes display dissimilar absorption spectra from that of $ZnTaO_2N$, and the absorption edge point is mainly lifted to the higher wavelengths compared with TaON. The BG of the fabricated TaON-based photoanodes was estimated via Kublenka-Munk (K-M) functions, and the data are presented in Table 1. The assessed optical band gaps decreased in the order TaON > $ZnTaO_2N$ > CoPi/$ZnTaO_2N$ > $LaTa_{0.3}Nb_{0.7}O_2N$ > $LaNbO_2N$ (Table 1), which is consistent with a similar effect for perovskite tantalum oxides [42].

Table 1. Structural properties and the band gap of various oxynitride based photoanodes.

Electrophoretic Deposited Oxynitrides on ITO	Particle Size (nm) [a]	Band Gap E_g (eV) [b]
TaON	29.5	2.95
$ZnTaO_2N$	22.7	2.75
CoPi/$ZnTaO_2N$	22.4	2.7, 2.47
$LaTa_{0.3}Nb_{0.7}O_2N$ [32]	18	1.84
$LaNbO_2N$	40	1.65

[a] Obtained using Scherer equation; [b] Determined from the diffuse reflectance spectroscopy (DRS) analysis via K-M function.

2.2. Morphological Characteristics of $ZnTaO_2N$ Photo-Anodes

Figure 3 displays SEM images of the surface morphology of the $ZnTaO_2N$ (Figure 3a,b) and CoPi/$ZnTaO_2N$ (Figure 3c) photoanodes. These display irregular shapes for both $ZnTaO_2N$ and CoPi/$ZnTaO_2N$ materials with the presence of some voids between the particles. The particle sizes generally range from 30 to 200 nm. The $ZnTaO_2N$ particles had clearer crystal edges and faces, indicating a superior crystallinity of these anodes. This is likely to favor the process of photohole generation, which can reach the active reaction sites at the boundary between the electrode/electrolyte [30]. However, the presence of large quantities of grain boundaries and loose inter-particle connections may lead to the incoherence of electron transport between the particles. The SEM image in Figure 3c for $ZnTaO_2N$ covered with CoPi clearly reveals that the loaded CoPi layer does not alter the morphology of $ZnTaO_2N$ particles. In order to prove the deposition of the CoPi layer at the $ZnTaO_2N$ surface, we examined the structure using energy-dispersive X-Ray analysis (EDX), and the results are shown in Figure 3d. The EDX characterization confirms the existence of Ta, Co, and P elements in the photoanodes at 46.25, 1.23, and 0.58 wt %, respectively.

Figure 3. Scanning electron microscopy (SEM) images of (**a**) $ZnTaO_2N$; (**b**) higher magnification of $ZnTaO_2N$; (**c**) CoPi/$ZnTaO_2N$ photo-anodes prepared via electrophoretic and photo-deposition deposition onto ITO; (**d**) energy-dispersive X-Ray analysis (EDX) spectrum of CoPi/$ZnTaO_2N$.

2.3. XPS Investigation of the CoPi/$ZnTaO_2N$ Photoanodes

Figure 4 shows the existence of Co, P, Zn, Ta, O, and N in the wide-scan X-ray photoelectron spectroscopy (XPS) spectrum of CoPi/$ZnTaO_2N$ photo-anodes. The molar ratio of Co and P in CoPi/$ZnTaO_2N$ was estimated to be 1:2.2. The major peaks of binding energies at 26.8 and 28.5 eV related to the spin-orbit separation of the Ta $4f_{5/2}$ and Ta $4f_{7/2}$ ingredients, respectively (Figure 4b), demonstrating the development of the Ta^{5+} [43]. The two dissimilar binding energies of the O element can be allocated to the characteristics of the Ta-O bond (530.1 eV) and oxygen in carbonate species or hydroxyl groups (531.8 eV) (Figure 4c) [44]. The major peaks for Ta $4P_{2/3}$ and N 1s in Figure 4d indicate that the N 1s region centered at 396.4 eV is associated with the binding energy of about 403.5 eV for Ta $4P_{2/3}$, further confirming the creation of various Ta-N bonds [44].

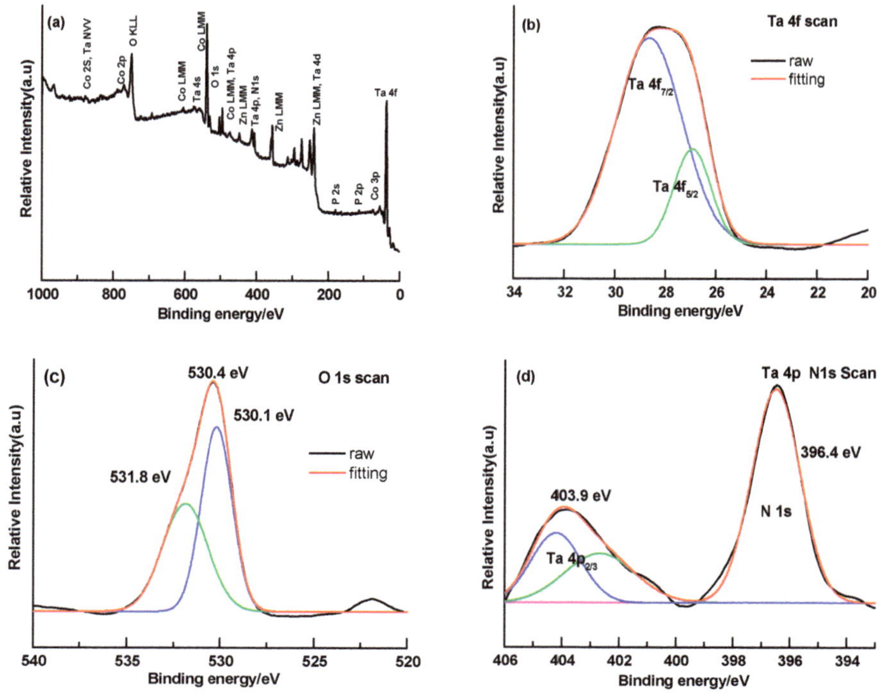

Figure 4. X-ray photoelectron spectroscopy (XPS) analysis of the CoPi/$ZnTaO_2N$ microsphere: (**a**) wide survey spectra scan for sample; narrow scan for (**b**) Ta element; (**c**) O element; (**d**) Ta and N element.

2.4. Photoelectrochemical (PEC) Properties of the $ZnTaO_2N$ Photoanodes

The PEC properties of $ZnTaO_2N$ photoanodes were examined by performing cyclic voltammetry (CV) and chronoamperometric (CA) measurements in an H-shaped cell. Figure 5a shows the CV measurements logged at 50 $mV \cdot s^{-1}$ in 1.0-M Na_2SO_4 for $ZnTaO_2N$/ITO photoanodes under AM 1.5 G simulated sunlight at various thicknesses of the film prepared by varying the EPD deposition time duration. The results show that the current density of the photoanode changes dramatically as the deposition time increases. As shown in Figure 5b, the maximum current density (measured at 2.0 V vs. RHE) was observed around 0.5 $\pm$ 0.05 mg of loaded $ZnTaO_2N$ on ITO substrate (equivalent to deposition for 4.0 min and a thickness of 2.0 μm), and at higher loading (>0.5 mg), the photoanode current density significantly decreased. This can be clarified by the catalytic water oxidation reaction at the CoPi/electrolyte interface, and when CoPi loading at higher loading (>0.5 mg), the photoholes need to be transferred in amid numerous CoPi molecules and CoPi/electrolyte interface, which affects the sluggish kinetics of the hole transfer, and subsequently, a smaller photocurrent is detected [32]. Moreover, as shown in Figure 5c (taken from Figure 5a), at a $ZnTaO_2N$ loading of around 0.5 $\pm$ 0.05 mg, the onset potential of water oxidation markedly shifted to a less positive value, indicating a more favorable process. To understand the effect of pH on the PEC behavior of the $ZnTaO_2N$ photoanode, Figure 5d shows the linear sweep voltammetry (LSV) plot for $ZnTaO_2N$ photoanode at different pH in 1.0-M Na_2SO_4 solution. It is evidenced that the current density of water oxidation was significantly enhanced at alkaline pH of 13, which is almost eight times superior than that at pH of 12. Additionally, the onset potential of water oxidation was shifted 120 mV more cathodic compared with the value obtained at pH of 12.

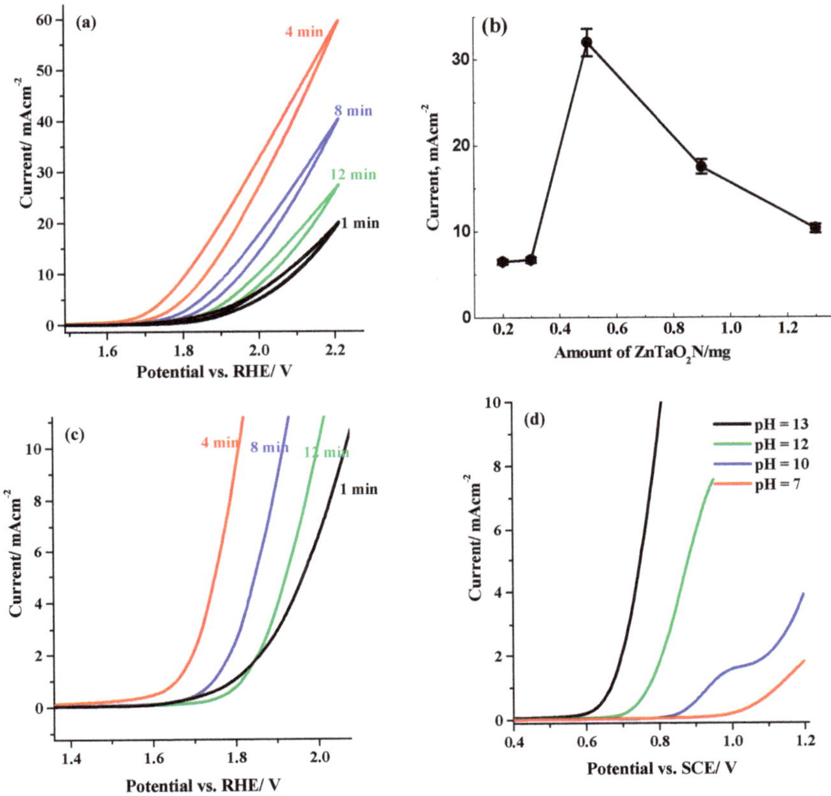

Figure 5. (**a**) Cyclic voltammetry (CV) at 50 $mV \cdot s^{-1}$ in 1.0 M Na_2SO_4 at pH 13 for photocurrent response for various amounts of $ZnTaO_2N$ loaded on ITO; (**b**) plot for photocurrent measured at 2.0 V vs. RHE and amount of loaded $ZnTaO_2N$ deposited by electrophoretic deposition (EPD) onto ITO; (**c**) CV in (**a**) zoomed in current less than 10 $mA \cdot cm^{-2}$; (**d**) Linear sweep voltammetry (LSV) recorded at various pH of electrolyte solution in the presence of 1.0 M Na_2SO_4 solution.

To further improve the PEC behavior of the $ZnTaO_2N$ photocatalyst for water oxidation reaction, the CoPi co-catalyst (OEC) was incorporated into the photoanodes using the photodeposition PD method [32,45]. Initially, the amount of CoPi loaded onto the $ZnTaO_2N$ film was optimized by varying the PD duration, followed by measurement of the photocurrent response using chronoamperometry at 1.7 V vs. RHE in 1.0-M Na_2SO_4 solution. Figure 6a shows the current time transients with a 10 s light pulse (1.5 AM) for a $ZnTaO_2N$ anode before and after the photodeposition of CoPi for a deposition time of 60 min. For clarity, the background current in the dark was subtracted for both photoanodes. Once the light pulse was applied, the current of both photoanodes increased significantly and the photocurrent density obtained at the $ZnTaO_2N$ anode was about 0.75 $mA \cdot cm^{-2}$, while the CoPi/$ZnTaO_2N$ photoanode displayed a much superior photocurrent density of 2.9 $mA \cdot cm^{-2}$ under the similar conditions. Clearly, the photocurrent obtained at CoPi/$ZnTaO_2N$ is more than three times higher than in the case of the $ZnTaO_2N$ electrode, which indicates that the separation and collection processes of the photogenerated electron/hole pairs are more efficient at CoPi/$ZnTaO_2N$ than at a simple $ZnTaO_2N$ photoanode.

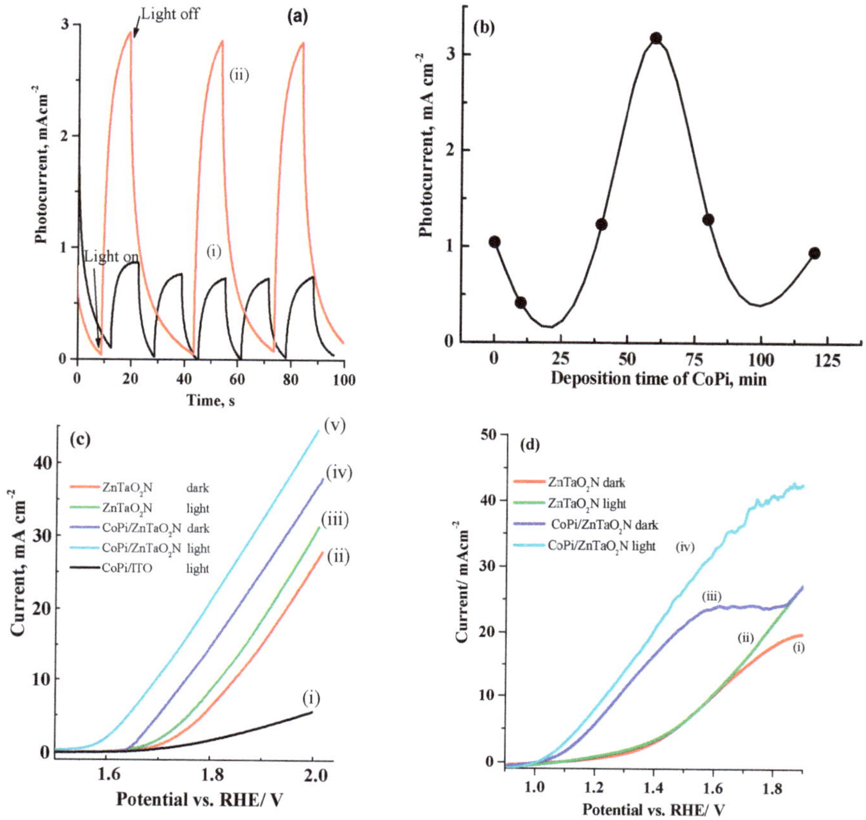

Figure 6. (**a**) Chronoamperometry diagram at 1.7 V vs. RHE during pulsing of light for 10 s in 1.0-M Na_2SO_4 solution (pH 13) for (i) $ZnTaO_2N$ and (ii) CoPi/$ZnTaO_2N$ photoanode, photodeposition time = 60 min; (**b**) plot for the photocurrent vs. deposition time of CoPi deposition on $ZnTaO_2N$ photoanode; (**c**) LSV at 50 mV·s^{-1} in dark and under 1.5 AM light illumination for (i) CoPi/ITO in light, (ii) $ZnTaO_2N$ in dark, (iii) $ZnTaO_2N$ in light, (iv) CoPi/$ZnTaO_2N$ in dark, and (v) CoPi/$ZnTaO_2N$ in light; photo-anodes in 1.0-M Na_2SO_4 solution (pH 13) and (**d**) LSV spectra of $ZnTaO_2N$-based photoanodes in the presence of H_2O_2 (i) $ZnTaO_2N$ in dark, (ii) $ZnTaO_2N$ in light, (iii) CoPi/$ZnTaO_2N$ in dark, and (iv) CoPi/$ZnTaO_2N$ in light.

Figure 6b shows the relationship between the CoPi PD time and the photocurrent obtained for the CoPi/$ZnTaO_2N$ anode at 1.7 V vs. RHE. Clearly, using our present PD methodology, the optimum period for CoPi photo-deposition on $ZnTaO_2N$ photo-anodes was found to be around 60 min. The PEC characteristics were compared for both $ZnTaO_2N$ and CoPi/$ZnTaO_2N$ anodes to examine the effect of the photodeposited CoPi co-catalyst on the photocurrent response of the $ZnTaO_2N$ photoanode. Figure 6c shows the LSV at 50 mV·s^{-1} in the dark and under 1.5 AM light illumination for CoPi/$ZnTaO_2N$ and $ZnTaO_2N$ photoanodes in 1.0-M Na_2SO_4 solution (pH 13). In comparison, the LSV of the control anode made by the PD of the CoPi catalyst directly onto the ITO substrate is also shown in Figure 6c (i). The LSV for the control anode of CoPi/ITO (i) shows a small level of the oxygen evolution current under light illumination. The CoPi/$ZnTaO_2N$ photoanode clearly shows negative potential shifts of about 120 mV in the onset potential of oxygen evolution, and an increase in the photocurrent of more than 2.3 mA·cm^{-2} at 1.23 V vs. RHE relative to its parent $ZnTaO_2N$ photoanode. For the CoPi/$ZnTaO_2N$ anode, the photocurrent was enhanced by a factor of three compared with the parent $ZnTaO_2N$ electrode. It is clear that the magnitude of the CoPi loading is influenced by the photoanode morphology, time, and the employed deposition procedure. A larger amount of CoPi deposited on $ZnTaO_2N$ (>60 min) results in the suppression of the photocurrent, as revealed in Figure 6b. This phenomenon can be explained by considering that the water catalytic oxidation reaction takes place at the CoPi/electrolyte interface. For a thick layer of CoPi, the photoholes have to transfer through a thick CoPi layer to reach the interface between the CoPi/electrolyte; consequently, the hole migration becomes very slow, and a decreased photocurrent is observed. On the contrary, for a thinner CoPi layer, the cobalt ions link more directly to the $ZnTaO_2N$ surface, and rapidly acquire the photoholes in the water oxidation reaction [46,47].

Hydrogen peroxide (H_2O_2) was introduced to the electrolyte (0.1 M) as an electron donor to evaluate the maximum photocurrent that might be acquired while injecting a photoexcited holes from the anode surface to the electrolyte solution was perfectly facilitated. It is believed that H_2O_2 captures the photoexcited holes efficiently without substantial recombination [27]. To observe these behaviors on the prepared photoanodes, LSV plots of CoPi/$ZnTaO_2N$ were logged in the presence and absence of H_2O_2 and the results are presented in Figure 6d. The LSV of $ZnTaO_2N$ and CoPi/$ZnTaO_2N$ photoanodes indicate that the photocurrent at applied potential values of more negative than 1.7 V vs. RHE was improved considerably in the presence of H_2O_2. This implies that the redox active Co species may be quickly oxidized at a lower applied potential (<1.4 V vs. RHE), while water could not. This results coincides with the results attained for CoPi/La(Ta,Nb)O_2N photoelectrodes [32]. These PEC performances show that the CoPi/$ZnTaO_2N$ photoelectrodes, photocurrent at <1.4 V vs. RHE was comparable, irrespective of the introduction of H_2O_2, signifying that the CoPi had improved the surface redox reactions effectively.

2.5. Band Positions of the $ZnTaO_2N$ Photoanodes

As previously discussed, the band-edge position at the electrolyte and the carrier mobility in the semiconductor photocatalyst are both significant factors for the characterization of the PEC performance of the photoanodes. The Mott–Schottky (MS) method was performed and the Nyquist plot was obtained to determine the characteristics of $ZnTaO_2N$ and CoPi/$ZnTaO_2N$ electrodes in 1-M Na_2SO_4 (pH 13) at a frequency of 100 Hz. The valence-band and conduction-band potentials of TaON, $LaTaO_2N$, and $ZnTaO_2N$ were calculated using the Butler and Ginley method, and the values are précised in Table S1.

The conduction-band edge (E_{CB}) for $ZnTaO_2N$ was obtained at 0.32 V vs. SCE by employing the method proposed by Butler and Ginley [48] using the optical BG from the K-M spectra. As presented in Figure 7a, the intercept on the potential axis of the $ZnTaO_2N$ photoanode reveals a flat band potential (E_{FB}) at 0.14 V, while the intercept of the CoPi/$ZnTaO_2N$ photoanodes displays an E_{FB} at 0.08 V vs. RHE at pH = 13. It is clear that CoPi/$ZnTaO_2N$ photoanodes have less positive E_{FB} values than the parent photoanodes. The addition of the CoPi co-catalyst as a hole acceptor promotes the charge-separation reaction, which leads to the enhancement of the photocurrent during the water oxidation reaction. In addition, the CoPi co-catalyst can cause the oxygen evolution reaction to occur at lower oxidation potential by altering the reaction pathway [39]. The CoPi layer on the photoanode has to be thin enough to complement the fast transfer of the photogenerated hole from the $ZnTaO_2N$ photoanode to the water oxidation reaction, which relieves the charge accumulation at the electrode/electrolyte interface.

Figure 7b shows Nyquist plots of the $ZnTaO_2N$ and CoPi/$ZnTaO_2N$ photo-anodes acquired in the light and in the dark. The equivalent circuit is shown in the inset of Figure 7b and the fitting results of the photoanodes are summarized in Table 2. The $ZnTaO_2N$ photo-anodes were perfect fitted to a RC circuit model which grasps a resistor and a RC circuit. RC circuit can be allocated to $ZnTaO_2N$/electrolyte interface. For the CoPi/$ZnTaO_2N$ photo-anode, capacitances increased, whereas the charge transfer resistances decreased, signifying that CoPi-loading sustained charge separation at the bulk photo-anode and, as a consequence, enriched PEC water oxidation ability. CoPi loading on photoanodes advances the PEC behaviour by endorsing the charge separation and water oxidation phenomenon on the $ZnTaO_2N$ photo-anodes surface which is very consistent with the behavior of CoPi/$LaTaNbO_2N$ [32], CoPi/TiO_2 [39] and CoPi/$BiVO_4$ structures [47].

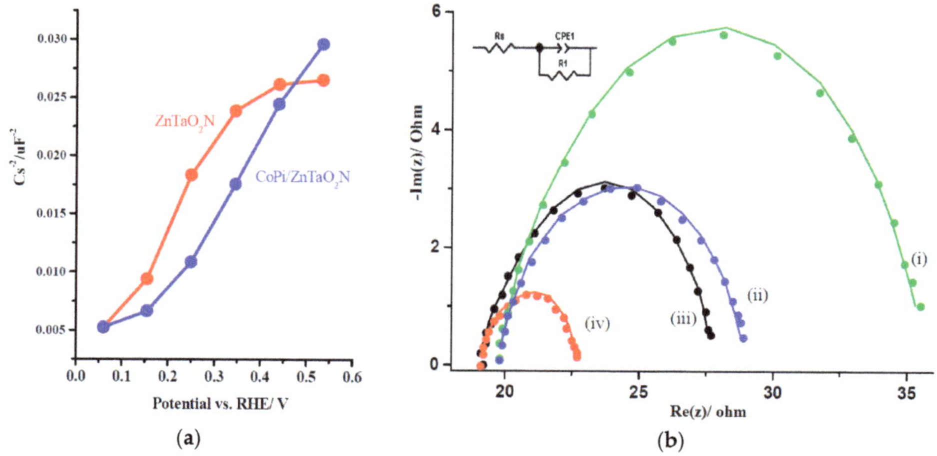

Figure 7. (**a**) Mott–Schottky plots for $ZnTaO_2N$ and CoPi/$ZnTaO_2N$ photo-anodes; (**b**) Nyquist plot for the photoanodes of (i) $ZnTaO_2N$ in dark, (ii) $ZnTaO_2N$ in light, (iii) CoPi/$ZnTaO_2N$ in dark, and (iv) CoPi/$ZnTaO_2N$ in light; illumination conditions are 1.5 AM and in 1.0-M Na_2SO_4 at pH = 13.

Table 2. Nyquist plot of equivalent circuit-model parameters obtained from the electrochemical impedance spectra of photoanodes.

Photoanode		Rs/Ω	Rct/Ω	CPE1 (F)
$ZnTaO_2N$	Dark	22.17	33.55	6.763×10^{-5}
	Light	22.17	27.23	5.714×10^{-5}
CoPi/$ZnTaO_2N$	Dark	19.17	25.81	1.15×10^{-4}
	Light	19.17	22.46	2.303×10^{-5}

2.6. Quantification of Dioxygen Evolution during Photoactivation

The irradiation of $ZnTaO_2N$ and CoPi/$ZnTaO_2N$ photoanodes with visible-light photons eventually increases the evolution of dioxygen. Figure 8a shows the oxygen-evolution measurements obtained at 1.7 V vs. RHE using an oxygen sensor oxysense system. The corresponding photocurrent responses were obtained and the results are presented in Figure 8b. As anticipated, under similar conditions, the CoPi/$ZnTaO_2N$ photoanodes and their parent photoanodes with visible-light photons ($\lambda > 420$ nm) caused an increase in the dioxygen evolution. To obtain an estimate, the precise oxygen concentration was logged before and after PEC reactions. Figure 8a confirms that by switching on the photoelectrolysis, the concentration of oxygen starts to increase linearly with time, and no substantial quantity of oxygen was observed before (<15 min) or after (>45 min) the photoelectrolysis was turned off. The CoPi/$ZnTaO_2N$ photoanodes showed increased oxygen generation relative to parent $ZnTaO_2N$, which is demonstrated by its oxygen-evolution rate. The oxygen-evolution rate is faster in the case of CoPi-coated $ZnTaO_2N$ compared with the parent photoanode (Figure 8a). Conversely, it is evident that the incorporation of CoPi as a co-catalyst enhanced the photocurrent response and stability under continued illumination, as shown in Figure 8b. The durability of the CoPi/$ZnTaO_2N$ photoanodes was examined, and it displayed in Figure S1. The chronoamperometric results indicate that the photoanode remains stable until 180 min, after which small changes are observed. The stability of the CoPi/$ZnTaO_2N$ photoanodes under light irradiation was investigated by relating their absorption spectrum before and after visible light irradiation, signifying that the photoanodes are stable under visible-light illumination (Figure S2). A more effective coupling of photoholes to reaction sites of the water-oxidation reaction can be anticipated, as discussed earlier, which validates the enhancement of oxygen-evolution performance as well as the stability of CoPi-loaded $ZnTaO_2N$ photoelectrodes.

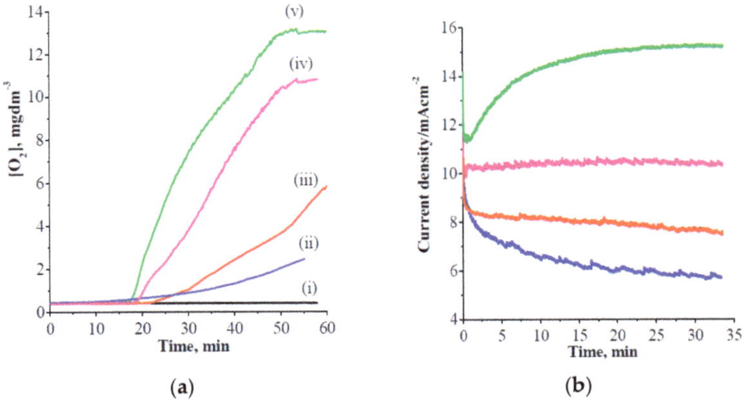

Figure 8. (**a**) Oxygen evolution and (**b**) its corresponding chronoamperometric measurements in a two-electrode setup in 1.0-M aqueous sulphate solution (pH = 13) during prolonged irradiation by visible light (λ > 420 nm), (i) ITO blank, (ii) $ZnTaO_2N$ dark, (iii) CoPi/$ZnTaO_2N$ dark, (iv) $ZnTaO_2N$ light, and (v) CoPi/$ZnTaO_2N$ light; the working electrode was biased at 1.7 V vs. RHE.

3. Experimental

3.1. Preparation of the $ZnTaO_2N$ Catalyst

The $ZnTaO_2N$ powder was fabricated via conventional solid-state reaction described in a previous work [49]. In the synthesis described here, stoichiometric quantities of $ZnCO_3$ and Ta_2O_5 (Aldrich, 99.9%, St. Louis, MO, USA) were well ground together in acetone in the presence of KCl (50% total weight), which is used as a mineralizer. The resulting mixture was heated at 850 °C under an ammonia flow (Air Products Electronic Grade, Riyadh, KSA) for 20 h at a flow rate of 7 $dm^3 \cdot h^{-1}$, and it was then permitted to cool to normal temperature under ammonia atmosphere. The mineralizer was leached from the products using excess de-ionized water, and the residual product was dried overnight at 80 °C. The TaON was prepared following similar procedure by heating pure Ta_2O_5 (Aldrich, 99.9%) at 850 °C under flowing ammonia for 18 h at a flow rate of 7 $dm^3 \cdot h^{-1}$.

3.2. Fabrication of the $ZnTaO_2N$ Photoanodes

$ZnTaO_2N$ film photoanodes were fabricated using the EPD process, as described in our previous work [32]. For instance, the $ZnTaO_2N$ (15 mg) and 3-mg iodine (Alfa-Aesar, Karlsruhe, Germany) powders were kept ultrasonically discrete in acetone (15 mL) to obtain a uniform suspension. The ITO substrates (174 nm, 1.0 × 1.0 cm, Asahi glass Co., Ltd., Tokyo, Japan) were submerged and held parallel at about 1.0 cm apart from each other in the solution. A +10 V bias was then applied to the two electrodes for 2 min using a potentiostat (Bio-Logic SAS, VSP 0478, Seyssinet-Pariset, France). This deposition process was repeated twice, and the electrode was then dried and annealed at 450 °C under a flow of N_2 gas at a rate of 500 mL/min for 1 h. This process resulted in the development of a $ZnTaO_2N$ layer (in total about 0.5 mg) with a comparatively uniform thickness of about 2 μm, as monitored using a surface profilometer. The cobalt phosphate (CoPi) co-catalyst layer was deposited on $ZnTaO_2N$ photoanodes using the PD method, as described in the literature [32].

CoPi/$ZnTaO_2N$ electrodes with various CoPi depositions were fabricated by altering the PD time (indicated as CoPi/$ZnTaO_2N$). Characterization and XRD measurements were carried out using a MiniFlex 600 (Rigaku, CuKα, 40 kV, 15 mA, Tokyo, Japan). The photoanodes were further characterized using an ultraviolet visible (UV-Vis) spectrophotometer (Shimadzu UV-2600, Tokyo, Japan) and EDAX (JED-2200 series, Tokyo, Japan). Electrochemical impedance spectroscopy (EIS) analysis was performed using a biologic (VSP-0478) potentiostat electrochemical workstation. A solar simulator (Asahi spectra, MAX 303, Torrance, CA, USA) provided visible-light irradiation to the fabricated photoanodes.

3.3. Photoelectrochemical Characterization

The PEC properties of the $ZnTaO_2N$ and CoPi/$ZnTaO_2N$ photoanodes were obtained in H-shaped two-compartment glass cells with a 2-cm diameter quartz window. The fabricated $ZnTaO_2N$-based photoanode film was taken as a working electrode (WE, 0.25 cm^2), the saturated calomel electrode as a reference electrode, and Pt foil was served as the counter electrode in the electrolyte solution containing 1-M Na_2SO_4, and the pH of the electrolyte solution was modified to 13 with the addition of KOH. Photoelectrochemical oxygen evolution was monitored by an oxygen analyzer (Oxysense Inc., Dallas, TX, USA, 300/5000 series). The compartment cell was sealed and argon gas was used for purging. Prior to the experiment, the electrolyte solution was bubbled with argon for 1 h, and the atmosphere above the electrolyte was maintained as argon throughout the measurements. Before starting the electrolysis, the cell setup was purged for 20 min to attain the equilibrium.

4. Conclusions

A photocatalyst of $ZnTaO_2N$ was initially synthesized via conventional solid state reaction and then photoanodes were fabricated by an electrophoretic deposition method into ITO. We studied the co-catalytic effect of a photo-assisted water oxidation reaction by photodeposition of CoPi oxygen electrocatalyst onto the $ZnTaO_2N$ photoanodes. An electrochemical investigation revealed that the PEC performance of $ZnTaO_2N$ was significantly enhanced after CoPi deposition and a 2.3 $mA \cdot cm^{-2}$ photocurrent density was reached at 1.23 V vs. RHE in a sulfate medium. Additionally, CoPi-loading assisted the PEC performance of $ZnTaO_2N$ film by reducing the charge recombination process and stabilizing the photo-anode performance for the oxygen evolution reaction under visible light illumination.

Supplementary Materials: The following are available online at www.mdpi.com/2079-4991/8/1/48/s1, Table S1: Photophysical and structural properties of TaO_2N based photoandodes, Figure S1: Chronoamperometric measurements in two-electrode setup in 1 M aqueous sulfate solution (pH = 13), Figure S2: Absorption spectra of CoPi/$ZnTaO_2N$ photoanodes before and after visible-light irradiation for 10 h.

Acknowledgments: The authors would like to extend their sincere appreciation to the Deanship of Scientific Research at King Saud University for funding this Research group NO: RG-1438-087.

Author Contributions: P.A. performed the experiments and wrote the manuscript. M.N.S. assisted in the photoelectrochemical characterization. M.A.G., A.M.A.-M. and M.T.W. provided suggestions and assistance in the experiment design, synthesis of materials and manuscript editing.

Conflicts of Interest: The authors declare no conflict of interest.

References

1. Kudo, A.; Miseki, Y. Heterogeneous photocatalyst materials for water splitting. *Chem. Soc. Rev.* **2009**, *38*, 253–278. [CrossRef] [PubMed]
2. Arunachalam, P.; Zhang, S.; Abe, T.; Komura, M.; Iyoda, T.; Nagai, K. Weak visible light (~mw/cm^2) organophotocatalysis for mineralization of amine, thiol and aldehyde by biphasic cobalt phthalocyanine/fullerene nanocomposites prepared by wet process. *Appl. Catal. B Environ.* **2016**, *193*, 240–247. [CrossRef]
3. Bockris, J. *Energy: The Solar-Hydrogen Alternative*; Halsted Press: New York, NY, USA, 1975; p. 381.

4. Osterloh, F.E. Inorganic materials as catalysts for photochemical splitting of water. *Chem. Mater.* **2008**, *20*, 35–54. [CrossRef]
5. Fujishima, A.; Honda, K. Electrochemical photolysis of water at a semiconductor electrode. *Nature* **1972**, *238*, 37–38. [CrossRef] [PubMed]
6. Sun, Y.; Zhao, Q.; Wang, G.; Yan, K. Influence of water content on the formation of TiO_2 nanotubes and photoelectrochemical hydrogen generation. *J. Alloys Compd.* **2017**, *711*, 514–520. [CrossRef]
7. Sun, M.; Fang, Y.; Kong, Y.; Yuan, X.; Shi, J.; Umar, A. Direct in situ synthesis of Fe_2O_3-codoped n-doped TiO_2 nanoparticles with enhanced photocatalytic and photo-electrochemical properties. *J. Alloys Compd.* **2017**, *705*, 89–97. [CrossRef]
8. Sun, W.; Wang, D.; Rahman, Z.U.; Wei, N.; Chen, S. 3D hierarchical WO_3 grown on TiO_2 nanotube arrays and their photoelectrochemical performance for water splitting. *J. Alloys Compd.* **2017**, *695*, 2154–2159. [CrossRef]
9. Watanabe, T.; Fujishima, A.; Honda, K.-I. Photoelectrochemical reactions at $SrTiO_3$ single crystal electrode. *Bull. Chem. Soc. Jpn.* **1976**, *49*, 355–358. [CrossRef]
10. Kudo, A. Development of photocatalyst materials for water splitting. *Int. J. Hydrogen Energy* **2006**, *31*, 197–202. [CrossRef]
11. Melián, E.P.; López, C.R.; Méndez, A.O.; Díaz, O.G.; Suárez, M.N.; Rodríguez, J.D.; Navío, J.; Hevia, D.F. Hydrogen production using Pt-loaded TiO_2 photocatalysts. *Int. J. Hydrogen Energy* **2013**, *38*, 11737–11748. [CrossRef]
12. Yuan, Y.; Zhang, X.; Liu, L.; Jiang, X.; Lv, J.; Li, Z.; Zou, Z. Synthesis and photocatalytic characterization of a new photocatalyst $BaZrO_3$. *Int. J. Hydrogen Energy* **2008**, *33*, 5941–5946. [CrossRef]
13. Hu, Y.-S.; Kleiman-Shwarsctein, A.; Stucky, G.D.; McFarland, E.W. Improved photoelectrochemical performance of Ti-doped α-Fe_2O_3 thin films by surface modification with fluoride. *Chem. Commun.* **2009**, *19*, 2652–2654. [CrossRef] [PubMed]
14. Amano, F.; Li, D.; Ohtani, B. Fabrication and photoelectrochemical property of tungsten (VI) oxide films with a flake-wall structure. *Chem. Commun.* **2010**, *46*, 2769–2771. [CrossRef] [PubMed]
15. Sayama, K.; Nomura, A.; Zou, Z.; Abe, R.; Abe, Y.; Arakawa, H. Photoelectrochemical decomposition of water on nanocrystalline $BiVO_4$ film electrodes under visible light. *Chem. Commun.* **2003**, 2908–2909. [CrossRef]
16. Scaife, D. Oxide semiconductors in photoelectrochemical conversion of solar energy. *Sol. Energy* **1980**, *25*, 41–54. [CrossRef]
17. Abe, R.; Takata, T.; Sugihara, H.; Domen, K. The use of $TiCl_4$ treatment to enhance the photocurrent in a TaON photoelectrode under visible light irradiation. *Chem. Lett.* **2005**, *34*, 1162–1163. [CrossRef]
18. Kato, H.; Asakura, K.; Kudo, A. Highly efficient water splitting into H_2 and O_2 over lanthanum-doped $NaTaO_3$ photocatalysts with high crystallinity and surface nanostructure. *J. Am. Chem. Soc.* **2003**, *125*, 3082–3089. [CrossRef] [PubMed]
19. Sasaki, Y.; Iwase, A.; Kato, H.; Kudo, A. The effect of co-catalyst for Z-scheme photocatalysis systems with an Fe^{3+}/Fe^{2+} electron mediator on overall water splitting under visible light irradiation. *J. Catal.* **2008**, *259*, 133–137. [CrossRef]
20. Si, W.; Pergolesi, D.; Haydous, F.; Fluri, A.; Wokaun, A.; Lippert, T. Investigating the behavior of various cocatalysts on $LaTaON_2$ photoanode for visible light water splitting. *Phys. Chem. Chem. Phys.* **2017**, *19*, 656–662. [CrossRef] [PubMed]
21. Feng, X.; LaTempa, T.J.; Basham, J.I.; Mor, G.K.; Varghese, O.K.; Grimes, C.A. Ta_3N_5 nanotube arrays for visible light water photoelectrolysis. *Nano Lett.* **2010**, *10*, 948–952. [CrossRef] [PubMed]
22. Nishimura, N.; Raphael, B.; Maeda, K.; Le Gendre, L.; Abe, R.; Kubota, J.; Domen, K. Effect of $TiCl_4$ treatment on the photoelectrochemical properties of $LaTiO_2N$ electrodes for water splitting under visible light. *Thin Solid Films* **2010**, *518*, 5855–5859. [CrossRef]
23. Maeda, K.; Teramura, K.; Lu, D.; Takata, T.; Saito, N.; Inoue, Y.; Domen, K. Photocatalyst releasing hydrogen from water. *Nature* **2006**, *440*, 295. [CrossRef] [PubMed]
24. Wang, X.; Maeda, K.; Lee, Y.; Domen, K. Enhancement of photocatalytic activity of $(Zn_{1+x}Ge)(N_2O_x)$ for visible-light-driven overall water splitting by calcination under nitrogen. *Chem. Phys. Lett.* **2008**, *457*, 134–136. [CrossRef]
25. Siritanaratkul, B.; Maeda, K.; Hisatomi, T.; Domen, K. Synthesis and photocatalytic activity of perovskite niobium oxynitrides with wide visible-light absorption bands. *ChemSusChem* **2011**, *4*, 74–78. [CrossRef] [PubMed]

26. Abe, R.; Higashi, M.; Domen, K. Facile fabrication of an efficient oxynitride TaON photoanode for overall water splitting into H_2 and O_2 under visible light irradiation. *J. Am. Chem. Soc.* **2010**, *132*, 11828–11829. [CrossRef] [PubMed]
27. Urabe, H.; Hisatomi, T.; Minegishi, T.; Kubota, J.; Domen, K. Photoelectrochemical properties of $SrNbO_2N$ photoanodes for water oxidation fabricated by the particle transfer method. *Faraday Discuss.* **2015**, *176*, 213–223. [CrossRef] [PubMed]
28. Kuno, Y.; Tassel, C.; Fujita, K.; Batuk, D.; Abakumov, A.M.; Shitara, K.; Kuwabara, A.; Moriwake, H.; Watabe, D.; Ritter, C. $ZnTaO_2N$: Stabilized high-temperature $LiNbO_3$-type structure. *J. Am. Chem. Soc.* **2016**, *138*, 15950–15955. [CrossRef] [PubMed]
29. Hisatomi, T.; Katayama, C.; Moriya, Y.; Minegishi, T.; Katayama, M.; Nishiyama, H.; Yamada, T.; Domen, K. Photocatalytic oxygen evolution using $BaNbO_2N$ modified with cobalt oxide under photoexcitation up to 740 nm. *Energy Environ. Sci.* **2013**, *6*, 3595–3599. [CrossRef]
30. Zhang, L.; Song, Y.; Feng, J.; Fang, T.; Zhong, Y.; Li, Z.; Zou, Z. Photoelectrochemical water oxidation of $LaTaON_2$ under visible-light irradiation. *Int. J. Hydrogen Energy* **2014**, *39*, 7697–7704. [CrossRef]
31. Landsmann, S.; Maegli, A.E.; Trottmann, M.; Battaglia, C.; Weidenkaff, A.; Pokrant, S. Design guidelines for high-performance particle-based photoanodes for water splitting: Lanthanum titanium oxynitride as a model. *ChemSusChem* **2015**, *8*, 3451–3458. [CrossRef] [PubMed]
32. Arunachalam, P.; Al-Mayouf, A.; Ghanem, M.A.; Shaddad, M.N.; Weller, M.T. Photoelectrochemical oxidation of water using $La(Ta,Nb)O_2N$ modified electrodes. *Int. J. Hydrogen Energy* **2016**, *41*, 11644–11652. [CrossRef]
33. Maeda, K.; Domen, K. Water oxidation using a particulate $BaZrO_3$-$BaTaO_2N$ solid-solution photocatalyst that operates under a wide range of visible light. *Angew. Chem. Int. Ed.* **2012**, *51*, 9865–9869. [CrossRef] [PubMed]
34. Wang, J.; Wang, X.; Liu, B.; Li, X.; Cao, M. Facile synthesis of $SrNbO_2$ n nanoparticles with excellent visible-light photocatalytic performances. *Mater. Lett.* **2015**, *152*, 131–134. [CrossRef]
35. Konta, R.; Ishii, T.; Kato, H.; Kudo, A. Photocatalytic activities of noble metal ion doped $SrTiO_3$ under visible light irradiation. *J. Phys. Chem. B* **2004**, *108*, 8992–8995. [CrossRef]
36. Arunachalam, P.; Ghanem, M.A.; Al-Mayouf, A.M.; Al-shalwi, M. Enhanced electrocatalytic performance of mesoporous nickel-cobalt oxide electrode for methanol oxidation in alkaline solution. *Mater. Lett.* **2017**, *196*, 365–368. [CrossRef]
37. Ghanem, M.A.; Al-Mayouf, A.M.; Arunachalam, P.; Abiti, T. Mesoporous cobalt hydroxide prepared using liquid crystal template for efficient oxygen evolution in alkaline media. *Electrochim. Acta* **2016**, *207*, 177–186. [CrossRef]
38. Theerthagiri, J.; Thiagarajan, K.; Senthilkumar, B.; Khan, Z.; Senthil, R.A.; Arunachalam, P.; Madhavan, J.; Ashokkumar, M. Synthesis of hierarchical cobalt phosphate nanoflakes and their enhanced electrochemical performances for supercapacitor applications. *Chem. Select.* **2017**, *2*, 201–210. [CrossRef]
39. Ai, G.; Mo, R.; Li, H.; Zhong, J. Cobalt phosphate modified TiO_2 nanowire arrays as co-catalysts for solar water splitting. *Nanoscale* **2015**, *7*, 6722–6728. [CrossRef] [PubMed]
40. Jo, W.J.; Jang, J.W.; Kong, K.J.; Kang, H.J.; Kim, J.Y.; Jun, H.; Parmar, K.; Lee, J.S. Phosphate doping into monoclinic $BiVO_4$ for enhanced photoelectrochemical water oxidation activity. *Angew. Chem. Int. Ed.* **2012**, *51*, 3147–3151. [CrossRef] [PubMed]
41. Shaddad, M.N.; Ghanem, M.A.; Al-Mayouf, A.M.; Gimenez, S.; Bisquert, J.; Herraiz-Cardona, I. Cooperative catalytic effect of ZrO_2 and α-Fe_2O_3 nanoparticles on $BiVO_4$ photoanodes for enhanced photoelectrochemical water splitting. *ChemSusChem* **2016**, *9*, 2779–2783. [CrossRef] [PubMed]
42. Kato, H.; Kudo, A. Water splitting into H_2 and O_2 on alkali tantalate photocatalysts $ATaO_3$ (A = Li, Na, and K). *J. Phys. Chem. B* **2001**, *105*, 4285–4292. [CrossRef]
43. Chen, Y.; Liang, S.; Wen, L.; Wu, W.; Yuan, R.; Wang, X.; Wu, L. A TaON nano-photocatalyst with low surface reduction defects for effective mineralization of chlorophenols under visible light irradiation. *Phys. Chem. Chem. Phys.* **2013**, *15*, 12742–12747. [CrossRef] [PubMed]
44. Ma, X.; Chen, Y.; Li, H.; Cui, X.; Lin, Y. Annealing-free synthesis of carbonaceous Nb_2O_5 microspheres by flame thermal method and enhanced photocatalytic activity for hydrogen evolution. *Mater. Res. Bull.* **2015**, *66*, 51–58. [CrossRef]
45. Li, P.; Jin, Z.; Xiao, D. A one-step synthesis of Co–P–B/rGO at room temperature with synergistically enhanced electrocatalytic activity in neutral solution. *J. Mater. Chem. A* **2014**, *2*, 18420–18427. [CrossRef]

46. Pilli, S.K.; Furtak, T.E.; Brown, L.D.; Deutsch, T.G.; Turner, J.A.; Herring, A.M. Cobalt-phosphate (Co-Pi) catalyst modified Mo-doped $BiVO_4$ photoelectrodes for solar water oxidation. *Energy Environ. Sci.* **2011**, *4*, 5028–5034. [CrossRef]
47. McAlpin, J.G.; Surendranath, Y.; Dinca, M.; Stich, T.A.; Stoian, S.A.; Casey, W.H.; Nocera, D.G.; Britt, R.D. EPR evidence for Co (IV) species produced during water oxidation at neutral pH. *J. Am. Chem. Soc.* **2010**, *132*, 6882–6883. [CrossRef] [PubMed]
48. Butler, M.; Ginley, D. Prediction of flatband potentials at semiconductor-electrolyte interfaces from atomic electronegativities. *J. Electrochem. Soc.* **1978**, *125*, 228–232. [CrossRef]
49. Rooke, J.; Weller, M. Synthesis and characterisation of perovskite-type oxynitrides. *Sol. State Phenom.* **2003**, *90*, 417–422. [CrossRef]

© 2018 by the authors. Licensee MDPI, Basel, Switzerland. This article is an open access article distributed under the terms and conditions of the Creative Commons Attribution (CC BY) license (http://creativecommons.org/licenses/by/4.0/).

nanomaterials

Article

In-Situ Synthesis of Hydrogen Titanate Nanotube/Graphene Composites with a Chemically Bonded Interface and Enhanced Visible Photocatalytic Activity

Juan Yang [1,2,*], Jun You [1], Jun Dai [1], Yumei Chen [2,*] and Yao Li [1]

[1] Institute of Chemical Safety, School of Safety Science and Engineering, Henan Polytechnic University, Jiaozuo 454003, China; youjunhpu@163.com (J.Y.); daijun@hpu.edu.cn (J.D.); leayao35@hpu.edu.cn (Y.L.)
[2] Institute of Applied Chemistry, College of Chemistry and Chemical Engineering, Henan Polytechnic University, Jiaozuo 454003, China
* Correspondence: yangjuanhpu@163.com (J.Y.); chenyumei@hpu.edu.cn (Y.C.); Tel.: +86-391-398-7881 (J.Y.)

Received: 21 March 2018; Accepted: 6 April 2018; Published: 8 April 2018

Abstract: Hydrogen titanate nanotube (HTT)/graphene nanocomposites are synthesized by hydrothermal reduction of graphene oxide (GO) and simultaneous preparation of nanotubular HTT via an alkaline hydrothermal process. By using this facile in-situ compositing strategy, HTT are densely supported upon the surface of graphene sheets with close interface contacts. The as-prepared HTT/graphene nanocomposites possess significantly enhanced visible light catalytic activity for the partial oxidation of benzylic alcohols. The amount of graphene has significant influence on catalytic activity and the optimal content of graphene is 1.0 wt %, giving a normalized rate constant k of 1.71×10^{-3} g/m^2·h, which exceeds that of pure HTT and HTT/graphene-1.0% mixed by a factor of 7.1 or 5.2. Other than the general role of graphene as a high-performance electron acceptor or transporter, the observed enhancement in photocatalytic activity over HTT/graphene can be ascribed to the improved interfacial charge migration from enhanced chemical bonding (Ti–C bonds) during the in-situ compositing process. The formation of Ti–C bonds is confirmed by XPS analysis and the resulting enhanced separation of photoinduced charge carriers is demonstrated by electrochemical impedance spectra and transient photocurrent response.

Keywords: photocatalytic selective oxidation; Ti–C bonds; hydrogen titanate; benzylic alcohols; interfacial charge transfer

1. Introduction

In the last several decades, the photocatalytic process in semiconductors has been attracting more and more attention as an advanced green technology for contaminant purification, energy production, and selective organic synthesis [1–3]. The effectiveness of photocatalysis is primarily dependent on three aspects: optical absorption, charge separation, and catalytic reaction upon the surface of the semiconductor catalyst. Further research shows only a small part of photoinduced charges can migrate successfully to the surface of catalysts and initiate catalytic reactions. As a result, the efficient charge transfer on the surface of a semiconductor photocatalyst is widely considered to be very critical, being an elementary process related to photocatalytic application.

Among various photocatalysts, TiO_2-derived nanosized materials appear to be a potential candidate, owing largely to their excellent performance, long term stability, low toxicity, and cost [4,5]. In the field of photocatalysis, hydrogen titanate (HT), especially nanotubular HT has been principally used to fabricate diverse TiO_2 nanostructures, such as TiO_2 nanotubes, nanowires, and nanofibers [6–8]. Their unique one-dimensional structural characteristics provide the preconditions necessary for suitable catalysts under ultraviolet or visible illumination [9,10]. For example, HT calcined at 300–500 °C, possessing TiO_2(B) or anatase phase, exhibits higher catalytic performance for pollutant degradation than P25-TiO_2 [11,12]. However, the inherently rapid recombination of photogenerated charge carriers within TiO_2 or titanate remains problematic. Numerous strategies have been studied to tailor the photocatalytic process and promote catalytic performance. Particularly, fabrication of carbon-titanate heterojunction composites has been identified as one of the most feasible strategies to enhance the separation of electron-hole pairs [13–15].

Owing to the great surface area of graphene and its excellent electronic transport properties, many studies have been carried out on fabricating composite catalysts of graphene and semiconductor particles in an attempt to promote photocatalytic activity [16,17]. Under light illumination, it is found that photoinduced electrons migrate from photocatalysts to the surface of graphene and suppress the recombination of electron-hole pairs, thereby enhancing the oxidative or reductive reactivity of the photocatalytic process [18]. Further research demonstrated that the interfacial action between the semiconductor and graphene can notably influence the interfacial electron migration, which is thought to be a critical step of heterogeneous photocatalysis [19–22]. Zhang et al. have reported the synthesis of graphene-Bi_2WO_6 composites with strong covalent bonding between graphene and Bi_2WO_6 nanoparticles. The improved photocatalytic performance for H_2 and O_2 production were ascribed to the efficient electron migration and separation of photoinduced charge carriers through the chemically bonded interaction [20]. Xie et al. have recently reported the existence of a chemically bonded interface, via the formation of C–Ti bonds, between graphene sheets and TiO_2 nanoparticles. This chemical bonding effect plays a pivotal role in the interfacial electron migration, which makes TiO_2/graphene composites possess an outstanding catalytic activity for the photo-oxidation of formaldehyde, compared to a mechanically mixed sample [22].

The selective oxidation of alcohol to carbonyl compounds is a significant chemical conversion process. Generally, the partial oxidations of alcohols are implemented with hazardous or toxic stoichiometric oxidants under high temperature and pressure conditions [23]. Deteriorating environmental problems caused by chemical production processes have compelled us to seek green and economic alternative reaction systems. Recently, photocatalysis-based organic synthesis reactions have attracted wide concerns because they can be carried out using green oxidants (such as molecular oxygen) and solar energy under ambient temperature or pressure [3,24,25]. Several types of catalysts have been studied to achieve photocatalytic conversion of alcohols to the corresponding carbonyl compounds, including g-C_3N_4, HNb_3O_8 nanosheets, CdS, and TiO_2-based composites [24,26–28]. What's more, the selective oxidation of alcohols can be performed over TiO_2 catalysts using visible light, although the photocatalysts themselves and the reactants can only absorb UV light [26,27]. In our previous report, hydrogen titanate (HT) nanotubes exhibited much higher visible light catalytic performance toward partial oxidation of benzylic alcohols, compared with HT nanosheets and nanofibers [28]. The superior photo-activity of HT nanotubes is primarily ascribed to their one-dimensional structural features, which include a high surface area, abundant unsaturated Ti sites, and efficient charge migration. Developing composite photocatalysts of HT nanotubes and graphene through an in-situ synthetic strategy can further improve catalytic performance for the selective oxidation of alcohols.

However, previously reported research on photocatalytic selective synthesis principally focused on reaction systems using TiO_2 or TiO_2 nanohybrids as photocatalysts [29–31]. The photocatalytic studies on titanate/graphene nanocomposites have largely focused on removing organic pollutants and splitting H_2O for hydrogen production [32,33], whereas the selective chemical conversion using titanate/graphene composite photocatalysts has been scarcely researched. Furthermore, studies on the effects of chemical bonding on interfacial charge transfer and catalytic activity of titanate/graphene nanocomposites for photocatalytic partial oxidation of alcohols are also lacking.

In this paper, HT nanotube (HTT)/graphene nanohybrids with numerous heterogeneous interfaces are prepared via a facile hydrothermal method under basic conditions. Due to their one-dimensional structural characteristics, HT nanotubes can be uniformly coupled onto graphene to form close interfacial contacts. The presence of chemical Ti–C bonds between nanotubular HT and graphene are confirmed by XPS analysis. The HTT/graphene nanocomposites with chemically bonded interfaces exhibit exceptional photocatalytic reactivity toward the partial oxidation of benzylic alcohols under visible illumination. The influence of graphene content on the catalytic performance is investigated and a possible mechanism based on the experimental results is also proposed.

2. Materials and Methods

2.1. Materials

Graphite was purchased from Nanjing XFNANO Materials Tech Co., Ltd. (Nanjing, China). $KMnO_4$, H_2SO_4, H_3PO_4, H_2O_2 (30%), NaOH, P25 TiO_2, and ethanol were obtained from Aladdin Industrial Inc. (Shanghai, China). Benzyl alcohol and its derivatives were purchased from J&K Scientific Ltd. (Beijing, China). 5,5-Dimethyl-1-pyrroline-N-oxide was obtained from TCI Development Ltd. (Shanghai, China). All the other chemicals were of analytical reagent grade and used as received without further purification. Deionized water was obtained using a Millipore Milli-Q system and used throughout the experiments.

2.2. Synthesis of HTT/Graphene Nanohybrids

To improve the hydrophilic and oxidation degree of graphite, a modified method was used to prepare graphite oxide [34]. In brief, a mixed solution of concentrated H_2SO_4/H_3PO_4 (volume ratio, 9:1) was added into a mixture of squama graphite and $KMnO_4$ (weight ratio, 1:6). Then, the obtained mixture was heated for 12 h at 323 K under magnetic stirring. After being cooled down to room temperature, the mixture was poured into an icy solution of 30% H_2O_2. The solid was recycled by filtration, washed by deionized water, and vacuum-dried at 298 K for 12 h. The obtained graphite oxide (GO) was dispersed in water (0.5 mg/mL) and subjected to sonication for 60 min to form an aqueous dispersion of exfoliated GO.

In a typical synthesis process for HTT/graphene nanohybrids, 1.3 g of P25 TiO_2 was added to 65 mL of concentrated NaOH solution (10.0 M) and was stirred for 30 min to form a uniform suspension. Then, a specific amount of GO dispersion was added to the above-obtained suspension under magnetic stirring. The weight ratios of GO to P25 TiO_2 were 0, 0.2, 0.4, 1.0, 2.0, and 4.0%, and the as-prepared samples are labeled as HTT/graphene-x%, where x refers to the mentioned weight ratios. After further stirring for 90 min, the mixture was transferred into a 100 mL stainless steel autoclave and heated at 160 °C for 12 h. The obtained precipitate was separated by centrifugation, washed by deionized water, and re-dispersed for several cycles until a pH value of nearly 8 was obtained. The precipitate was finally washed with absolute ethanol and dried at 90 °C for 10 h to obtain HTT/graphene-x% nanocomposites.

For comparison, bare HTT and graphene samples were prepared using a similar process without GO or P25 TiO_2 and NaOH solution. About 10 mg of graphene was added into 50 mL of deionized water and sonicated for 1 h. Then 1.0 g of as-made HTT was mixed into the graphene suspension under magnetic stirring. After being stirred for 2 h, the homogeneous suspension was treated by centrifugation. The precipitate was washed with deionized water and dried in a vacuum oven at 90 °C for 10 h. The as-obtained product is denoted as HTT/graphene-1.0%-mixed. For the preparation of the P25 TiO_2/graphene-1.0% composite, 1.0 g of P25 TiO_2 was added to 65 mL of deionized water and stirred for 30 min. 10 mL of GO dispersion (1.0 mg/mL) was added into P25 TiO_2 suspension under magnetic stirring. After further stirring for 90 min, the mixture was transferred into a 100 mL Teflon-lined autoclave and heated at 160 °C for 12 h. The obtained precipitate was washed with deionized water and dried at 90 °C overnight. The as-synthesized sample is named as P25 TiO_2/graphene-1.0%.

2.3. Characterization

X-ray diffraction measurements were conducted using a D8 Advance diffractometer (Bruker) with Cu Kα radiation (λ = 0.15418 nm). Raman spectra were recorded on a Labram HR 800 Raman spectrometer (Horiba Jobin Yvon, Paris, France) using 532 nm excitation light at room temperature. The morphological features of HTT/graphene nanohybrids were observed by a transmission electron microscope (TEM, Tecnai G20, FEI, Hillsboro, WA, USA) operating at 200 kV. FTIR spectra were acquired with a Nicolet 750 FTIR spectrometer in the range 400–4000 cm^{-1}. X-ray photoelectron spectra were recorded using a Thermo Scientific ECALAB 250xi system with a monochromated Al Ka X-ray source. UV-visible diffuse reflectance spectra were collected on a UV-vis spectrophotometer (UV-2550, Shimadzu, Kyoto, Japan) using $BaSO_4$ as the reflectance standard. The specific surface area of as-synthesized samples were measured by N_2 adsorption-desorption with a Micrometrics ASAP 2020 apparatus.

Photoelectrochemical experiments were performed in a standard three-electrode system. Platinum foil and Ag/AgCl were used as the counter and reference electrodes, respectively. The working electrode was prepared by the following method: 40 mg of powder sample was dispersed in 0.2 mL of polyethylene glycol solution and ultrasonicated for 45 min to obtain a uniform slurry. The resulting slurry was then spread on a fluorine doped tin oxide (FTO) conducting glass and was dried at 90 °C to get HTT/graphene/FTO electrodes. The photocurrent was recorded at a bias voltage of 0.6 V on an electrochemical workstation (CHI-660C, Shanghai, China). The electrolyte was 0.1 M $LiClO_4$ and nitrogen gas was continuously purged through the electrolyte solution. After being immersed in a benzyl alcohol solution for 1.5 h, the working electrode was irradiated using a 300 W Xe lamp with a 400 nm cut-off filter. Electrochemical impedance spectra (EIS) were measured using the above-mentioned three-electrode cell at 0.0 V with 0.1M Na_2SO_4 aqueous solution. Electron spin resonance (ESR) signals of the radicals trapped by DMPO were collected on a Bruker ER200-SRC spectrometer at ambient temperature.

2.4. Photocatalytic Experiment of Alcohol Oxidation

The partial conversion of benzyl alcohol was performed in a 20 mL round-bottom quartz glass flask. Benzyl alcohol (0.25 mmol) and the photocatalyst (50 mg) were dispersed into 5 mL of acetonitrile and the obtained mixture was bubbled with O_2 for 30 min. This suspension was stirred magnetically for another 60 min to reach the adsorption-desorption equilibrium of the reactant alcohols and catalysts. The mixture was then irradiated using a 300 W Xe lamp with 400 nm cut-off filter under ambient temperature. At certain time intervals, a 0.5 mL mixture was withdrawn and filtered using a Millipore filter (pore size 0.45 μm) to remove the catalyst particles. The concentration changes of reactants in the remaining filtrate were quantitatively analyzed on a gas chromatograph (Agilent 7890) equipped with a standard FID detector and a DM-5 amine capillary column. Oxidation products were identified by comparing the retention times with the standards and further confirmed with GC-MS.

2.5. Radical Scavenging Experiments

In order to elucidate the photocatalytic mechanism, control experiments using different radical scavengers (*tert*-butyl alcohol, triethanolamine, and benzoquinone as the scavenger for the hydroxyl, photogenerated hole, and superoxide radicals, respectively) were performed via an analogous process of benzyl alcohol photo-oxidation, except that the radical scavenger (0.2 mmol) was included. The primary role of the active species were revealed by analyzing the changes of the benzyl alcohol conversion yield.

3. Results and Discussion

3.1. XRD, FTIR and Raman Spectra

Figure 1 shows the XRD patterns of as-synthesized samples including blank HTT and nanocomposites of HTT/graphene with different contents of graphene. Broad diffraction peaks are observed at $2\theta = 10.7°$, $24.7°$, $29.0°$, and $48.5°$, which can be assigned to the characteristic crystalline phase peaks of layered hydrogen titanate [13,35]. For the sample of pure GO (inset of Figure 1), the diffraction peak at $2\theta = 10.5°$ corresponds to the (002) reflection of stacked GO sheets. Additionally, pure reduced GO obtained in the absence of HTT indicates a broad (002) graphene reflection at 26.5°, which clearly demonstrates the conversion of GO to graphene [36]. However, in comparison with the XRD patterns of GO and graphene (inset of Figure 1), no diffraction peaks belonging to GO or graphene are observed in the XRD spectra of HTT/graphene nanocomposites. It is hard to distinguish the characteristic peak at $2\theta = 10.5°$ of pure GO and the broad peak at $2\theta = 10.7°$ assigned to (200) crystallographic planes of hydrogen titanate. The absence of these peaks might also be due to the low weight content and weak diffraction intensity of GO and graphene.

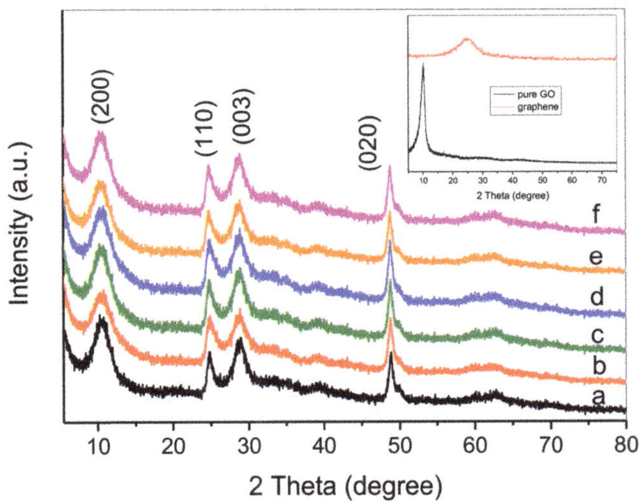

Figure 1. XRD patterns of hydrogen titanate nanotube (HTT)/graphene nanocomposites: (a) pure HTT, (b) HTT/graphene-0.2%, (c) HTT/graphene-0.4%, (d) HTT/graphene-1.0%, (e) HTT/graphene-2.0%, (f) HTT/graphene-4.0%. Inset: XRD patterns of pure GO and graphene.

The presence of graphene in the HTT/graphene nanocomposites could be confirmed using FTIR spectra (Figure 2). Bare GO exhibits several characteristic peaks belonging to oxygen functional groups, such as C=O stretching at 1723 cm^{-1}, O–H deformation at 1389 cm^{-1}, C–OH stretching at 1220 cm^{-1}, and C–O stretching at 1053 cm^{-1} [20,34]. Compared with that of bare GO, the spectrum of HTT/graphene-1.0% shows a significant decline in intensity for peaks associated with oxygen-containing groups (i.e., C=O, O–H, C–OH, and C–O), indicating that GO in the composites is reduced. Additionally, the strong absorption band centered at 490 cm^{-1} can be assigned to the Ti–O–Ti stretching vibration in the pure HTT sample. A slight blue shift can be observed in the FTIR spectrum of HTT/graphene-1.0% (centered at 502 cm^{-1}), suggesting an interfacial interaction between graphene and HTT in the nanocomposites.

Raman spectroscopy is also widely employed to characterize the electronic structure of carbon products. Two bands are observed in the 850–2000 cm^{-1} range for GO and HTT/graphene nanocomposites, as shown in Figure 3. They are the D band and G band of carbonaceous materials [36]. The former denotes a common characteristic of carbon sp^3 defects, and the latter gives information about the in-plane vibrations of sp^2 bonded carbon [21,34]. The intensity ratio between D band and G band (I_D/I_G) represents the relative concentration of defects in GO or graphene. Compared with bare GO, the calculated I_D/I_G of the HTT/graphene composites are lower, demonstrating a lower concentration of defects in graphene. The decreased I_D/I_G ratios in HTT/graphene nanohybrids imply the employed hydrothermal process facilitates the formation of more graphene without reducing the average size of sp^2 domains. Furthermore, it is worth noting that a G band up-shift from 1588 to 1596 cm^{-1} is observed for HTT/graphene compared with GO and graphene, which gives evidence for chemical bonding of carbon materials [37,38]. The results indicate significant interaction between graphene and HTT, consistent with the results of the FTIR investigation.

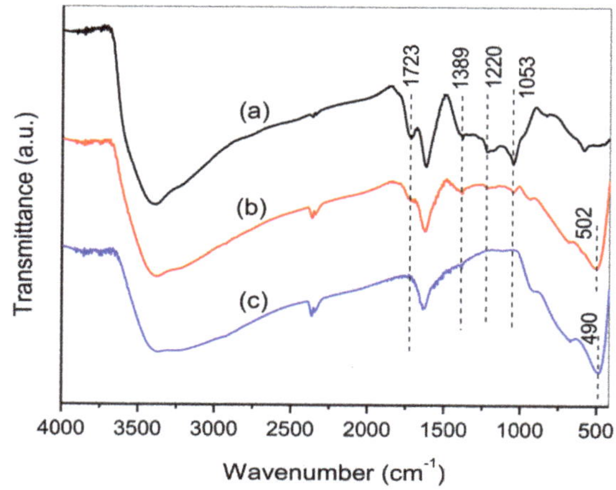

Figure 2. FTIR spectra of samples (a) GO, (b) HTT/graphene-1.0%, and (c) pure HTT.

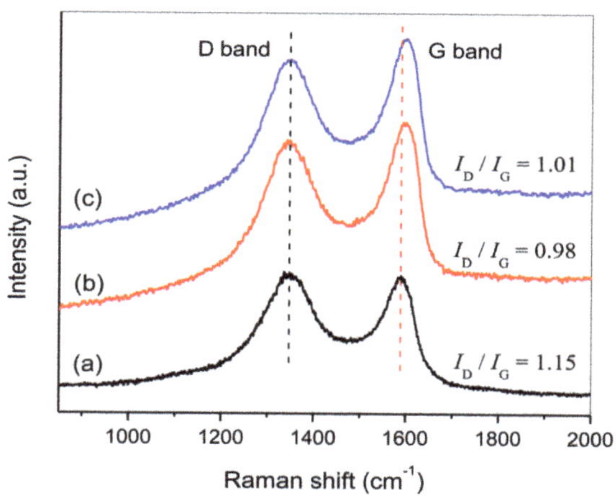

Figure 3. Raman spectra of (a) bare GO, (b) HTT/graphene-1.0%, and (c) HTT/graphene-4.0% composites.

3.2. *TEM and Brunauer–Emmett–Teller (BET) Surface Area Analyses*

The morphologies of HTT/graphene composites can be observed by TEM, as shown in Figure 4. Pure HTT exhibit a tubular framework structure, although, occasionally the aggregation of HTT can be seen (Figure 4a). Upon introducing GO, HTT are anchored onto the graphene sheets after hydrothermal treatment (Figure 4b). The surfaces of graphene sheets are covered by interconnected networks of HT nanotubes. The introduced GO acts as a substrate of the tightly packed HTT. It is noted from Figure 4b, that HTT occupy most of the available surface of graphene, leading to high loading amounts of HTT in as-synthesized nanocomposites. Compared to HTT/graphene-1.0% obtained via the in-situ synthesis strategy, the HTT/graphene-1.0%-mixed sample, prepared by the mechanically mixing of HTT and graphene, exhibits an inhomogeneous distribution of HTT on graphene, as shown in Figure 4c. These results demonstrate that the oxygen-containing functional groups of GO facilitate the nucleation and growth of HTT upon the graphene surface. It is beneficial for HTT to anchor directly on graphene sheets, consequently forming a uniform distribution and close interfacial contact between HTT and graphene.

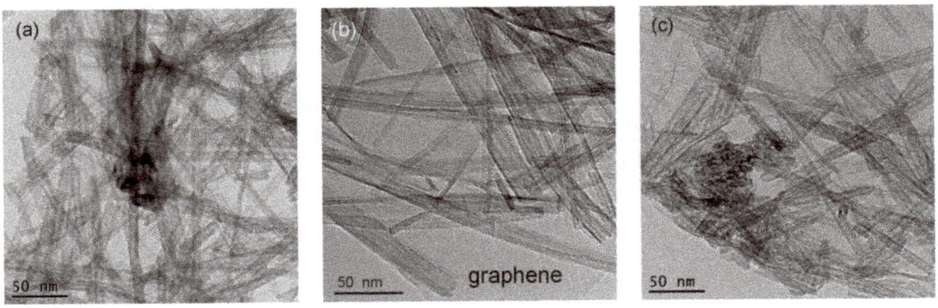

Figure 4. Typical TEM images of (**a**) pure HTT, (**b**) HTT/graphene-1.0%, and (**c**) HTT/graphene-1.0%-mixed nanocomposites.

N_2 adsorption-desorption experiments were used to investigate the textual properties of the as-synthesized nanocomposites. The specific surface area, S_{BET} was calculated according to the classical Brunauer–Emmett–Teller (BET) theory and the corresponding data are summarized in Table 1. Table 1 indicates that all of the HTT/graphene composite samples possess larger S_{BET} than bare HTT, which is due to the huge surface area of graphene [20]. Additionally, the BET surface area of HTT/graphene hybrids increases with an increasing amount of graphene, reconfirming the influence of graphene in improving the total surface area of HTT/graphene samples. For instance, S_{BET} increases from 228.9 to 248.0 $m^2 \cdot g^{-1}$ when the content of graphene increases from 0.2% to 4.0% for as-synthesized HTT/graphene composites. There is no doubt that the enhanced S_{BET} can provide more surface active sites and facilitate the adsorption of the reactants, thus promoting heterogeneous catalytic reactions.

Table 1. Brunauer–Emmett–Teller (BET) surface area of as-prepared HTT/graphene composites and S_{BET} normalized rate constants *k* for the selective oxidation of benzyl alcohol.

Catalyst	BET Surface Area/$m^2 \cdot g^{-1}$	Conv./%	Rate Constant k/h^{-1}	S_{BET} Normalized Rate Constants k
pure HTT	225.1	27.6	0.053	0.24×10^{-3}
HTT/graphene-0.2%	228.9	57.1	0.141	0.61×10^{-3}
HTT/graphene-0.4%	234.4	78.0	0.252	1.08×10^{-3}

Table 1. *Cont.*

Catalyst	BET Surface Area/m^2·g^{-1}	Conv./%	Rate Constant k/h^{-1}	S_{BET} Normalized Rate Constants k
HTT/graphene-1.0%	240.5	91.8	0.411	1.71×10^{-3}
HTT/graphene-2.0%	244.7	69.2	0.195	0.80×10^{-3}
HTT/graphene-4.0%	248.0	43.8	0.097	0.39×10^{-3}
HTT/graphene-1.0%-mixed	233.6	32.3	0.078	0.33×10^{-3}

Reaction conditions: catalyst 50 mg, benzylic alcohol 0.25 mmol, acetonitrile 5 mL, reaction temperature 298 K, visible light irradiation 6 h.

3.3. *XPS Analysis*

The interactions between HTT and graphene are further studied by XPS. The high-resolution C 1s XPS spectrum of bare GO is shown in Figure 5a. Two characteristic peaks can be observed at 285.0 and 287.1 eV, which correspond to adventitious carbon or sp^2-hybridized carbon of GO and the oxygen-containing carbonaceous bands (C–OH) [20,39]. These oxygen-containing bands can provide active anchoring sites for HTT growth. The C 1s spectrum of the HTT/graphene-1.0%-mixed sample is deconvoluted into three peaks at binding energies of 284.6, 286.5, and 288.6 eV. The C 1s peaks located at 284.6 and 286.5 eV can be assigned to C–O or C=O groups (Figure 5b). The relative weak peak at 288.6 eV belongs to the O=C–O group. These results indicate that the –OH groups of HTT are likely to interact with the surface –COOH groups of GO via forming O=C–O bonds. Additionally, it can be seen the peak at 287.1 eV, belonging to the C–OH groups of GO, decreases significantly or disappears in the C 1s XPS spectra of HTT/graphene composites. This further demonstrates the successful reduction of GO to graphene through the hydrothermal process. The C 1s spectrum of in-situ synthesized HTT/graphene-1.0% can be deconvoluted into four peaks (Figure 5c). Besides the three peaks at 284.6, 286.5, and 288.6 eV— belonging to the C–C, C–O, and O=C–O groups—another shoulder peak centered at 283.2 eV is observed, which is attributed to forming chemical bonds (Ti–C) between carbon atoms and titanium atoms in the lattice of HTT. This is the first report on the formation of Ti–C bonds in nanocomposites of hydrogen titanate and graphene. The intimate interfacial contacts originated from the chemical bonding would benefit the migration of photogenerated charges and thus improve the photocatalytic activity of HTT.

Forming Ti–C bonds is also investigated and identified by analyzing the Ti 2p XPS spectra of the HTT/graphene nanocomposites. As depicted in Figure 6a, in the Ti 2p spectrum of HTT/graphene-1.0%-mixed, two peaks are observed at binding energies of 464.4 and 458.7 eV. These peaks can be attributable to the Ti $2p_{1/2}$ and Ti $2p_{3/2}$ spin-orbital splitting photoelectrons of Ti^{4+}. Two bands at binding energies of 463.3 and 457.7 eV usually correspond to Ti^{3+} [21,36]. In Figure 6b, besides the above-mentioned four characteristic peaks belonging to Ti^{4+} and Ti^{3+} in HTT, two other weak peaks located at 465.3 and 459.4 eV are found and are likely due to the forming of Ti–C bonds in the HTT/graphene nanohybrids. This suggests that a chemical interaction between HTT and graphene is present in as-synthesized HTT/graphene composites and absent in the mixed samples.

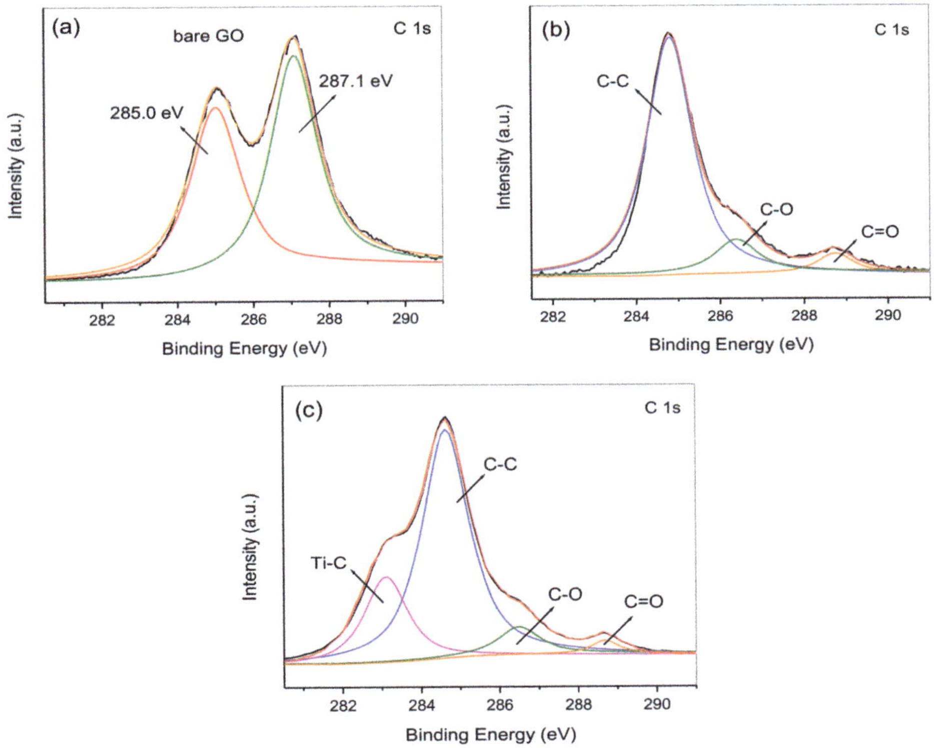

Figure 5. C 1s XPS spectra of (**a**) bare GO, (**b**) HTT/graphene-1.0%-mixed, and (**c**) HTT/graphene-1.0% chemically bonded nanocomposites.

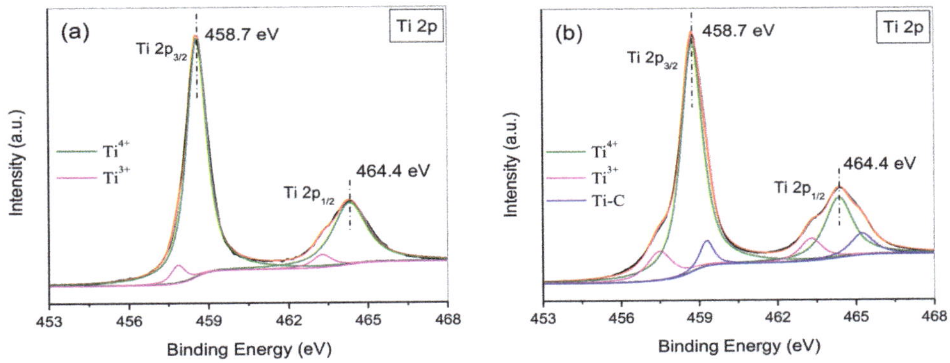

Figure 6. Ti 2p XPS spectra of (**a**) HTT/graphene-1.0%-mixed and (**b**) HTT/graphene-1.0% chemically bonded nanocomposites.

3.4. Enhanced Visible Light Catalytic Activities

To investigate the photocatalytic performance of HTT/graphene nanohybrids, the selective oxidation of benzylic alcohols was chosen as a model reaction. No product of aldehyde was obtained without visible radiation or the photocatalyst, indicating that the selective oxidation of benzyl alcohol is conducted through the photocatalytic reaction. Figure 7 shows a comparison of photocatalytic oxidation activity of pure HTT, HTT/graphene-0.2%, HTT/graphene-0.4%, HTT/graphene-1.0%, HTT/graphene-2.0%, HTT/graphene-4.0%, HTT/graphene-1.0%-mixed, and P25 TiO_2/graphene -1.0%. As depicted in Figure 7, without graphene, pure HTT shows moderate photocatalytic activity and 27.6% of benzyl alcohol was converted into benzaldehyde after 6 h of visible light irradiation. Even when the weight content of graphene is low (0~4.0 wt %), HTT/graphene nanohybrids synthesized using the in-situ compositing approach exhibit remarkably enhanced photocatalytic activities, and the prepared photocatalysts possess high selectivity (>95%) for the conversion of benzyl alcohol to benzaldehyde. The photo-oxidation activity of HTT/graphene increases with graphene amount from 0.2 to 1.0 wt %. The optimal conversion yield of benzyl alcohol, obtained in the sample HTT/graphene-1.0%, is 91.8%, which exceeds by 3.3 times that of pure HTT. To our knowledge, this is the first report demonstrating that hydrogen titanate/graphene composites serve as efficient visible light catalysts toward the partial oxidation of benzylic alcohols. When the content of graphene is above 1.0%, further increasing the graphene content results in a reduction of photo-activity. For instance, the conversion yield of benzyl alcohol over HTT/graphene-4.0% has a sharp decline. This may be due to the increased turbidity and light scattering of HTT/graphene-4.0%. Additionally, high loading of graphene shields HTT from absorbing light irradiation, which leads to an obvious reduction of illumination passing through the reaction suspension. The results are analogous to the previous studies of TiO_2/graphene nanohybrids for contaminant degradation and hydrogen production [19,20].

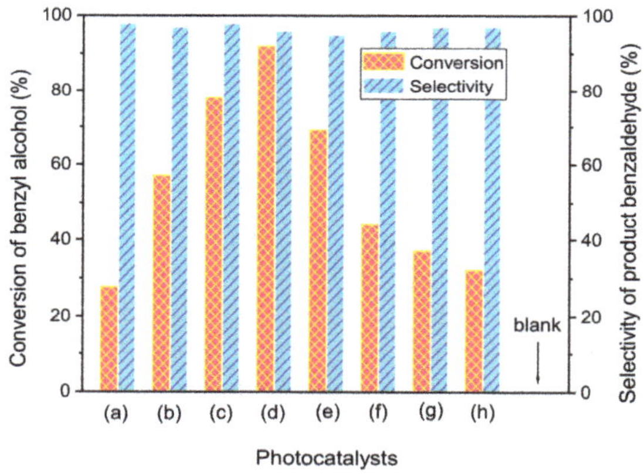

Figure 7. The selective oxidation of benzyl alcohol over HTT/graphene nanocomposites under visible light irradiation: (a) pure HTT, (b) HTT/graphene-0.2%, (c) HTT/graphene-0.4%, (d) HTT/graphene-1.0%, (e) HTT/graphene-2.0%, (f) HTT/graphene-4.0%, (g) P25 TiO_2/graphene-1.0%, and (h) HTT/graphene-1.0%-mixed. The blank column represents the experimental results in the absence of visible irradiation or photocatalyst.

For comparison, the selective conversion of benzyl alcohol in P25 TiO_2/graphene-1.0% and HTT/graphene-1.0%-mixed have also been tested and analyzed under the same experimental conditions. The sample of P25 TiO_2/graphene-1.0% shows dramatically lower activity (37.0% of alcohol is oxidized) than the HTT/graphene-1.0% composites, suggesting that more direct contact and the resulting strong interactions between HTT and graphene sheets are critical for the enhancement of photocatalytic activity because of improved electron/hole separation and migration. The physical mixed sample (HTT/graphene-1.0%-mixed) shows a significantly lower conversion yield than that of HTT/graphene-1.0% synthesized via a one-step hydrothermal process. This suggests that simple mechanical-mixing with graphene has a small effect on improving the photocatalytic activity of HTT, and is due to the absence of a chemical-bonding interface between HTT and graphene. This also indicates that the interaction state between HTT and graphene has crucial influence on photocatalytic performance of HTT/graphene nanohybrids.

Additionally, the reaction kinetics of benzyl alcohol oxidation were investigated and the corresponding results are shown in Figure 8. As the reaction time is prolonged, slight decreases in the conversion yields of benzyl alcohol occur. Derived from the experimental data in Figure 8A, the kinetics of benzyl alcohol oxidation are shown in Figure 8B. A linear correlation between $\ln(C_0/C_t)$ (C_0 and C_t are the concentrations of benzyl alcohol at the initial time and at specific time intervals) and the reaction time is displayed, indicating the oxidation is pseudo-first order. The pseudo-first-order rate constant (k) of the HTT/graphene photocatalysts can be obtained from the slope. The rate constant (k) of benzyl alcohol oxidation over the optimal photocatalyst, HTT/graphene-1.0%, is 0.411 h^{-1}, which is 5.3 and 6.4 times higher than those of the P25 TiO_2/graphene-1.0% and HTT/graphene-1.0% -mixed samples.

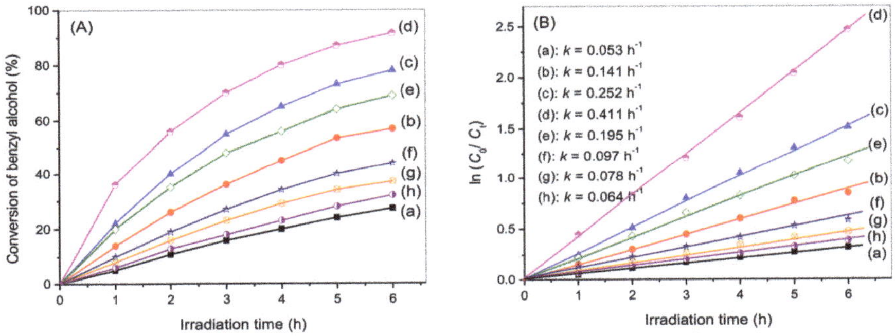

Figure 8. (**A**) Time involved photocatalytic conversion of benzyl alcohol over as-synthesized HTT/graphene nanocomposites: (a) pure HTT, (b) HTT/graphene-0.2%, (c) HTT/graphene-0.4%, (d) HTT/graphene-1.0%, (e) HTT/graphene-2.0%, (f) HTT/graphene -4.0%, (g) P25 TiO_2/graphene-1.0%, and (h) HTT/graphene-1.0%-mixed; (**B**) The corresponding first-order kinetics plots over these photocatalysts.

To investigate the general applicability of HTT/graphene composites, the oxidation of various benzylic alcohols over the optimal HTT/graphene-1.0% were examined and the corresponding results are displayed in Table 2. It can be seen from Table 2 that the selective conversion of various benzylic alcohol derivatives proceeds, and the desirable aldehyde products are obtained in high selectivity (entries 1–5) under visible radiation. Additionally, it can be noted that the electron-donating substituents (CH_3– and CH_3O–) facilitate the formation of aldehyde products; for example, 97.6% of 4-methoxybenzyl alcohol can be converted in 4.5 h with 96% selectivity. Whereas, the electron-withdrawing substituents (such as F– and Cl–) retard the conversion of alcohols. This suggests that the existence of an electron-donating group at the *p*-position of benzyl alcohol benefits the selective conversion over an HTT/graphene photocatalyst.

Table 2. Photocatalytic oxidation of various benzylic alcohols over HTT/graphene-1.0%.

Entry	Substrate	Product	Time (h)	Conv. (%)	Sel. (%)
1	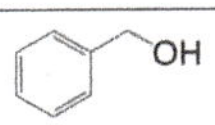	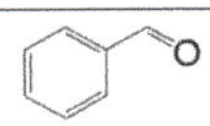	6	91.8	96
2	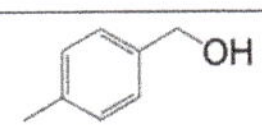	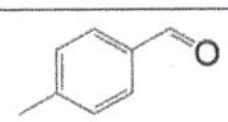	5	92.5	97

Table 2. *Cont.*

Entry	Substrate	Product	Time (h)	Conv. (%)	Sel. (%)
3	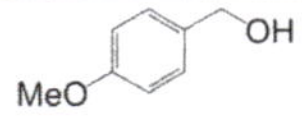	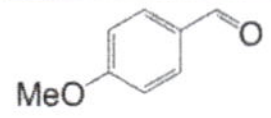	4.5	97.6	96
4	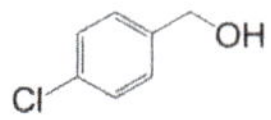	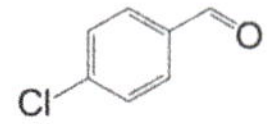	7	89.2	97
5	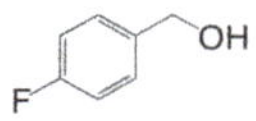	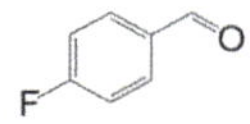	9	84.9	98

Reaction conditions: catalyst 50 mg, benzylic alcohol 0.25 mmol, acetonitrile 5 mL, reaction temperature 298 K, visible light irradiation (λ > 400 nm).

3.5. Stability of Composite Photocatalysts

The stability or recyclability of catalysts is another important practical issue. In order to study the stability and durability of HTT/graphene composites, cycle experiments are performed. After the catalytic reactions, HTT/graphene-1.0% photocatalysts are separated and recovered by washing, and are reused. As depicted in Figure 9, there is only a small loss of activity and no obvious change of selectivity after five consecutive runs for partial photo-oxidation of benzyl alcohol. These results indicate that HTT/graphene is an efficient and stable photocatalyst for selective conversion of benzyl alcohol under the applied experimental conditions.

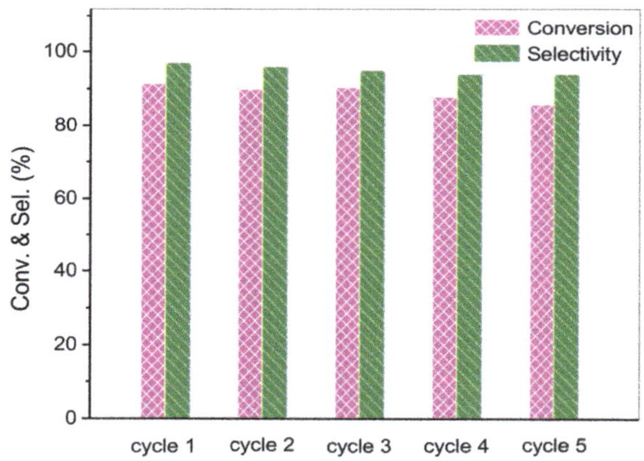

Figure 9. Recycled photoactivity test for five operational runs over the optimal HTT/graphene-1.0%, on selective oxidation of benzyl alcohol.

3.6. Visible Light Absorption Generated by Benzyl Alcohol Surface Adsorption

Surface adsorption generally plays a critical role toward photocatalysis-driven selective oxidation of alcohols. The surface adsorption of benzyl alcohol on HTT/graphene composites is first investigated using FTIR spectra (Figure 10A). In comparison with free benzyl alcohol molecules, the characteristic peaks assigned to C–O and O–H groups shift to lower wavenumbers (at 1036 and 1363 cm^{-1}) after being adsorbed over HTT/graphene-1.0%. On the other hand, the characteristic peaks belonging to skeletal vibrations of the phenyl ring (at 1495 and 1451 cm^{-1}) and the C–C stretching vibration (at 1207 cm^{-1}) show no changes [26,28]. These results imply that benzyl alcohol is adsorbed on the surface of HTT/graphene and the –CH_2OH groups of the alcohol molecules interact with the unsaturated Ti sites of HTT, forming alcoholate species –CH_2–O(H)–Ti.

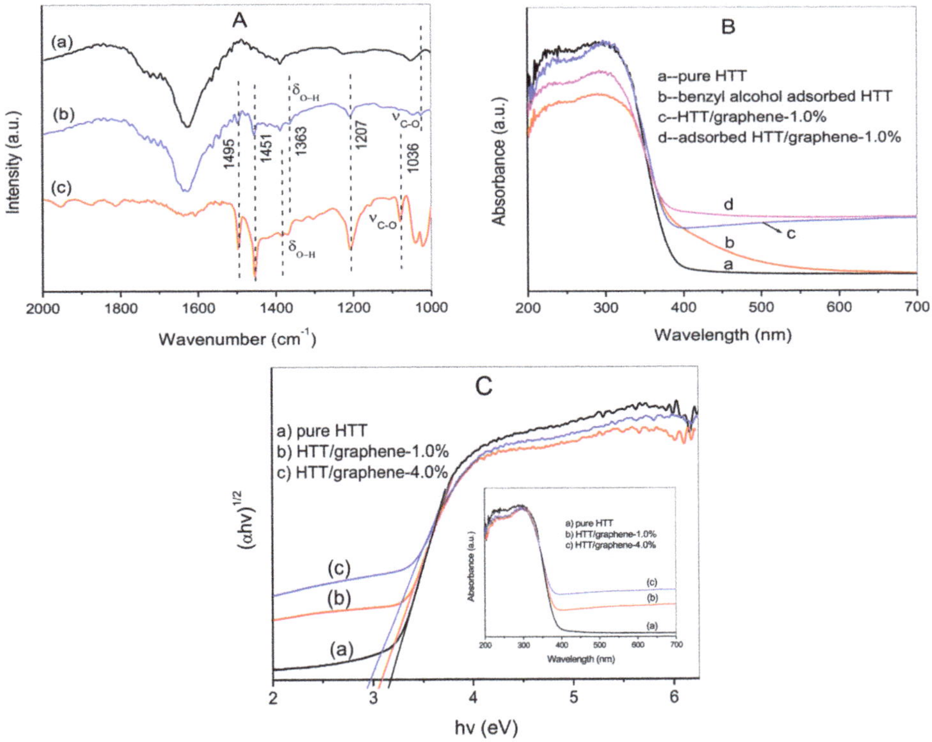

Figure 10. (**A**) FTIR spectra of (a) HTT/graphene-1.0%, (b) benzyl alcohol adsorbed HTT/graphene-1.0% and (c) free benzyl alcohol; (**B**) UV-vis DRS of pure HTT, benzyl alcohol adsorbed HTT, HTT/graphene-1.0%, and benzyl alcohol adsorbed HTT/graphene-1.0%; (**C**) Plot of transformed Kubelka–Munk function versus the energy of light (Inset: the corresponding UV-vis DRS).

The formation of surface complexes –CH_2–O(H)–Ti can produce visible absorption, which provides a prior condition for the selective conversion of benzylic alcohol to product aldehyde. This has been previously shown in research on photocatalytic reactions over TiO_2 or Nb_2O_5 [40,41]. Based on the literature, surface-unsaturated metals (such as Ti or Nb) can serve as Lewis acid sites and the alcohol molecules are adsorbed via Lewis acid-base interactions. The visible light absorption can be attributed to the ligand-to-metal charge transfer (LMCT) of in-situ formed surface complexes. The visible light absorption caused by the surface adsorption of benzyl alcohol over pure HTT and HTT/graphene-1.0% is indicated in Figure 10B. As presented in Figure 10B, the pure HTT sample shows no visible light response, whereas the benzyl alcohol adsorbed HTT exhibits obvious visible light absorption. This absorption can be ascribed to the formation of surface complexes between adsorbed benzyl alcohol and the unsaturated surface Ti atoms of HTT. It can be concluded from Figure 10B that the incorporation of graphene results in continuous absorption bands in visible light range for HTT/graphene composites, which is analogous to results on TiO_2/graphene nanohybrids [42]. Although the HTT/graphene nanocomposites extend the wide background absorption in visible light region, additional visible light absorption can be observed for benzyl alcohol adsorbed composites. This demonstrates that the introduction of graphene has no influence on the formation of surface complexes.

The band gap energy of the photocatalyst is an important factor affecting the catalytic performance, especially for a visible light photocatalytic system. The band gap energies of pure HTT, HTT/graphene-1.0%, and HTT/graphene-4.0% are estimated through a Tauc plot of the modified

Kubelka–Munk function with linear extrapolations (Figure 10C). The calculated band gap energies are 3.16, 3.09, 2.98 eV for pure HTT, HTT/graphene-1.0%, and HTT/graphene-4.0%, respectively. The reduction in band gap energy can be ascribed to chemical bonding between HTT and the graphene support. As shown in the inset of Figure 10C, a slight red-shift in the absorption edge of the HTT/graphene composites is observed. For instance, the absorption edge of HTT/graphene-1.0% is about 401 nm, which implies that HTT/graphene-1.0% can hardly absorb visible light. That is, the red-shift originating from the chemical bonding between HTT and graphene is insufficient to bring forth excellent catalytic activity for the selective oxidation of benzylic alcohols. Hence, the visible light oxidation of benzylic alcohols over HTT/graphene nanocomposites is mainly driven by the formation of surface complexes.

3.7. Improved Charge Transfer via the Chemically Bonded Interface between HTT and Graphene

To investigate visible light induced charge transfer during the photo-oxidation of benzylic alcohols over HTT/graphene catalysts, the transient photocurrent responses were recorded over working electrodes made of pure HTT, HTT/graphene-1.0%-mixed and in-situ compositing HTT/graphene-1.0%. Figure 11A presents the *I-t* curves of the above-mentioned three working electrodes in a benzyl alcohol containing electrolyte solution under visible light illumination. The photocurrent is primarily determined by the charge transfer from the adsorbed benzyl alcohols to the HTT conduction band and the subsequent graphene surface. The photocurrent density of the HTT/graphene-1.0% electrode is enhanced ~4.6 times that of pure HTT, indicating the separation efficiency of photoinduced charges can be improved by the interfacial electronic interactions between HTT and graphene. Moreover, the photocurrent density of the in-situ composited HTT/graphene-1.0% is enhanced ~2.3 times that of the mechanical-mixed hybrid (HTT/graphene-1.0%-mixed), which is primarily ascribed to the chemical interactions between HTT and the graphene sheets. Additionally, for the HTT/graphene-1.0% electrode, the photocurrent density in the presence of benzyl alcohol is predominantly higher than that in its absence. This result implies the generation of anodic photocurrent is closely related to the surface adsorption of benzyl alcohol upon the HTT/graphene-1.0% electrode. The photocurrent experiments offer direct evidence for electron migration from the adsorbed benzyl alcohol to the conduction band of HTT and subsequently to graphene surface under visible light.

Based on the experimental data in Figure 11A and Table 1, the photocurrent results are in accordance with the S_{BET} normalized rate constants of benzyl alcohol oxidation. This suggests that visible light induced charge transfer plays a pivotal role in the photocatalytic conversion of benzylic alcohol. For graphene-based photocatalytic systems, the enhanced charge separation efficiency is dominated by the high spatial charge transportation properties of graphene [43,44]. The extent of interaction between graphene and semiconductor catalysts determines the electron migration, further influencing the photocatalytic activity [19–22]. The formation of a chemically bonded interface between catalyst nanoparticles and graphene can improve the photocatalytic performance of semiconductor catalysts [19–22]. Based on the above characterization results (Figures 5, 6 and 11A) and photocatalytic experimental data (Figure 8 and Table 1), an efficient interfacial charge transfer based on a chemical-bonding effect is proposed to clarify the promoted catalytic activity of HTT/graphene nanocomposites, especially compared to mechanically mixed samples. The improved charge separation caused by the interaction extent between HTT and graphene sheets can be further investigated by EIS.

Figure 11B compares the EIS Nyquist curves of the in situ-synthesized and mechanically mixed HTT/graphene composites under irradiation by a xenon lamp. Figure 11B shows only one semicircle on the EIS plan, which suggests that charge migration is occurring, and the supposed equivalent circuit is given in the inset of Figure 11B. The charge transfer resistances of the pure HTT, HTT/graphene-1.0%-mixed, and HTT/graphene-1.0% samples are 4.9×10^5, 2.2×10^5, and 1.0×10^5 ohm, determined by fitting with the equivalent circuit. This demonstrates that the introduction of graphene facilitates the charge transfer of HTT. Furthermore, the arc radius of the EIS Nyquist curve for HTT/graphene-1.0% is smaller than that of both the physically mixed

(HTT/graphene-1.0%-mixed) and pure HTT samples. Generally, the arc radius of an EIS spectrum reflects the resistance of the interface layer at the surface of the electrodes [37,45]. The smaller semicircle size represents more efficient separation of electrons/holes and more rapid interfacial charge transfer [37,45]. These results clearly reveal that the chemical bonding of HTT and graphene can efficiently promote the migration and separation of photogenerated charges, contributing to improve the catalytic performance for selective oxidation of benzylic alcohols.

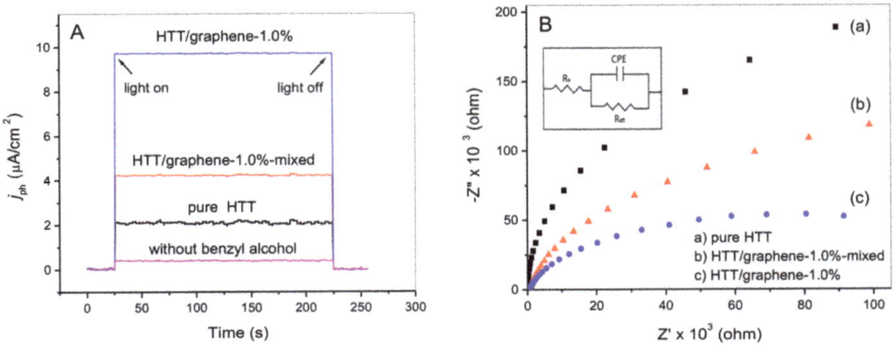

Figure 11. (**A**) Photocurrent curves of pure HTT, HTT/graphene-1.0%-mixed, and HTT/graphene-1.0% electrodes under visible light irradiation; the violet curve represents the photocurrent response over HTT/graphene-1.0% electrode without benzyl alcohol in the electrolyte solution. (**B**) Electrochemical impedance spectra (EIS) Nyquist plots of the pure HTT, HTT/graphene-1.0%-mixed, and HTT/graphene-1.0% samples under light irradiation with a Xe lamp. Inset: Electrical equivalent circuit supposed for fitting of impedance spectra. R_S, CPE, and R_{ct} represent the electrolyte resistance, space charge capacitance, and the charge transfer resistance, respectively.

3.8. Radical Trapping Experiments

The involved catalytic mechanism of alcohol photo-oxidation on HTT/graphene composites is further studied. Control experiments on radical trapping were carried out and the obtained results are illustrated in Figure 12A. When adding *tert*-butyl alcohol (TBA) to the reaction solution [46], the conversion of benzyl alcohol is almost unchanged, indicating ·OH has little effect on the selective conversion of benzyl alcohol. Removing O_2 by sparging with N_2 results in a low conversion of benzyl alcohol (ca. 15%), indicating that O_2 plays a critical role in photocatalytic oxidation of benzyl alcohol. When triethanolamine (TEA, a scavenger of photohole h^+) is put into the reaction system, the oxidation of benzyl alcohol is remarkably inhibited, implying that partial oxidation of benzyl alcohol is primarily driven by photo-holes. Additionally, the introduction of benzoquinone (BQ) results in an obvious decrease in the conversion of benzyl alcohol, suggesting that $O_2^{\cdot-}$ is also involved during the photo-oxidation of benzyl alcohol. To confirm the formation of $O_2^{\cdot-}$, active radical species generated over HTT/graphene-1.0% are detected by ESR using DMPO as the spin-trapping agent. As depicted in Figure 12B, after irradiation of the HTT suspension for 5 min, the characteristic spectrum of the adduct DMPO/·OOH formed between DMPO and superoxide radicals can be clearly observed. In the presence of HTT/graphene-1.0%, the signal intensity of DMPO/·OOH increased. These results indicate that superoxide radicals can be formed upon pure HTT or HTT/graphene under light illumination, and that introduction of graphene significantly enhances the formation of $O_2^{\cdot-}$.

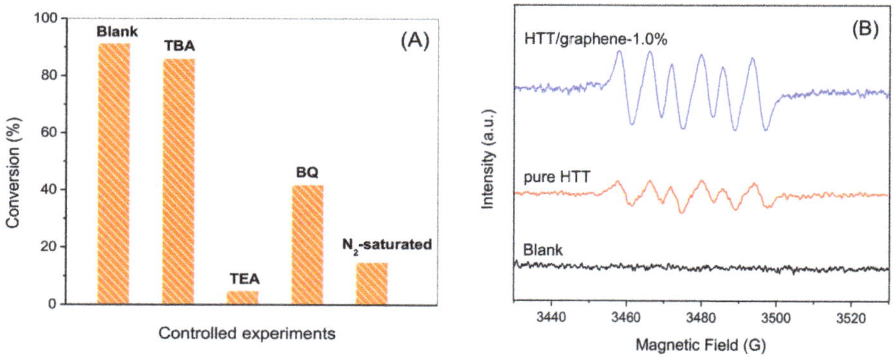

Figure 12. (**A**) Control experiments of photocatalytic selective oxidation of benzyl alcohol in the presence of various scavengers or in the absence of O_2 over HTT/graphene-1.0% under visible light irradiation for 6 h. (**B**) Electron spin resonance (ESR) spectra of $O_2^{\cdot-}$ trapped by DMPO over pure HTT and HTT/graphene-1.0% suspensions in methanol solution, blank represents both the sample containing DMPO alone under light illumination and the sample containing DMPO and catalysts in dark.

3.9. Mechanism of Improved Visible Light Catalytic Activity

According to these results, a possible mechanism for the improved visible light catalytic activity over as-prepared HTT/graphene composites is proposed and depicted in Figure 13. Reactant benzylic alcohol is first adsorbed on unsaturated Ti site of HTT and generates the surface complex. A visible light induced LMCT process occurs, which produces an electron and a hole upon the in-situ formed surface complex. The photogenerated electron transfers into the conduction band of neighbor HTT and subsequently the surface of the graphene sheet via the in-situ formed Ti–C bonds, which can be trapped by adsorbed O_2 to produce a superoxide anion radical. Meanwhile, the adsorbed benzylic alcohol transforms into an intermediate with a positive charge. Subsequently, the photogenerated hole drives the α-deprotonation of the intermediate and gives rise to a neutral carbon radical. Then the reaction between the substrate radical and superoxide species yields the product aldehyde. Compared to the HTT/graphene-mixed samples, the in-situ composited HTT/graphene catalysts possess significantly enhanced photocatalytic activity for the selective oxidation of benzylic alcohols under visible light illumination, as indicated in Figure 8 and Table 1. It can be mainly ascribed to the chemical-bonding effect and the resulting intimate interactions between HTT and graphene sheets, which facilitate the interfacial electron transfer from adsorbed alcohols to HTT and finally to graphene. The close interfacial contacts between graphene and HTT are not being able to form in the mechanical mixed composites. In brief, the improved oxidation efficiency of as-synthesized HTT/graphene composites is primarily ascribed to the interfacial charge transfer effect, which inhibits the recombination of photoinduced electron-hole pairs and thus enhances the photoactivity of alcohol oxidation.

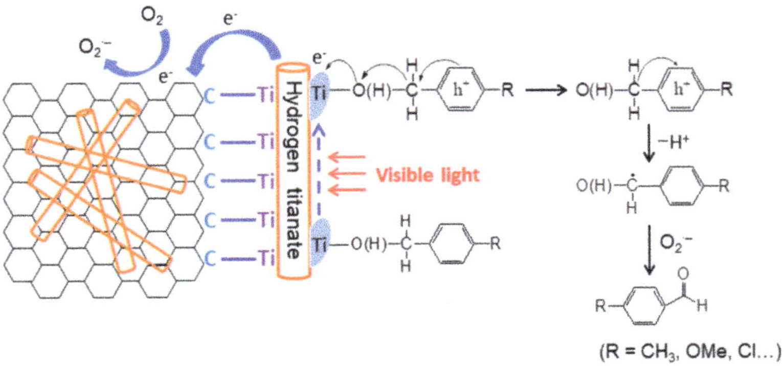

Figure 13. Schematic illustrations for the efficient interfacial charge transfer effect over HTT/graphene composites toward the selective oxidation of benzylic alcohols under visible light irradiation.

4. Conclusions

In summary, HTT/graphene nanohybrids with superior photocatalytic activity were fabricated through a facile hydrothermal process. Nanotubular HTT are densely supported upon the surface of graphene with close interfacial contacts. Introducing graphene can markedly improve the catalytic performance for the selective conversion of benzylic alcohols under visible illumination and the contents of graphene have significant effects on the catalytic activity of HTT. The optimal content of graphene is 1.0 wt % and the corresponding normalized rate constant k of benzyl alcohol oxidation is 1.71×10^{-3} g/m^2·h, which exceeds that of pure HTT by more than seven times. Further characterization results indicate that HTT are chemically bonded with graphene surfaces via Ti–C bonds, which is supported by XPS analysis. More importantly, the chemical-bonded nanocomposite exhibits a remarkable enhancement of photocatalytic activity, which is 5.2 times that of a mechanical-mixed sample. Based on the results of transient photocurrent and EIS, this enhancement can be ascribed to the interfacial charge transfer effect, which provides favorable spatial conditions for electrons migration from in-situ generated surface complexes to neighboring HTT and finally to graphene. This effect inhibits the recombination of electrons and holes, and thus yields superior photocatalytic performance. Our study on the efficient interfacial charge transfer via a chemically bonded interface provides new insights for guiding the design of novel photocatalytic reaction systems, particularly for solar-light driven selective organic synthesis.

Acknowledgments: This works has been supported by the National Nature Science Foundation of China (21307027), the Natural Science Foundation of Henan Province (14A440007), the Funding Scheme for the Young Backbone Teachers of Higher Education Institutions in Henan Province (2015GGJS-071) and the Foundation for Distinguished Young Scientists of Henan Polytechnic University (J2016-4).

Author Contributions: Juan Yang proposed and planned the research, and supervised the experiments. Yumei Chen designed the synthesis of photocatalysts. Jun You and Jun Dai prepared the photocatalysts, and preformed the characterization and photocatalytic measurements. Jun Dai and Yao Li improved the data analysis. All the authors collaborated and participated in the writing of the manuscript.

Conflicts of Interest: The authors declare no conflict of interest.

References

1. Chen, C.C.; Ma, W.H.; Zhao, J.C. Semiconductor-mediated photodegradation of pollutants under visible-light irradiation. *Chem. Soc. Rev.* **2010**, *39*, 4206–4219. [CrossRef]
2. Kamat, P.V. Meeting the clean energy demand: Nanostructure architectures for solar energy conversion. *J. Phys. Chem. C* **2007**, *111*, 2834–2860. [CrossRef]

3. Lang, X.J.; Ma, W.H.; Chen, C.C.; Ji, H.W.; Zhao, J.C. Selective aerobic oxidation mediated by TiO_2 photocatalysis. *Acc. Chem. Res.* **2014**, *47*, 355–363. [CrossRef] [PubMed]
4. Humayun, M.; Li, Z.J.; Sun, L.Q.; Zhang, X.L.; Raziq, F.; Zada, A.; Qu, Y.; Jing, L.Q. Coupling of nanocrystalline anatase TiO_2 to aporous nanosized $LaFeO_3$ for efficient visible-light photocatalytic degradation of pollutants. *Nanomaterials* **2016**, *6*, 22. [CrossRef] [PubMed]
5. Yurdakal, S.; Palmisano, G.; Loddo, V.; Augugliaro, V.; Palmisano, L. Nanostructured rutile TiO_2 for selective photocatalytic oxidation of aromatic alcohols to aldehydes in water. *J. Am. Chem. Soc.* **2008**, *130*, 1568–1569. [CrossRef] [PubMed]
6. Yang, D.J.; Liu, H.W.; Zheng, Z.F.; Yuan, Y.; Zhao, J.C.; Waclawik, E.R.; Ke, X.B.; Zhu, H.Y. An efficient photocatalyst structure: TiO_2(B) nanofibers with a shell of anatase nanocrystals. *J. Am. Chem. Soc.* **2009**, *131*, 17885–17893. [CrossRef] [PubMed]
7. Liu, Y.X.; Wang, Z.L.; Wang, W.D.; Huang, W.X. Engineering highly active TiO_2 photocatalysts via the surface-phase junction strategy employing a titanate nanotube precursor. *J. Catal.* **2014**, *310*, 16–23. [CrossRef]
8. Zheng, Z.K.; Huang, B.B.; Lu, J.B.; Wang, Z.Y.; Qin, X.Y.; Zhang, X.Y.; Dai, Y.; Whangbo, M.H. Hydrogenated titania: Synergy of surface modification and morphology improvement for enhanced photocatalytic activity. *Chem. Commun.* **2012**, *28*, 5733–5735. [CrossRef]
9. Chatterjee, S.; Bhattacharyya, K.; Ayyub, P.; Tyagi, A.K. Photocatalytic properties of one-dimensional nanostructured titanates. *J. Phys. Chem. C* **2010**, *114*, 9424–9430. [CrossRef]
10. Yu, J.G.; Xiang, Q.J.; Zhou, M.H. Preparation, characterization and visible-light-driven photocatalytic activity of Fe-doped titania nanorods and first-principles study for electronic structures. *Appl. Catal. B-Environ.* **2009**, *90*, 595–602. [CrossRef]
11. Zheng, Z.F.; Liu, H.W.; Ye, J.P.; Zhao, J.C.; Waclawika, E.R.; Zhu, H.Y. Structure and contribution to photocatalytic activity of the interfaces in nanofibers with mixed anatase and TiO_2(B) phases. *J. Mol. Catal. A-Chem.* **2010**, *316*, 75–82. [CrossRef]
12. Turki, A.; Kochkar, H.; Guillard, C.; Berhault, G.; Ghorbel, A. Effect of Na content and thermal treatment of titanate nanotubes on the photocatalytic degradation of formic acid. *Appl. Catal. B-Environ.* **2013**, *138*, 401–415. [CrossRef]
13. Tang, Z.R.; Li, F.; Zhang, Y.H.; Fu, X.Z.; Xu, Y.-J. Composites of titanate nanotube and carbon nanotube as photocatalyst with high mineralization ratio for gas-phase degradation of volatile aromatic pollutant. *J. Phys. Chem. C* **2011**, *115*, 7880–7886. [CrossRef]
14. Kim, I.Y.; Lee, J.M.; Kim, T.W.; Kim, H.N.; Kim, H.; Choi, W.; Hwang, S.-J. A strong electronic coupling between graphene nanosheets and layered titanate nanoplates: A soft-chemical route to highly porous nanocomposites with improved photocatalytic activity. *Small* **2012**, *8*, 1038–1048. [CrossRef] [PubMed]
15. Liu, G.G.; Han, K.; Ye, H.Q.; Zhu, C.Y.; Gao, Y.P.; Liu, Y.; Zhou, Y.H. Graphene oxide/triethanolamine modified titanate nanowires as photocatalytic membrane for water treatment. *Chem. Eng. J.* **2017**, *320*, 74–80. [CrossRef]
16. Kim, S.; Han, K.I.; Lee, I.G.; Park, W.K.; Yoon, Y.; Yoo, C.S.; Yang, W.S.; Hwang, W.S. A gallium oxide-graphene oxide hybrid composite for enhanced photocatalytic reaction. *Nanomaterials* **2016**, *6*, 127. [CrossRef] [PubMed]
17. Chen, Y.; Sun, H.Q.; Peng, W.C. 2D transition metal dichalcogenides and graphene-based ternary composites for photocatalytic hydrogen evolution and pollutants degradation. *Nanomaterials* **2017**, *7*, 62. [CrossRef] [PubMed]
18. Zhang, N.; Yang, M.-Q.; Tang, Z.-R.; Xu, Y.-J. CdS-graphene nanocomposites as visible light photocatalyst for redox reactions in water: A green route for selective transformation and environmental remediation. *J. Catal.* **2013**, *303*, 60–69. [CrossRef]
19. Zhang, H.; Lv, X.J.; Li, Y.M.; Wang, Y.; Li, J.H. P25-graphene composite as a high performance photocatalyst. *ACS Nano* **2010**, *4*, 380–386. [CrossRef] [PubMed]
20. Sun, Z.H.; Guo, J.J.; Zhu, S.M.; Mao, L.; Ma, J.; Zhang, D. A high-performance Bi_2WO_6-graphene photocatalyst for visible light-induced H_2 and O_2 generation. *Nanoscale* **2014**, *6*, 2186–2193. [CrossRef] [PubMed]
21. Gao, F.D.; Zeng, D.W.; Huang, Q.W.; Tian, S.Q.; Xie, C.S. Chemically bonded graphene/BiOCl nanocomposites as high-performance photocatalysts. *Phys. Chem. Chem. Phys.* **2012**, *14*, 10572–10578. [CrossRef] [PubMed]

22. Huang, Q.W.; Tian, S.Q.; Zeng, D.W.; Wang, X.X.; Song, W.L.; Li, Y.Y.; Xiao, W.; Xie, C.S. Enhanced photocatalytic activity of chemically bonded TiO_2/graphene composites based on the effective interfacial charge transfer through the C-Ti bond. *ACS Catal.* **2013**, *3*, 1477–1485. [CrossRef]
23. Palmisano, G.; Augugliaro, V.; Pagliaro, M.; Palmisano, L. Photocatalysis: A promising route for 21st century organic chemistry. *Chem. Commun.* **2007**, *43*, 3425–3437. [CrossRef] [PubMed]
24. Zhang, M.; Wang, Q.; Chen, C.; Zang, L.; Ma, W.; Zhao, J. Oxygen atom transfer in the photocatalytic oxidation of alcohols by TiO_2: Oxygen isotope studies. *Angew. Chem. Int. Ed.* **2009**, *48*, 6081–6084. [CrossRef] [PubMed]
25. Tanaka, A.; Hashimoto, K.J.; Kominami, H. Preparation of Au/CeO_2 exhibiting strong surface plasmon resonance effective for selective or chemoselective oxidation of alcohols to aldehydes or ketones in aqueous suspensions under irradiation by green light. *J. Am. Chem. Soc.* **2012**, *134*, 14526–14533. [CrossRef] [PubMed]
26. Liang, S.J.; Wen, L.R.; Lin, S.; Bi, J.H.; Feng, P.Y.; Fu, X.Z.; Wu, L. Monolayer HNb_3O_8 for selective photocatalytic oxidation of benzylic alcohols with visible light response. *Angew. Chem. Int. Ed.* **2014**, *53*, 2951–2955. [CrossRef] [PubMed]
27. Higashimoto, S.; Kitao, N.; Yoshida, N.; Sakura, T.; Azuma, M.; Ohue, H.; Sakata, Y. Selective photocatalytic oxidation of benzyl alcohol and its derivatives into corresponding aldehydes by molecular oxygen on titanium dioxide under visible light irradiation. *J. Catal.* **2009**, *266*, 279–285. [CrossRef]
28. Yang, J.; Shen, X.X.; Wei, J.P.; Zhang, L.N.; Zhao, D.; Yao, B.H. Selective oxidation of alcohols on hydrogen titanate nanotubes under visible light irradiation: Relationship between nanostructure and catalytic activity. *Catal. Sci. Technol.* **2016**, *6*, 7604–7614. [CrossRef]
29. Yu, L.H.; Lin, Y.M.; Li, D.Z. Visible-light-induced aerobic oxidation of alcohols in a green catalytic system of carbonate-like species doped TiO_2. *Appl. Catal. B-Environ.* **2017**, *216*, 88–94. [CrossRef]
30. Andryushina, N.S.; Stroyuk, O.L. Influence of colloidal graphene oxide on photocatalytic activity of nanocrystalline TiO_2 in gas-phase ethanol and benzene oxidation. *Appl. Catal. B-Environ.* **2014**, *148*, 543–549. [CrossRef]
31. Yang, X.; Zhao, H.; Feng, J.F.; Chen, Y.M.; Gao, S.Y.; Cao, R. Visible-light-driven selective oxidation of alcohols using a dye-sensitized TiO_2-polyoxometalate catalyst. *J. Catal.* **2017**, *351*, 59–66. [CrossRef]
32. Zhai, Q.Q.; Bo, T.; Hu, G.X. High photoactive and visible-light responsive graphene/titanate nanotubes photocatalysts: Preparation and characterization. *J. Hazard. Mater.* **2011**, *198*, 78–86.
33. Dang, H.F.; Dong, X.F.; Dong, Y.C.; Huang, J.S. Facile and green synthesis of titanate nanotube/graphene nanocomposites for photocatalytic H_2 generation from water. *Int. J. Hydrog. Energy* **2013**, *38*, 9178–9185. [CrossRef]
34. Zou, J.P.; Ma, J.; Huang, Q.; Luo, S.L.; Yu, J.; Luo, X.B.; Dai, W.L.; Sun, J.; Guo, G.C.; Au, C.T.; et al. Graphene oxide as structure-directing and morphology-controlling agent for the syntheses of heterostructured graphene-Bi_2MoO_6/$Bi_{3.64}Mo_{0.36}O_{6.55}$ composites with high photocatalytic activity. *Appl. Catal. B-Environ.* **2014**, *156*, 447–455. [CrossRef]
35. Sun, X.; Li, Y. Synthesis and characterization of ion-exchangeable titanate nanotubes. *Chem.-Eur. J.* **2003**, *9*, 2229–2238. [CrossRef]
36. Perera, S.D.; Mariano, R.G.; Vu, K.; Nour, N.; Seitz, O.; Chabal, Y.; Balkus, K.J. Hydrothermal synthesis of graphene-TiO_2 nanotube composites with enhanced photocatalytic activity. *ACS Catal.* **2012**, *2*, 949–956. [CrossRef]
37. Yang, J.; Wang, X.H.; Zhao, X.L.; Dai, J.; Mo, S.R. Synthesis of uniform Bi_2WO_6-reduced graphene oxide nanocomposites with significantly enhanced photocatalytic reduction activity. *J. Phys. Chem. C* **2015**, *119*, 3068–3078. [CrossRef]
38. Manna, A.K.; Pati, S.K. Tuning the electronic structure of graphene by molecular charge transfer: A computational study. *Chem.-Asian J.* **2009**, *4*, 855–860. [CrossRef] [PubMed]
39. Akhavan, O.; Ghaderi, E. Photocatalytic reduction of graphene oxide nanosheets on TiO_2 thin Film for photoinactivation of bacteria in solar light irradiation. *J. Phys. Chem. C* **2009**, *113*, 20214–20220. [CrossRef]
40. Shishido, T.; Teramura, K.; Tanaka, T. A unique photo-activation mechanism by "in situ doping" for photo-assisted selective NO reduction with ammonia over TiO_2 and photooxidation of alcohols over Nb_2O_5. *Catal. Sci. Technol.* **2011**, *1*, 541–551. [CrossRef]
41. Lang, X.; Ma, W.; Zhao, Y.; Chen, C.; Ji, H.; Zhao, J. Visible-light-induced selective photocatalytic aerobic oxidation of amines into imines on TiO_2. *Chem.-Eur. J.* **2012**, *18*, 2624–2631. [CrossRef]

42. Yang, M.Q.; Zhang, N.; Xu, Y.J. Synthesis of fullerene-, carbon nanotube-, and graphene-TiO_2 nanocomposite photocatalysts for selective oxidation: A comparative study. *ACS Appl. Mater. Inter.* **2013**, *5*, 1156–1164. [CrossRef] [PubMed]
43. Chen, D.; Zhang, H.; Liu, Y.; Li, J.H. Graphene and its derivatives for the development of solar cells, photoelectrochemical, and photocatalytic applications. *Energy Environ. Sci.* **2013**, *6*, 1362–1387. [CrossRef]
44. Xiang, Q.J.; Yu, J.G. Graphene-based photocatalysts for hydrogen generation. *J. Phys. Chem. Lett.* **2013**, *4*, 753–759. [CrossRef] [PubMed]
45. Xiao, F.X.; Wang, F.C.; Fu, X.Z.; Zheng, Y. A green and facile self-assembly preparation of gold nanoparticles/ZnO nanocomposite for photocatalytic and photoelectrochemical applications. *J. Mater. Chem.* **2012**, *22*, 2868–2877. [CrossRef]
46. Xiao, X.Y.; Jiang, J.; Zhang, L.Z. Selective oxidation of benzyl alcohol into benzaldehyde over semiconductors under visible light: The case of $Bi_{12}O_{17}Cl_2$ nanobelts. *Appl. Catal. B-Environ.* **2013**, *142*, 487–493. [CrossRef]

© 2018 by the authors. Licensee MDPI, Basel, Switzerland. This article is an open access article distributed under the terms and conditions of the Creative Commons Attribution (CC BY) license (http://creativecommons.org/licenses/by/4.0/).

nanomaterials

MDPI

Article

A Facile Approach to Prepare Black TiO_2 with Oxygen Vacancy for Enhancing Photocatalytic Activity

Shihao Chen [1], Yang Xiao [1], Yinhai Wang [1,*], Zhengfa Hu [1], Hui Zhao [1] and Wei Xie [2]

1 School of Physics &Optoelectronic Engineering, Guangdong University of Technology, Guangzhou 510006, China; shchenbhx@163.com (S.C.); xiaoyxy1023@163.com (Y.X.); zhfhu@gdut.edu.cn (Z.H.); kkhui@gdut.edu.cn (H.Z.)

2 School of Physics Science and Technology, Lingnan Normal University, Zhanjiang 524048, China; xiewei@lingnan.edu.cn

* Correspondence: yhwang@gdut.edu.cn; Tel.: 86-020-39322265

Received: 14 March 2018; Accepted: 13 April 2018; Published: 16 April 2018

Abstract: Black TiO_2 has triggered worldwide research interest due to its excellent photocatalytic properties. However, the understanding of its structure–property relationships and a more effective, facile and versatile method to produce it remain great challenges. We have developed a facile approach to synthesize black TiO_2 nanoparticles with significantly improved light absorption in the visible and infrared regions. The experimental results show that oxygen vacancies are the major factors responsible for black coloration. More importantly, our black TiO_2 nanoparticles have no Ti^{3+} ions. These oxygen vacancies could introduce localized states in the bandgap and act as trap centers, significantly decreasing the electron–hole recombination. The photocatalytic decomposition of both rhodamine B and methylene blue demonstrated that, under ultraviolet light irradiation, better photocatalytic performance is achieved with our black TiO_2 nanoparticles than with commercial TiO_2 nanoparticles.

Keywords: photocatalysis; black TiO_2; oxygen vacancies; photodegradation

1. Introduction

Since the pioneering work of Fujishima and Honda in 1972 [1], titanium dioxide (TiO_2) has attracted extensive interest as a widely used semiconductor photocatalyst in the fields of hydrogen production, photocatalytic water-splitting [2], environmental protection technologies [3] and photocatalytic reduction of carbon dioxide [4].Especially in the area of environmental protection, textile dyes and other industrial dye stuffs constitute one of the largest group of organic compounds that represent an increasing environmental danger. Therefore, improving the efficiency of photocatalysis is a current research hotspot. Absorption of light to generate electron-hole pairs and effective charge separation are of great significance for enhancing the efficiency of the photocatalytic reaction in TiO_2 semiconductor materials. In general, the more light TiO_2 can absorb, the more excited charges are likely to be on its surface, improving its photocatalytic efficiency [5]. However, the large band gap of anatase TiO_2 (~3.2 eV) greatly limits its optical absorption in the UV region of the solar spectrum, resulting in poor efficiency for solar-driven photocatalysis.

Therefore, great efforts have been devoted to modifying the band structure to enhance the visible-light absorption, such as doping TiO_2 with metal [6–9] or nonmetal [10,11] elements and co-doping with metal and non-metal elements[12–14]. All these efforts have enhanced its visible light absorption and photocatalytic activity. However, the results are not yet satisfactory.

Recently, a hydrogenated black TiO_2 (B-TiO_2) material was reported by Mao and co-workers, having a narrow bandgap of ~1.5 eV with increased light harvesting efficiency in the visible and infrared regions and enhanced photocatalytic activity [15]. Unsurprisingly, this discovery has triggered worldwide research interest in black TiO_2 nanomaterials, which represent a major breakthrough for TiO_2 in photocatalysis. Since then, different methods for the synthesis of black TiO_2 have been developed, which have allowed the preparation of hydrogenated or reduced TiO_2 of different colors like yellow, blue, gray and black, by the use of different initial TiO_2 materials and reaction conditions [16–18]. Most of these approaches can be divided into two types: reduction from TiO_2 and incomplete oxidation from low-valence-state Ti species. To reduce white TiO_2 to black TiO_2, Al, Mg, Zn, H_2, $NaBH_4$, imidazole and ascorbic acid have been used as reducing agents [19,20]. For example, Liu et al. reported black TiO_2 nanotubes obtained by pressurized H_2 reduction at high temperature (500 °C, 20 bar, for 1 h), with a high photocatalytic hydrogen production rate [21]. Wang et al. reported a mass-production approach to synthesize black TiO_2 with a unique crystalline-core amorphous-shell structure ($TiO_2@TiO_{2-x}$), by using aluminum instead of hydrogen as the reducing agent [22]. Sinhamahapatra et al. developed a controlled magnesiothermic reduction to synthesize reduced black TiO_2 under a 5% H_2/Ar atmosphere, with an optimum bandgap and band position, oxygen vacancies, surface defects and charge recombination centers and showing significantly improved optical absorption in the visible and infrared regions [23]. For the second approach, the Ti sources TiH_2, Ti, TiO, Ti_2O_3, $TiCl_3$ and TiN have been incompletely oxidized to synthesize black or hydrogenated TiO_2 [18,24]. For example, Liu et al. synthesized rice-shaped Ti^{3+} self-doped TiO_{2-x} nanoparticles by mild hydrothermal treatment of TiH_2 in a H_2O_2 aqueous solution; the particles showed a strong absorption from the UV to the visible light regions and retained their light-blue color upon storage under ambient atmosphere or water, for one month at 40 °C [25].Pei et al. prepared gray TiO_{2-x} with enhanced visible light photocatalytic activity by a facile hydrothermal treatment of TiO in HCl solution [26].All these synthetic approaches resulted in changes in the TiO_2 color and enhancement of its light absorption. Surface lattice disorders, high-concentration of oxygen vacancies, Ti^{3+} ions, Ti-OH groups and Ti-H bands have been found in some black TiO_2 and explain their color change, enhancement of light absorption and photocatalytic activity. These highly defective TiO_2, referred to as Ti^{3+} ions/oxygen-vacancy self-doping (TiO_{2-x}) and H-doping ($TiO_{2-x}H_x$), always present black color and possess strong visible light absorption. Some papers have also reported prepared black TiO_2 that have only oxygen vacancies but no Ti^{3+} ions. For example, Wang et al. suggested the absence of Ti^{3+} in hydrogenated black TiO_2 nanowires treated at 450 °C [27]. Dong et al. synthesized defective black TiO_{2-x} with a remarkable photocatalytic activity by a facile anodization technique and indicated that there were oxygen vacancies present but no Ti^{3+} [28]. This may be attributed to their lower preparation temperatures (≤450 °C). Furthermore, the defect structure of black TiO_2 is the major factor that determines its various properties, including color change, light absorption and photocatalytic activity [29]. According to previous papers, black TiO_2 with disordered shell/hydroxyl groups can induce valence-band/conduction-band tails, that reduce their bandgap [15,30]. They can also induce oxygen vacancies, Ti^{3+} ions and H-doping defects that reduce the bandgap and introduce mid-gap/localized-donor states below the conduction band, in addition to an upshift of the Fermi level [31,32],which renders color changes. For example, for Ti^{3+} self-doping black TiO_2, the localized excitation of the Ti^{3+} ions via 3d transitions from the gap state to the empty excited state, significantly increases its light absorption and extends it to the visible region [22].

However, some questions still exist regarding black TiO_2. For example, there is no consensus on what are the most essential modifications and the most effective synthetic method for their preparation. Some black TiO_2 do not show the expected efficiency in visible light. Furthermore, their preparation is cumbersome and costly. The oxygen vacancies and Ti^{3+} species created during their synthesis are usually not stable enough because the TiO_2 reduction occurs mainly on its surface. These factors severely limit their wide applicability. Therefore, developing a simple and economic strategy to synthesize a higher stability, higher performance black TiO_2 photocatalyst is still a great challenge.

Inspired by the above-mentioned considerations, we have developed a facile technique for synthesizing black TiO_2 nanoparticles, for the first time by low-temperature annealing in a vacuum-tube furnace. The obtained samples reveal very strong visible and infrared light absorption.

The B-TiO_2 nanoparticles have oxygen vacancies and no Ti^{3+} ions. We evaluated the effectiveness of the as-prepared B-TiO_2 nanoparticles and the feasibility of applying them to photocatalytic degradation tests. Very promisingly, the resultant B-TiO_2 nanoparticles showed superior photocatalytic activity, far beyond that of commercial white TiO_2 nanoparticles.

2. Experimental

2.1. Materials

Commercial catalyst white titanium dioxide nanoparticles (20–30 nm) and absolute ethanol were purchased from Macklin Biochemical. All the chemicals used in this study were analytical grade and were used as received. All the solutions were made using ultrapure Milli-Q (Millipore Corporation, Burlington, MA, USA) water.

2.2. Synthesis of B-TiO_2 Nanoparticles

In a simple synthetic procedure, commercially available (W-TiO_2) nanoparticles were used as precursors. W-TiO_2 nanoparticles (0.5 g) and absolute ethanol (10 mL) were mixed in a 50 mL quartz tube using a magnetic stirrer to form a milky-white suspension. The milky-white suspension was stirred for 10 min. Then the suspension was poured into a sintering boat (length 6 cm, width 3 cm, height 1.5 cm) and transferred to a vacuum tube furnace (SK-G06163, $\Phi 60/50 \times 1000$ mm). The suspension was annealed at 400 °C for 3 h. The detailed annealing procedure is as follows: The initial temperature is set to 50 °C and the temperature is raised at a rate of 5°C/min until reaches 400 °C and the temperature is maintained at 400 °C for three hours. Then, the temperature is lowered at a rate of 5 °C/min until reaches 50 °C. Finally, the sample was allowed to cool to room temperature and a black mass of B-TiO_2 nanoparticles was obtained.

2.3. Characterization

The X-ray diffraction patterns (XRD) of the samples were obtained at room temperature using Cu Kα irradiation (λ = 1.5406 Å) operating at 36 kV tube voltage and 20 mA tube current. Data were collected between 10° and 70° (2θ) with a 0.02° step size. Raman spectra were recorded on a UV laser Raman spectrometer at λ = 785 nm. UV-vis diffuse reflectance (DRUV-vis) absorption spectra were recorded on an UV-2700 Spectrophotometer in the range of 200–800 nm. X-ray photoelectron spectra (XPS) were measured on an ESCALAB 250Xi using the reference of C1s (284.8 eV) with an excitation source of 150 W Al Kα (hv = 1486.6 eV) X-rays. The electron paramagnetic resonance (EPR) spectra were obtained using a JES-FA200 spectrometer (JEOL, Tokyo, Japan) at room temperature and an FEI Tecnai G2 F20 (FEI, Hillsboro, OR, USA) model was used to determine the size and morphologies of the nanoparticles. Sample specimens for the transmission electron microscopy (TEM) measurement were prepared by briefly ultrasonicating the sample powders in ethanol, followed by placing a drop of the suspension onto lacey support films that were dried before imaging. Fourier transformed infra-red (FTIR, 80/80v Bruker TENSOR27, BRUKER OPTICS, Karlsruhe, Germany) was used to analyze the functional groups in the catalyst from 500 to 4000 cm^{-1}. Photoluminescence (PL) emission spectrum was recorded on an FLS980 fluorescence spectrophotometer equipped with a 450 W xenon lamp as the excitation source, at room temperature (excitation wavelength λ_{ex} = 300 nm).

2.4. Photocatalytic Test

Rhodamine B (RhB) and methylene blue (MB) are organic dye with a bright red color and blue color, respectively. They are widely applied as a test model pollutant in photocatalytic processes. Therefore, the photocatalytic activity of the two prepared samples was evaluated by monitoring the decomposition of (RhB)/(MB) in an aqueous solution under UV irradiation from a 500 W Hg lamp. The lamp was positioned in a cylindrical Pyrex vessel and cooled by circulating water to control the reaction temperature at about 27 °C. A quartz tube was used as the photo catalytic reactor. The sample of TiO_2 (0.01 g) was mixed with an aqueous solution of RhB/MB (40 mL, 4 × 10^{-5} M). After stirring for 30 min in the dark, to reach an adsorption equilibrium between the photocatalyst and the RhB/MB solution, the mixture was exposed to UV light. Vigorous magnetic stirring was maintained in order to keep the B-TiO_2 and W-TiO_2 nanoparticles suspended in the RhB/MB solution. The concentration of aqueous RhB/ MB was determined with a UV-Vis spectrophotometer by measuring the peak intensity at 553/668 nm, respectively. The percentage of degradation was recorded as C/C_0, where C and C_0 refer to the absorbance of the RhB/MB solution after a certain time interval (20 min) and the initial absorbance, respectively.

The photocatalytic activity of B-TiO_2 and W-TiO_2 under visible light was also evaluated by monitoring the decomposition of RhB/MB. The apparatus for studying the photocatalytic decomposition of RhB/MB was identical under visible and ultraviolet light, except that an 800W Xe lamp was used instead of a Hg lamp and ultraviolet cut off filter (>420 nm) was applied to cut off the UV light. The amount of catalyst has also been increased to 0.02 g.

3. Results and Discussion

3.1. Characterization

The prepared B-TiO_2 nanoparticles showed an outstanding stability. No color change was observed for the B-TiO_2 nanoparticles over one year after they were synthesized and stayed black even when annealed at 200 °C in air for 3 days. Figure 1 shows the UV-visible absorption spectra of the B-TiO_2 and W-TiO_2 nanoparticles. A large absorption peak is observed, for both, at wavelengths shorter than 400 nm (~3.1 eV), which can be attributed to the intrinsic bandgap absorption of crystalline anatase TiO_2. Compared with pristine W-TiO_2, the B-TiO_2 extended the absorption from UV light to the visible and infrared light regions. The extended absorption is consistent with the changes of sample color from white to black (photograph shown in the inset of Figure 1). Furthermore, for the B-TiO_2 nanoparticles, a step appeared at approximately 740 nm (~1.67 eV). This step could be due to oxygen vacancies with a trapped electron and it is discussed in detail in the EPR-analysis section.

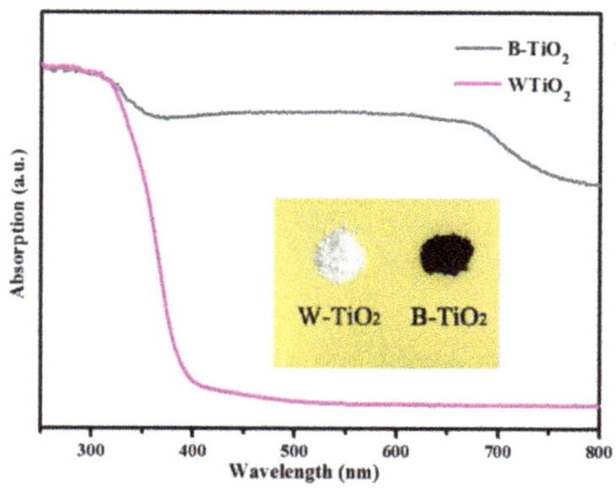

Figure 1. UV-Vis absorption spectra of black B-TiO_2 and W-TiO_2 nanoparticles. The insert is a photograph of B-TiO_2 and W-TiO_2 nanoparticles.

The crystalline structures of the B-TiO_2 and W-TiO_2 nanoparticles were determined by XRD analysis. The powder XRD patterns of the two samples, shown in Figure 2a, exhibited characteristic diffraction peaks matching the (101), (004), (200), (105) and (204) planes of anatase TiO_2 (JCPDS Card No. 21-1272). It could also be seen that there was almost no phase change in the B-TiO_2 nanoparticles and no other diffraction peaks could be observed after the low temperature annealing. However, from the local enlargement of diffraction peaks (inset of Figure 2a), the diffraction peaks of the B-TiO_2

nanoparticles moved slightly to a higher angle. According to the Bragg equation (2dsinθ = λ, where d is the crystal spacing, θ the diffraction angle and λ the X-ray wavelength), the shift of peaks towards higher angles suggests that the lattice parameters decrease. This result can be explained in terms of the oxygen vacancies produced by the annealing of the samples [33].

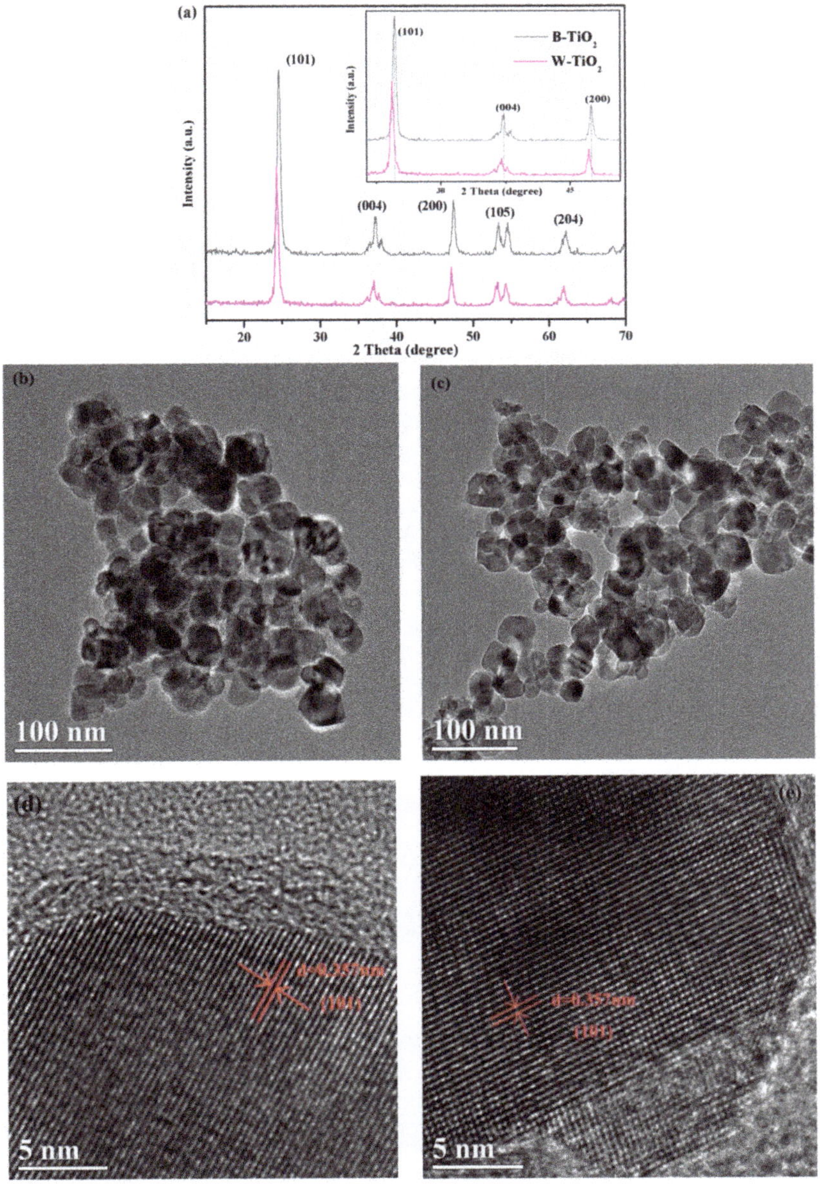

Figure 2. The XRD patterns of the as-prepared samples (**a**); TEM images of TiO_2 nanoparticles before (**b**) W-TiO_2 and after (**c**) B-TiO_2 low temperature annealing; HRTEM images of TiO_2 nanoparticles before (**d**) W-TiO_2 and after (**e**) B-TiO_2 low temperature annealing.

The morphology and crystal structure of the B-TiO_2 and W-TiO_2 nanoparticles were examined by high-resolution transmittance microscopy (HR-TEM) analysis, as shown in Figure 2b–e. The TEM images showed similar particle sizes and morphologies for B-TiO_2 (Figure 2b) and W-TiO_2 (Figure 2c), which proves that the low temperature annealing did not change its morphology. The lattice fringe pattern with a spacing of ~0.35 nm, shown in the high resolution TEM (HRTEM) images (Figure 2d,e), confirms the (101) plane of anatase TiO_2 for the W-TiO_2 and B-TiO_2 nanoparticles (in agreement with the XRD analysis). Both the W-TiO_2 and B-TiO_2 nanoparticles were highly crystallized, as seen from the well resolved lattice features. However, other researchers have reported a disordered surface layer surrounding the crystalline core in the hydrogenated/aluminum-reduced black TiO_2 nanoparticles obtained under high temperature (>450 °C) and high pressure (>5 bar) [22,34]. In contrast, the B-TiO_2 nanoparticles prepared in this work used low temperature (400 °C) and atmospheric pressure conditions that did not lead to the specific core/shell structure consisting of a disordered surface layer and a crystalline core. The specific surface areas using the Barrett-Emmett-Teller (BET) technique were 52.8 and 61.9 m^2/g for the B-TiO_2 and W-TiO_2 nanoparticles, respectively. The similar BET specific surface areas confirmed that the crystal size of the B-TiO_2 nanoparticles did not increase with the low temperature annealing, consistent with the above-discussed TEM results.

Raman spectroscopy is very sensitive to short-range distortions arising from microstructural defects. The structural characteristics of the B-TiO_2 nanoparticles were further examined by Raman scattering, as shown in Figure 3. For the W-TiO_2 nanoparticles, five characteristic Raman peaks at 143.6, 199.8, 397.2, 516.3 and 639.6 cm^{-1} were assigned as the E_g, E_g, B_{1g}, A_{1g} + B_{1g} and E_g modes of the anatase phase, respectively [35]. However, compared to the W-TiO_2 nanoparticles, it is clear that the frequency of the strongest E_g mode (at 143.6 cm^{-1}) in the B-TiO_2 nanoparticles, which results from the external vibration of the Ti-O bonds, had a large blue shift to 152.5 cm^{-1}, accompanied by peak broadening. As reported in previous studies, ruling out grain size effects, the blue shift and peak broadening of the anatase phase TiO_2 can be ascribed to the oxygen stoichiometry [36]. Since the B-TiO_2 and W-TiO_2 nanoparticles had similar particle sizes, the blue shift and broadening of the strongest mode at 143.6 cm^{-1}, can be attributed to a shortening of the correlation length due to the presence of oxygen vacancies (nonstoichiometric). Thus, Raman spectroscopy unambiguously supported the oxygen deficiency of the B-TiO_2 nanoparticles.

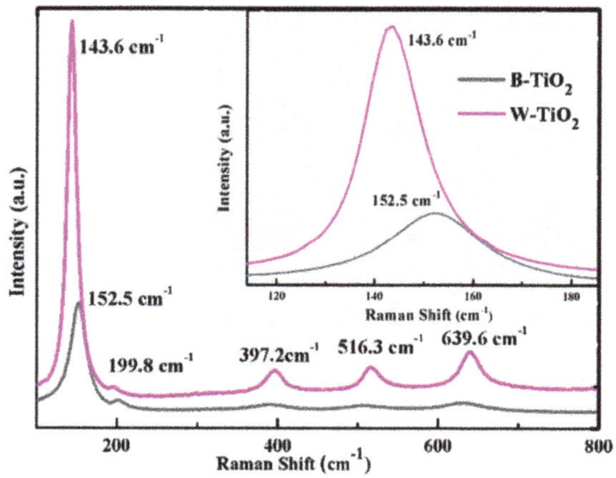

Figure 3. Raman spectra of B-TiO_2 nanoparticles and W-TiO_2, inset is the magnification of E_g peak.

The existence of oxygen vacancies in the samples was also verified by electron paramagnetic resonance (EPR) at room temperature. EPR is widely-used to examine unpaired spins. In Figure 4, it is easily seen a strong EPR signal in the B-TiO_2 nanoparticles at g = 2.003, which is characteristic of oxygen vacancies with a trapped electron in TiO_2 and not of Ti^{3+} ions (g = 1.96–1.99) and consistent with the reported oxygen-deficient TiO_{2-x} [37,38]. Upon removal of an oxygen atom, one or two electrons will be localized in an oxygen vacancy site (reaction 1) and form color centers. The color centers associated with oxygen vacancies in TiO_2 are the F, F^+ and F^{++} centers (reactions 2–4) [39]. In our samples, one electron is localized in an oxygen vacancy state (F^+-centers).

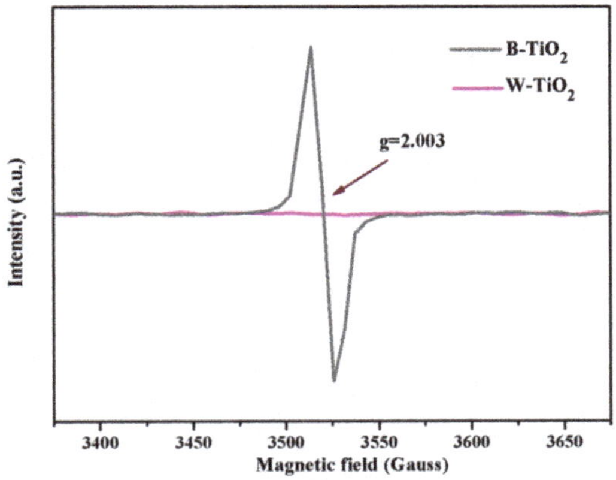

Figure 4. EPR spectra of B-TiO_2 and W-TiO_2 nanoparticles.

$$-O^{2-} \rightarrow O + V_O + 2e^- \quad (1)$$

$$V_O + 2e^- \rightarrow F \quad (2)$$

$$V_O + 1e^- \rightarrow F^+ \quad (3)$$

$$V_O + 0e^- \rightarrow F^{++} \quad (4)$$

Furthermore, Chen et al. have proved that electrons transit from the color center energy level to the conduction band bottom. The calculated excited energy of the F^+-center electron is 1.67 eV [40], which is in reasonable agreement with the step value which appeared at approximately 740 nm (~1.67 eV) in the UV-Vis absorption spectra (Figure 1). Thus, the absorption spectral features in the 1.67 eV region (740 nm) are assigned to a transition from the ground state of the F^+ center to its corresponding excited F^{+*} state. By contrast, the W-TiO_2 nanoparticles did not contain any paramagnetic site, as a flat line was observed for this sample. Thus, the strong EPR signal peak and also the blue shift of the Raman peak, indicated that B-TiO_2 nanoparticles had a high concentration of oxygen vacancies and that no Ti^{3+} ions existed.

X-ray photo-electron spectroscopy (XPS) analysis was conducted to investigate the chemical states of the Ti and O elements, to gain further insight into the oxygen vacancies in the B-TiO_2 nanoparticles and the valence band position on the sample surface. Figure 5a shows the Ti 2p XPS of the B-TiO_2 and W-TiO_2 nanoparticles. For the W-TiO_2 nanoparticles, the Ti 2p3/2 and 2p1/2 XPS peaks were centered at binding energies (BE) 458.6 and 464.36 eV and the calculated ΔBE value between Ti 2p1/2 and Ti 2p3/2 (ΔBE = BE(Ti 2p1/2) − BE(Ti 2p3/2)) was 5.76 eV. These features are typical for the Ti^{4+}-O bonds in TiO_2 [18]. The Ti 2p3/2 and 2p1/2 XPS peaks of the B-TiO_2 nanoparticles were located at 459.25 and 465.0 eV, respectively, showing a BE positive shift of about 0.65 eV compared with the peaks of the W-TiO_2 nanoparticles. However, the ΔBE value between Ti 2p1/2 and Ti 2p3/2 was 5.75 eV, indicating a normal state of Ti^{4+} in the B-TiO_2 nanoparticles. According to previous papers on black TiO_2, when the valence of Ti changes to Ti^{3+}, additional peaks at lower binding energies can be observed, or the peaks will show a negative shift in binding energy [41,42].Our results confirmed that the charge states of the Ti atoms at the surface of the B-TiO_2 sample were Ti^{4+} and that no Ti^{3+} existed at the surface. Additionally, the same phenomenon appeared in the O 1s XPS spectra. Figure 5b shows the high-resolution O 1s XPS spectra of the W-TiO_2 nanoparticles: there are two peaks located at about 529.88 and 531.66 eV, which were attributed to the Ti^{4+}-O bond and to surface hydroxyl groups (Ti-OH bonds), respectively. Compared with the W-TiO_2 nanoparticles, the O 1s XPS peak of the B-TiO_2 nanoparticles (Figure 5c) located at ~530.54 eV, also showed a clear positive shift in binding energy of about 0.66 eV, which is consistent with the Ti 2p. The extra peak existing in the O 1s spectra of the B-TiO_2 nanoparticles at higher BE (~532.91 eV) was ascribed to surface free −OH species [43]. Furthermore, the calculated value $\Delta BE = BE(O\ 1s) - BE(Ti\ 2p3/2) = 530.54\ eV - 459.25\ eV = 71.29\ eV$, is close to that of anatase (71.4 eV) and quite different from that of the characteristic Ti^{3+}-containing oxides (72.9–73.1 eV) [44]. These features indicated that the valence state of Ti did not change in the B-TiO_2 nanoparticles. Reasonably, the clear positive shifts of binding energies of Ti 2p and O 1s were ascribed to the strong interaction between Ti^{4+} and oxygen vacancies, revealing lattice distortions on the B-TiO_2 nanoparticles. It is possible that the oxygen vacancies, sustaining a positive charge, rejected the Ti atoms toward the adjacent oxygen atoms in the crystal structure of B-TiO_2, consequently reducing the Ti-O bond length as well as increasing the BE of Ti 2p and O 1s [45,46]. Therefore, the positive shift of the XPS peak is due to the shortening of the correlation length because of the presence of oxygen vacancies. The valence-band maxima were estimated by a linear extrapolation of the peaks to the baselines, which gave a band edge position of 2.7 eV below the Fermi energy for both the B-TiO_2 and the W-TiO_2 nanoparticles, as shown in Figure 5d. Thus, the oxygen vacancies had a negligible effect on the valence-band position of the B-TiO_2 nanoparticles surface.

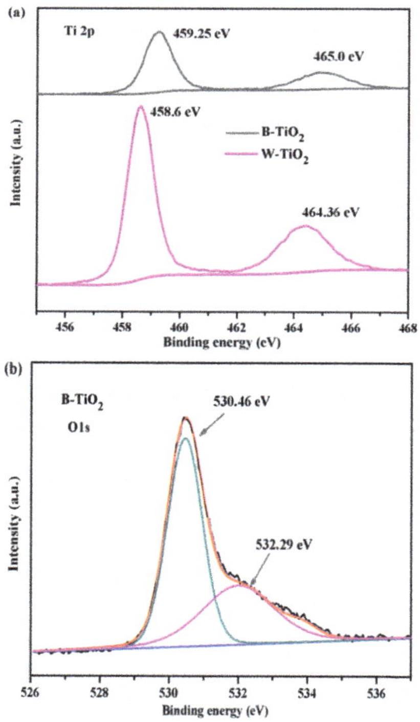

Figure 5. *Cont.*

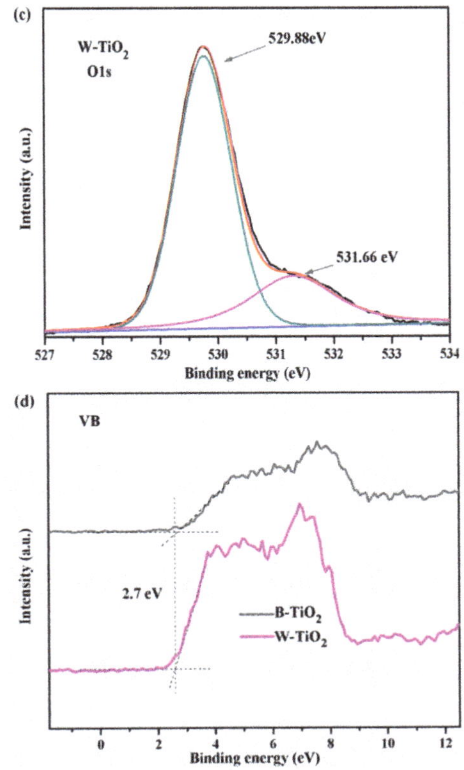

Figure 5. (**a**) The Ti 2p XPS spectra for the as-prepared B-TiO_2 and W-TiO_2 nanoparticles; (**b**)The O 1s XPS spectra for W-TiO_2 nanoparticles and (**c**) the O 1s XPS spectra as-prepared B-TiO_2 nanoparticles along with the Gaussian fits; (**d**) Valence-band XPS spectra of B-TiO_2 and W-TiO_2 nanoparticles.

To gain further insights on the chemical changes produced in the B-TiO_2 nanoparticles, Fourier transform infrared (FTIR) spectroscopic measurements were performed, as shown in Figure 6a. Both the B-TiO_2 and W-TiO_2 nanoparticles showed similar absorption features from 500 cm^{-1} to 4000 cm^{-1}. Characteristic features of the spectrum are the presence of a band at ~710 cm^{-1}, due to the symmetric stretching vibrations of the Ti-O-Ti bonds of TiO_2 and of a peak at about ~3418 cm^{-1}, attributed to the stretching vibration mode of hydroxyls and adsorbed water [47]. Compared with the W-TiO_2 nanoparticles, the infrared absorption of the B-TiO_2 nanoparticles decreased greatly at ~710 cm^{-1}. This change can be attributed to the presence of a high concentration of oxygen vacancies in the B-TiO_2 nanoparticles [48].

The photoluminescence (PL) spectra of the B-TiO_2 and W-TiO_2 nanoparticles were assessed to investigate the behavior of the electron-hole recombination, which are very important for the photocatalytic activity. Their emission spectra, in the wavelength range of 320–560 nm with excitation at 300 nm, are shown in Figure 6b and were very similar. The main emission peaks of the W-TiO_2 nanoparticles appeared at 395 nm (3.1 eV), 436 nm (2.8 eV), 451 nm (2.75 eV) and 459 nm (2.7 eV), respectively. The first one is attributed to the bandgap transition, corresponding to the bandgap energy of anatase. The peaks at 451 nm and 459 nm were attributed to band-edge free excitons [47]. The PL peak intensities of the B-TiO_2 nanoparticles showed a significant decrease compared with those of the W-TiO_2 nanoparticles. Furthermore, the main emission peak at 395 nm (3.1 eV), could not be observed, which indicated that the recombination rate of photo-generated electrons and holes had been considerably inhibited in the B-TiO_2 nanoparticles [49]. Given the fact that oxygen vacancies usually serve as electron traps, the weaker intensity of emission and disappearance of the emission peak at 395 nm should be due to the increased oxygen vacancies. All the results indicated that the B-TiO_2 nanoparticles had a relative low recombination rate of electrons and holes. Normally, the low recombination rate of electrons and holes favors high photocatalytic activity.

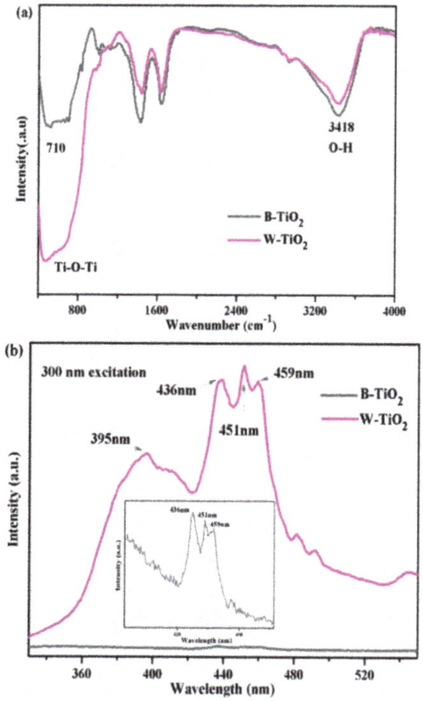

Figure 6. (**a**) FT-IR spectra of B-TiO_2 and W-TiO_2 nanoparticles; (**b**) Photoluminescence spectra of B-TiO_2 and W-TiO_2 nanoparticles.

3.2. *Photocatalytic Activity*

The photocatalytic activities of the as-prepared samples were evaluated by monitoring the decomposition of RhB in an aqueous solution under UV light irradiation. Commercial W-TiO_2 nanoparticles were used as the reference photocatalyst. Before the irradiation of these samples, an adsorption experiment was performed in the dark in order to ensure the adsorption equilibrium of RhB and MB on the catalyst surface. As shown in Figure 7a, the B-TiO_2 nanoparticles distinctly showed much better photocatalytic activity than the W-TiO_2 nanoparticles. After 20 min of UV-light irradiation, the B-TiO_2 nanoparticles could decompose about 50% of the RhB dye, while the value for the W-TiO_2 nanoparticles was very low (~10%). When the irradiation time was prolonged to 60 min, the RhB dye was almost completely decomposed (96.2%) by the B-TiO_2 nanoparticles but the W-TiO_2 nanoparticles only decomposed about 40%. The Figure 7a also shows that the photocatalytic decomposition of RhB, for the two samples, followed a pseudo-first-order reaction [50]; the pseudo-first-order kinetics of the decompositions are illustrated in Figure 7b. The B-TiO_2 nanoparticles showed a first-order rate constant of k_{app} = 0.036 min^{-1}, 4.5 times greater than that of the W-TiO_2 nanoparticles (k_{app} = 0.008 min^{-1}), under UV light irradiation.

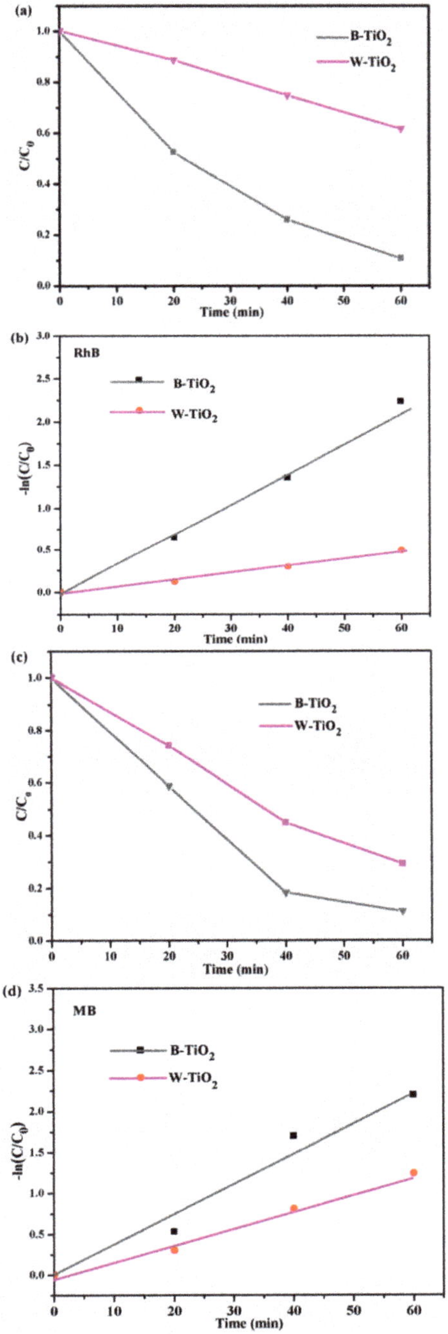

Figure 7. *Cont.*

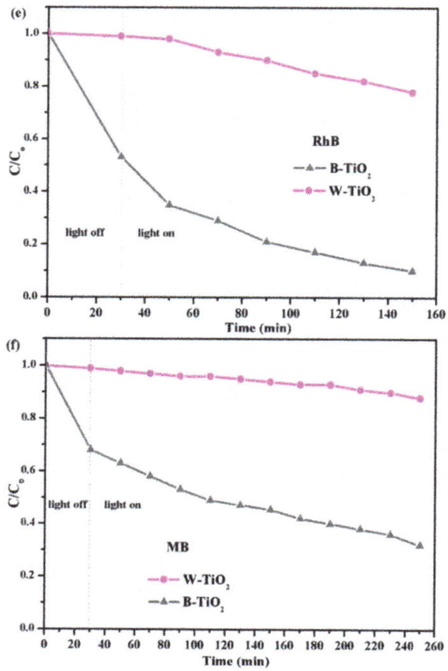

Figure 7. Comparison of photodecomposition of RhB (**a**) and MB (**c**) with different photocatalysts under UV light. $-\ln(C/C_0)$ of the RhB (**b**) and MB (**d**) concentration as a function of UV light irradiation time. Photodecomposition of RhB (**e**) and MB (**f**) with different photocatalysts under visible light.

The UV-light photocatalytic activity of the B-TiO_2 nanoparticles was further evaluated by measuring the photo-catalytic-degradation efficiency of the MB dye in aqueous solution. As shown in Figure 7c, under UV-light irradiation, in the short period of 20 min, the B-TiO_2 nanoparticles could decomposed about 40% of the original organic MB dye. After 60 min irradiation, the MB dye had been almost completely decomposed (95.2%) by the B-TiO_2 nanoparticles. In contrast, the W-TiO_2 nanoparticles photocatalyst were able to decompose only about 70% of the original MB after 60 min irradiation. Figure 7d shows that the first-order rate constants of the B-TiO_2 and W-TiO_2 nanoparticles were 0.036 min^{-1} and 0.02 min^{-1}, respectively. The as-prepared B-TiO_2 nanoparticles had about 1.8 times better MB-decomposing photocatalytic activity than the W-TiO_2 nanoparticles, under the same conditions.

The photocatalytic activity of the B-TiO_2 sample was also investigated under visible light ($\lambda > 420$ nm) illumination, as shown in Figure 7e,f. Figure 7e shows the photodecomposition of RhB by B-TiO_2 and W-TiO_2 samples. Under visible light illumination, the RhB solutions containing the B-TiO_2 undergo significant decomposition and became nearly transparent within 150 min. In contrast, W-TiO_2 exhibits limited activity on photodecomposition of RhB solution. We also evaluated the decomposition of MB dye solution as a model reaction under visible-light irradiation, shown in Figure 7f. The photocatalytic efficiency of B-TiO_2 under these conditions is 34%, which is higher than that of W-TiO_2 (88%) after an irradiation time of 250 min. Thus, the B-TiO_2 samples clearly exhibit improved photocatalytic activity.

3.3. Mechanism

In this work, the B-TiO_2 nanoparticles greatly extended their absorption range from the UV- to visible- to infrared-light regions and their photocatalytic activity was greatly improved. All the characterization results proved that this B-TiO_2 behavior is attributable to its high concentration of oxygen vacancies. The oxygen vacancies have been demonstrated to be electron donors in TiO_2 [51] and have been considered to contribute to the enhanced donor density in B-TiO_2 nanoparticles [27]. The electrons located on the oxygen vacancy states have a direct effect on the electronic structure of TiO_2 by forming a donor level below the conduction band. The oxygen vacancies introduced localized states at 0.75–1.18 eV below the TiO_2 conduction band minimum [46]. Consequently, the visible and near infrared-light absorptions are associated with transitions from the B-TiO_2 valence band to the oxygen vacancy levels or from the oxygen vacancies to the TiO_2 conduction band or with transitions from the ground state of the F^+ center to its corresponding excited state F^{+*}. In order to further understand our B-TiO_2 nanoparticles, the energy band diagram of the B-TiO_2 nanoparticles is schematized in Figure 8. The valence band maximum (VBM) and conduction band minimum (CBM) of TiO_2 are derived mainly from the O 2p orbitals and the Ti 3d orbitals, respectively. According to the UV-vis absorbance spectra, the bandgap of both the B-TiO_2 and the W-TiO_2 nanoparticles was 3.1 eV. Furthermore, the increased oxygen vacancies can improve charge transport in TiO_2 and shift its Fermi level toward the conduction band, facilitating the charge separation at the semiconductor/electrolyte interface [52,53]. The PL results also indicated that, in our experiments, the B-TiO_2 nanoparticles had a low recombination rate of electrons and holes. A low recombination rate leads to a large increase in photocatalytic activity.

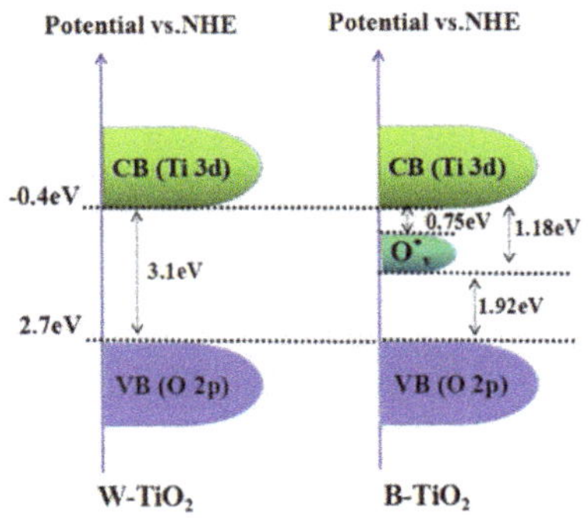

Figure 8. The band energy diagram of B-TiO_2 and W-TiO_2 nanoparticles.

4. Conclusions

The present work deals with the development of a facile method for the preparation of a highly active black TiO_2 photocatalyst. The black TiO_2 was prepared via low temperature annealing of commercial W-TiO_2 catalyst nanoparticles. The prepared B-TiO_2 nanoparticles showed remarkable photocatalytic activity for the degradation of the RhB and MB dyes. An appropriate amount of oxygen vacancies was introduced into the B-TiO_2 nanoparticles and Ti^{3+} ions were not found. The oxygen vacancies are responsible for the increased visible-and infrared-light absorption as they can introduce localized states into the bandgap. At the same time, the oxygen vacancies act as traps for reducing the recombination of electrons and holes and significantly improve the e–h separation efficiency, thus greatly enhancing the photocatalytic activity.

Acknowledgments: This work is supported by the National Nature Science Foundation of China (No. 21271048), Natural Science Foundation of China (11747074), Guangdong province science and technology plan project public welfare fund and ability construction project (2016A010103041, 2017A010103025), Doctoral Program of Lingnan Normal University (ZL1503), China Spark Program (2015GA780058).

Author Contributions: Shihao Chen: performed photocatalysts synthesis, XRD, Raman, TEM, UV-visible absorption spectra analysis, photocatalytic activity tests and wrote the manuscript, Yang Xiao: supervised rhodamine B and methylene blue products analysis in cooperation with Shihao Chen, Yinhai Wang: conceived the concept, designed the experiments, analyzed the data and revised the manuscript, Zhengfa Hu: performed EPR analysis, Hui Zhao: performed XPS analysis, Wei Xie: performed FTIR and PL analysis .

Conflicts of Interest: The authors declare no conflict of interest.

References

1. Fujishima, A.; Honda, K. Electrochemical Photolysis of Water at a Semiconductor Electrode. *Nature* **1972**, *238*, 37–38. [CrossRef] [PubMed]
2. Zuo, F.; Wang, L.; Wu, T.; Zhang, Z.Y.; Borchardt, D.; Feng, P.Y. Self-Doped Ti^{3+} Enhanced Photocatalyst for Hydrogen Production under Visible Light. *J. Am. Chem. Soc.* **2010**, *132*, 11856–11857. [CrossRef] [PubMed]
3. Shao, J.; Sheng, W.C.; Wang, M.S.; Lia, S.J.; Chen, J.R.; Zhang, Y.; Cao, S.S. In situ Synthesis of Carbon-doped TiO_2 Single-crystal Nanorods with Aremarkably Photocatalytic Efficiency. *Appl. Catal. B Environ.* **2017**, *209*, 311–319. [CrossRef]
4. Yu, J.G.; Low, J.X.; Xiao, W.; Zhou, P.; Jaroniec, M. Enhanced Photocatalytic CO_2-Reduction Activity of Anatase TiO_2 by Coexposed {001} and {101} Facets. *J. Am. Chem. Soc.* **2014**, *136*, 8839–8842. [CrossRef] [PubMed]
5. Wang, W.; Ni, Y.R.; Lu, C.H.; Xu, Z.Z. Hydrogenation Temperature Related Inner Structures and Visible-light-driven Photocatalysis of N–F co-doped TiO_2 Nanosheets. *Appl. Surf. Sci.* **2014**, *290*, 125–130. [CrossRef]
6. Liu, B.; Chen, H.M.; Liu, C.; Andrews, S.C.; Hahn, C.; Yang, P.D. Large-Scale Synthesis of Transition-Metal-Doped TiO_2 Nanowires with Controllable Overpotential. *J. Am. Chem. Soc.* **2013**, *135*, 9995–9998. [CrossRef] [PubMed]
7. Inturi, S.N.R.; Boningari, T.; Suidan, M.; Smirniotis, P.G. Flame Aerosol Synthesized Cr Incorporated TiO_2 for Visible Light Photodegradation of Gas Phase Acetonitrile. *J. Phys. Chem. C* **2014**, *118*, 231–242. [CrossRef]
8. Chen, Z.Y.; Fang, L.; Dong, W.; Zheng, F.G.; Shen, M.R.; Wang, J.L. Inverse Opal Structured Ag/TiO_2 Plasmonic Photocatalyst Prepared by Pulsed Current Deposition and its Enhanced Visible Light Photocatalytic Activity. *J. Mater. Chem. A* **2014**, *2*, 824–832. [CrossRef]
9. Kamegawa, T.; Matsuura, S.; Seto, H.; Yamashita, H. A Visible-Light-Harvesting Assembly with a Sulfocalixarene Linker between Dyes and a Pt-TiO_2 Photocatalyst. *Angew. Chem. Int. Ed.* **2013**, *52*, 916–919. [CrossRef] [PubMed]
10. Pan, J.H.; Cai, Z.Y.; Yu, Y.; Zhao, X.S. Controllable Synthesis of Mesoporous F-TiO_2 Spheres for Effective Photocatalysis. *J. Mater. Chem.* **2011**, *21*, 11430–11438. [CrossRef]
11. Senna, M.; Myers, N.; Aimable, A.; Laporte, V.; Pulgarin, C.; Baghriche, O.; Bowen, P. Modification of titania nanoparticles for photocatalytic antibacterial activity via a colloidal route with glycine and subsequent annealing. *J. Mater. Res.* **2013**, *28*, 354–361. [CrossRef]
12. Milosevic, I.; Jayaprakash, A.; Greenwood, B.; van Driel, B.; Rtimi, S.; Bowen, P. Synergistic Effect of Fluorinated and N Doped TiO_2 Nanoparticles Leading to Different Microstructure and Enhanced Photocatalytic Bacterial Inactivation. *Nanomaterials* **2017**, *7*, 391. [CrossRef] [PubMed]
13. Zhang, Q.Y.; Gao, T.T.; Andino, J.M.; Li, Y. Copper and Iodine Co-modified TiO_2 Nanoparticles for Improved Activity of CO_2 Photo Reduction with Water Vapor. *Appl. Catal. B Environ.* **2012**, *123*, 257–264. [CrossRef]
14. Zhang, M.; Wu, J.; Hou, J.; Yang, J.J. Molybdenum and Nitrogen Co-doped Titanium Dioxide Nanotube Arrays with Enhanced Visible Light Photocatalytic Activity. *Sci. Adv. Mater.* **2013**, *5*, 535–541. [CrossRef]
15. Chen, X.B.; Liu, L.; Yu, P.Y.; Mao, S.S. Increasing Solar Absorption for Photocatalysis with Black Hydrogenated Titanium Dioxide Nanocrystals. *Science* **2011**, *331*, 746–750. [CrossRef] [PubMed]
16. Shin, J.Y.; Joo, J.H.; Samuelis, D.; Maier, J. Oxygen-Deficient $TiO_{2-\delta}$ Nanoparticles via Hydrogen Reduction for High Rate Capability Lithium Batteries. *Chem. Mater.* **2012**, *24*, 543–551. [CrossRef]
17. Samsudina, E.M.; Hamida, S.B.A.; Juana, J.C.; Basiruna, W.J.; Kandjani, A.E. Surface Modification of Mixed-phase Hydrogenated TiO_2 and Corresponding Photocatalytic Response. *Appl. Surf. Sci.* **2015**, *359*, 883–896. [CrossRef]
18. Zhu, Q.; Peng, Y.; Lin, L.; Fan, C.M.; Gao, G.Q.; Wang, R.X.; Xu, A.W. Stable Blue TiO_{2-x} Nanoparticles for Efficient Visible Light Photocatalysts. *J. Mater. Chem. A* **2014**, *2*, 4429–4437. [CrossRef]

19. Zou, X.X.; Liu, J.K.; Su, J.; Zuo, F.; Chen, J.S.; Feng, P.Y. Facile Synthesis of Thermal-and Photostable Titania with Paramagnetic Oxygen Vacancies for Visible-Light Photocatalysis. *Chem. Eur. J.* **2013**, *19*, 2866–2873. [CrossRef] [PubMed]
20. Kang, Q.; Cao, J.Y.; Zhang, Y.J.; Liu, L.Q.; Xu, H.; Ye, J.H. Reduced TiO_2 Nanotube Arrays for Photoelectrochemical Water Splitting. *J. Mater. Chem. A* **2013**, *1*, 5766–5774. [CrossRef]
21. Liu, N.; Schneider, C.; Freitag, D.; Hartmann, M.; Venkatesan, U.; Müller, J.; Spiecker, E.; Schmuki, P. Black TiO_2 Nanotubes: Cocatalyst-Free Open-Circuit Hydrogen Generation. *Nano Lett.* **2014**, *14*, 3309–3313. [CrossRef] [PubMed]
22. Wang, Z.; Yang, C.Y.; Lin, T.Q.; Yin, H.; Chen, P.; Wan, D.Y.; Xu, F.F.; Huang, F.Q.; Lin, J.H.; Xie, X.M.; et al. Visible-light Photocatalytic, Solar Thermal and Photoelectrochemical Properties of Aluminium-reduced Black Titania. *Energy Environ. Sci.* **2013**, *6*, 3007–3014. [CrossRef]
23. Sinhamahapatra, A.; Jeon, J.P.; Yu, J.S. A New Approach to Prepare Highly Active and Stable Black Titania for Visible Light-assisted Hydrogen Production. *Energy Environ. Sci.* **2015**, *8*, 3539–3544. [CrossRef]
24. Zhou, X.M.; Zolnhofer, E.M.; Nguyen, N.T.; Liu, N.; Meyer, K.; Schmuki, P. Stable Co-Catalyst-Free Photocatalytic H_2 Evolution From Oxidized Titanium Nitride Nanopowders. *Angew. Chem. Int. Ed.* **2015**, *54*, 13385–13389. [CrossRef] [PubMed]
25. Liu, X.; Gao, S.; Xu, H.; Lou, Z.Z.; Wang, W.J.; Huang, B.B.; Dai, Y. Green Synthetic Approach for Ti^{3+} Self-doped TiO_{2-x} Nanoparticles with Efficient Visible Light Photocatalytic Activity. *Nanoscale* **2013**, *5*, 1870–1875. [CrossRef] [PubMed]
26. Pei, Z.X.; Ding, L.Y.; Lin, H.; Weng, S.X.; Zheng, Z.Y.; Hou, Y.D.; Liu, P. Facile Synthesis of Defect-mediated TiO_{2-x} with Enhanced Visible Light Pphotocatalytic Activity. *J. Mater. Chem. A* **2013**, *1*, 10099–10102. [CrossRef]
27. Wang, G.M.; Wang, H.Y.; Ling, Y.C.; Tang, Y.C.; Yang, X.Y.; Fitzmorris, R.C.; Wang, C.C.; Zhang, J.Z.; Li, Y. Hydrogen-Treated TiO_2 Nanowire Arrays for Photoelectrochemical Water Splitting. *Nano Lett.* **2011**, *11*, 3026–3033. [CrossRef] [PubMed]
28. Dong, J.Y.; Han, J.; Liu, Y.S.; Nakajima, A.; Matsushita, S.; Wei, S.H.; Gao, W. Defective Black TiO_2 Synthesized via Anodization for Visible-Light Photocatalysis. *ACS Appl. Mater. Interfaces* **2014**, *6*, 1385–1388. [CrossRef] [PubMed]
29. Liu, X.Y.; Zhu, G.L.; Wang, X.; Yuan, X.T.; Lin, T.Q.; Huang, F.Q. Progress in Black Titania: A New Material for Advanced Photocatalysis. *Adv. Energy Mater.* **2016**, *6*, 1600452. [CrossRef]
30. Fan, C.Y.; Chen, C.; Wang, J.; Fu, X.X.; Ren, Z.M.; Qian, G.D.; Wang, Z.Y. Black Hydroxylated Titanium Dioxide Prepared via Ultrasonication with Enhanced Photocatalytic Activity. *Sci. Rep.* **2015**, *5*, 11712. [CrossRef] [PubMed]
31. Pan, H.; Zhang, Y.W.; Shenoy, V.B.; Gao, H.J. Effects of H-, N-, and (H, N)-Doping on the Photocatalytic Activity of TiO_2. *J. Phys. Chem. C* **2011**, *115*, 12224–12231. [CrossRef]
32. Wang, Z.Q.; Wen, B.; Hao, Q.Q.; Liu, L.M.; Zhou, C.Y.; Mao, X.C.; Lang, X.F.; Yin, W.J.; Dai, D.X.; Selloni, A.; et al. Localized Excitation of Ti^{3+} Ions in the Photoabsorption and Photocatalytic Activity of Reduced Rutile TiO_2. *J. Am. Chem. Soc.* **2015**, *137*, 9146–9152. [CrossRef] [PubMed]
33. Wang, W.; Lu, C.H.; Ni, Y.R.; Song, J.B.; Su, M.X.; Xu, Z.Z. Enhanced Visible-light Photoactivity of {001} Facets Dominated TiO_2 Nanosheets with Even Distributed Bulk Oxygen Vacancy and Ti^{3+}. *Catal. Commun.* **2012**, *22*, 19–23. [CrossRef]
34. Chen, X.B.; Liu, L.; Liu, Z.; Marcus, M.A.; Wang, W.C.; Oyler, N.A.; Grass, M.E.; Mao, B.H.; Glans, P.A.; Yu, P.Y. Properties of Disorder-engineered Black Titanium Dioxide Nanoparticles Through Hydrogenation. *Sci. Rep.* **2013**, *3*, 1510. [CrossRef] [PubMed]
35. Gupta, S.K.; Desai, R.; Jha, P.K.; Sahoob, S.; Kirinc, D. Titanium Dioxide Synthesized Using Titanium Chloride: Size Effect Study Using Raman Spectroscopy and Photoluminescence. *J. Raman Spectrosc.* **2010**, *41*, 350–355. [CrossRef]
36. Parker, J.C.; Siegel, R.W. Calibration of the Raman Spectrum to the Oxygen Stoichiometry of Nanophase TiO_2. *Appl. Phys. Lett.* **1990**, *57*, 943–945. [CrossRef]
37. Liu, H.; Ma, H.T.; Li, X.Z.; Li, W.Z.; Wu, M.; Bao, X.H. The Enhancement of TiO_2 Photocatalytic Activity by Hydrogen Thermal Treatment. *Chemosphere* **2003**, *50*, 39–46. [CrossRef]
38. Mao, L.; Cai, X.Y.; Gao, H.; Diao, X.G.; Zhang, J.Y. A Newly Designed Porous Oxynitride Photoanode with Enhanced Charge Carrier Mobility. *Nano Energy* **2017**, *39*, 172–182. [CrossRef]

39. Serpone, N. Is the Band Gap of Pristine TiO_2 Narrowed by Anion-and Cation-Doping of Titanium Dioxide in Second-Generation Photocatalysts? *J. Phys. Chem. B* **2006**, *110*, 24287–24293. [CrossRef] [PubMed]
40. Chen, J.; Lin, L.B.; Jing, F.Q. Theoretical Study of F-type Color Center in Rutile TiO_2. *J. Phys. Chem. Solids* **2001**, *62*, 1257–1262. [CrossRef]
41. Chen, J.; Song, W.X.; Hou, H.S.; Zhang, Y.; Jing, M.J.; Jia, X.N.; Ji, X.B. Ti^{3+} Self-Doped Dark Rutile TiO_2 Ultrafine Nanorods with Durable High-Rate Capability for Lithium-Ion Batteries. *Adv. Funct. Mater.* **2015**, *25*, 6793–6801. [CrossRef]
42. Huo, J.C.; Hu, Y.J.; Jiang, H.; Li, C.Z. In Situ Surface Hydrogenation Synthesis of Ti^{3+} Self-doped TiO_2 with Enhanced Visible Light Photoactivity. *Nanoscale* **2014**, *6*, 9078–9084. [CrossRef] [PubMed]
43. Ullattila, S.G.; Periyat, P. A 'One Pot' Gel Combustion Strategy towards Ti^{3+} Self-doped 'black' Anatase TiO_{2-x} Solar Photocatalyst. *J. Mater. Chem. A* **2016**, *4*, 5854–5858. [CrossRef]
44. Atuchin, V.V. Comment on "A 'one pot' Gel Combustion Strategy towards Ti^{3+} Self-doped 'black' Anatase TiO_{2-x} Solar Photocatalyst," by S.G. Ullattil and P. Periyat, *J. Mater. Chem. A*, 2016, 4, 5854. *J. Mater. Chem. A* **2017**, *5*, 426–427. [CrossRef]
45. Pan, X.Y.; Yang, M.Q.; Fu, X.Z.; Zhang, N.; Xu, Y.J. Defective TiO_2 with Oxygen Vacancies: Synthesis, Properties and Photocatalytic Applications. *Nanoscale* **2013**, *5*, 3601–3614. [CrossRef] [PubMed]
46. Moya, A.; Cherevan, A.; Marchesan, S.; Gebhardt, P.; Prato, M.; Eder, D.; Vilatela, J.J. Oxygen Vacancies and Interfaces Enhancing Photocatalytic Hydrogen Production in Mesoporous CNT/TiO_2 Hybrids. *Appl. Catal. B Environ.* **2015**, *179*, 574–582. [CrossRef]
47. Wang, Z.; Yang, C.Y.; Lin, T.Q.; Yin, H.; Chen, P.; Wan, D.Y.; Xu, F.F.; Huang, F.Q.; Lin, J.H.; Xie, X.M.; et al. H-Doped Black Titania with Very High Solar Absorption and Excellent Photocatalysis Enhanced by Localized Surface Plasmon Resonance. *Adv. Funct. Mater.* **2013**, *23*, 5444–5450. [CrossRef]
48. Sharma, P.K.; Pandey, A.C.; Zolnierkiewicz, G.; Guskos, N.; Rudowicz, C. Relationship between Oxygen Defects and the Photoluminescence Property of ZnO Nanoparticles: A Spectroscopic View. *J. Appl. Phys.* **2009**, *106*, 094314. [CrossRef]
49. Tian, Z.L.; Cui, H.L.; Zhu, G.L.; Zhao, W.L.; Xu, J.J.; Shao, F.; He, J.Q.; Huang, F.Q. Hydrogen Plasma Reduced Black TiO_2-B Nanowires for Enhanced Photoelectrochemical Water-Splitting. *J. Power Sources* **2016**, *325*, 697–705. [CrossRef]
50. Komaguchi, K.; Maruoka, T.; Nakano, H.; Imae, I.; Ooyama, Y.; Harima, Y. Electron-Transfer Reaction of Oxygen Species on TiO_2 Nanoparticles Induced by Sub-band-gap Illumination. *J. Phys. Chem. C* **2010**, *114*, 1240–1245. [CrossRef]
51. Janotti, A.; Varley, J.B.; Rinke, P.; Umezawa, N.; Kresse, G.; Van de Walle, C.G. Hybrid Functional Studies of the Oxygen Vacancy in TiO_2. *Phys. Rev. B* **2010**, *81*, 085212. [CrossRef]
52. Cronemeyer, D.C. Infrared Absorption of Reduced Rutile TiO_2 Single Crystals. *Phys. Rev.* **1959**, *113*, 1222–1226. [CrossRef]
53. Hu, Y.H. A Highly Efficient Photocatalyst-Hydrogenated Black TiO_2 for the Photocatalytic Splitting of Water. *Angew. Chem. Int. Ed.* **2012**, *51*, 12410–12412. [CrossRef] [PubMed]

© 2018 by the authors. Licensee MDPI, Basel, Switzerland. This article is an open access article distributed under the terms and conditions of the Creative Commons Attribution (CC BY) license (http://creativecommons.org/licenses/by/4.0/).

nanomaterials

MDPI

Article

A Facile Method for the Preparation of Colored $Bi_4Ti_3O_{12-x}$ Nanosheets with Enhanced Visible-Light Photocatalytic Hydrogen Evolution Activity

Yizeng Zhang [1], Zhiwu Chen [1,2,*] and Zhenya Lu [1]

[1] College of Materials Science and Engineering, South China University of Technology, Guangzhou 510640, China; zyzdd286@163.com (Y.Z.); zhylu@scut.edu.cn (Z.L.)

[2] State Key Laboratory of Pulp and Paper Engineering, South China University of Technology, Guangzhou 510640, China

* Correspondence: chenzw@scut.edu.cn; Tel.: +86-20-22236602

Received: 15 March 2018; Accepted: 16 April 2018; Published: 21 April 2018

Abstract: $Bi_4Ti_3O_{12-x}$ nanosheet photocatalysts with abundant oxygen vacancies are fabricated by a facile solid-state chemical reduction method for the first time. This method is simple in operation, has short reaction time, and can be conducted at mild temperatures (300~400 °C). The electron paramagnetic resonance, thermogravimetric analysis, X-ray photoelectron spectrometer, and positron annihilation lifetime spectra results indicate that oxygen vacancies are produced in $Bi_4Ti_3O_{12-x}$, and they can be adjusted by tuning the reduction reaction conditions. Control experiments show that the reduction time and temperature have great influences on the photocatalytic activities of $Bi_4Ti_3O_{12-x}$. The optimal $Bi_4Ti_3O_{12-x}$ is the sample undergoing the reduction treatment at 350 °C for 60 min and it affords a hydrogen evolution rate of 129 $\mu mol \cdot g^{-1} \cdot h^{-1}$ under visible-light irradiation, which is about 3.4 times that of the pristine $Bi_4Ti_3O_{12}$. The $Bi_4Ti_3O_{12-x}$ photocatalysts have good reusability and storage stability and can be used to decompose formaldehyde and formic acid for hydrogen production. The surface oxygen vacancies states result in the broadening of the valence band and the narrowing of the band gap. Such energy level structure variation helps promote the separation of photo-generated electron-hole pairs thus leading to enhancement in the visible-light photocatalytic hydrogen evolution. Meanwhile, the narrowing of the band gap leads to a broader visible light absorption of $Bi_4Ti_3O_{12-x}$.

Keywords: $Bi_4Ti_3O_{12}$ nanosheets; photocatalytic hydrogen evolution; solid-state chemical reduction; oxygen vacancy

1. Introduction

The development of green energy has become one of the most prominent research fields. Among many energy sources, hydrogen gas has been considered one of the prime candidates for solving the emerging worldwide energy crisis, owing to the fact that it does not produce pollution and it has high energy density [1,2]. In recent years, hydrogen generation through photocatalytic water splitting utilizing solar energy has become an extremely active research area [2]. An ideal photocatalyst should be stable, non-toxic, easily available from nature, able to function under visible light, and highly efficient in separating photo-generated electron-hole pairs. So far, various types of photocatalysts, such as TiO_2 [3], ZnO [4], and CdS [5], have been developed. However, none of the photocatalysts can meet all the above requirements simultaneously and most of them suffer from a narrow photo response wavelength range because of large band gaps, which lead to very low utilization efficiency of solar energy. Therefore, the exploration of novel, efficient visible-light photocatalysts for splitting water to produce H_2 is of utmost importance [6,7].

Bismuth titanate ($Bi_4Ti_3O_{12}$) is well-known as a ferroelectric agent that has a special Aurivillius architecture with unique, electro-optic-converting performance. The Aurivillius phases of $Bi_4Ti_3O_{12}$ have structures that are intergrown layers of $[Bi_2O_2]^{2+}$ alternating with perovskite-like $[Bi_2Ti_3O_{10}]^{2-}$ blocks [8–10]. The advantage of such a layered structure is its ability to efficiently assist in diffusion and separation of the electron-hole pairs generated by light irradiation. Such separation increases the lifetime of the associated charge carriers, thus improving the quantum efficiency of photo-degradation [8,9,11]. Moreover, the hybridized $6s^2$ of Bi^{3+} and O2p generate a new valence band, which reduces the band gap of $Bi_4Ti_3O_{12}$ [8,9]. Since the photocatalytic activity of $Bi_4Ti_3O_{12}$ for water splitting was reported by Kudo et al., its photocatalytic properties have been receiving immense attention [12]. Various types of $Bi_4Ti_3O_{12}$ photocatalysts, such as nanofibers [8,11], particles [12,13], platelets [14], and films [15] have been developed for processes involving alternative energy development or destruction of different pollutants. However, these $Bi_4Ti_3O_{12}$ nanomaterials are not very efficient as visible-light photocatalysts because of the high recombination rate of photo-induced electron-hole pairs [16–18]. A widely adopted methodology to overcome this recombination problem is to couple $Bi_4Ti_3O_{12}$ with semiconductors having narrower band gaps, such as Bi_2MoO_6 [11], BiOI [16], $Bi_2Ti_2O_7$ [17], BiOCl [18], Ag_3PO_4 [19], and g-C_3N_4 [20]. However, the synthesis of composite compounds generally requires sophisticated synthesis techniques. Additionally, such synthesis increases the possibilities of introducing thermodynamic and structural instability due to the additionally introduced layer. Lastly, electrons would go through a multi-step transport process in such layered structures, which may diminish the efficient charge separation. Thus, catalysts comprised of single-phase metal oxides are highly desirable because they can provide both reliable stability and efficient electron-hole separation [21]. It is therefore very important to explore facile and economic techniques to prepare a single-phase $Bi_4Ti_3O_{12}$ photocatalyst with increased absorbance of visible light and a low carrier recombination rate.

It has been found recently that the oxygen vacancy defects in TiO_2 [22,23], ZnO [24,25], and Fe_2O_3 [26] are able to enhance photocatalytic performance. In particular, surface oxygen vacancies can first capture photo-generated electrons and then promote the reaction between these electrons with the adsorbed species, thus effectively preventing the recombination of photo-generated electron-hole pairs and improving the photocatalytic performance [22–26]. Bulk oxygen vacancies, on the contrary, are recombination centers of photo-generated electron-hole pairs and will reduce the photocatalytic performance [25]. The surface and bulk oxygen vacancies play different roles in the photocatalytic reaction. However, the effect of oxygen vacancies on the photocatalytic activity of $Bi_4Ti_3O_{12}$ has not been thoroughly explored. Therefore, the development of cost-effective synthesis procedures for the production of $Bi_4Ti_3O_{12-x}$ with oxygen vacancies and the in-depth understanding of its catalytic behavior are of profound importance in order to realize fully the great potential of $Bi_4Ti_3O_{12-x}$ in water splitting for H_2 production.

We recently reported a sol-gel hydrothermal technique to prepare highly crystalline $Bi_4Ti_3O_{12}$ nanosheets with enhanced catalytic activity towards the photodegradation of Rhodamine B, especially when comparing with the calcined sample [9]. In the present work, to improve the photocatalytic activity even further, a $Bi_4Ti_3O_{12}$ nanosheet photocatalyst with oxygen vacancies ($Bi_4Ti_3O_{12-x}$) was fabricated for the first time by the solid-state chemical reduction method using $NaBH_4$ and $Bi_4Ti_3O_{12}$ nanosheets. The reported methods on oxygen vacancy creation include heating the sample under an oxygen-deficient atmosphere (e.g., vacuum) or reducing conditions (e.g., H_2) [25,26], chemical vapor deposition, high-energy particle (laser, electron, or Ar^+) bombardment [27], combustion method [23], high pressure [28], high temperature aluminum vapor reduction [29], etc. For practical application, these strategies have a number of limitations, such as multiple steps, harsh synthesis conditions (high temperature (>500 °C) or high-pressure hydrogen (20 bar)), or expensive facilities. Compared with the above traditional methods, the solid-state chemical reduction method has many advantages, such as simple operation, moderate reaction temperature (350 °C), short reaction time (less than 80 min), simple equipment, etc. Using this method, several $Bi_4Ti_3O_{12-x}$ samples with different reduction degrees and, as a consequence, with different colors (several shades of blue, as well as black) have been synthesized. The effects of reduction time and temperature on the visible-light photocatalytic properties of the as-prepared $Bi_4Ti_3O_{12-x}$ nanosheets have been investigated systematically. The electron paramagnetic resonance (EPR), thermogravimetric analysis (TGA), and positron annihilation lifetime spectra (PALS) indicate that oxygen vacancies are produced on $Bi_4Ti_3O_{12}$ nanosheets during the reduction process. It has been found that the aggregation of oxygen vacancies raises the valence band maximum (VBM), thus decreasing the band gap and extending the photo response wavelength range. Moreover, the energy level variation induced by oxygen vacancy can facilitate the separation efficiency of the photo-generated electron-hole pairs, which contributes significantly to the improvement of the photocatalytic performance of $Bi_4Ti_3O_{12-x}$. In this paper, we also propose a mechanism for the decrease of the band gap of $Bi_4Ti_3O_{12-x}$ and its photocatalytic activity improvement.

2. Experimental Section

2.1. Synthesis of the $Bi_4Ti_3O_{12-x}$ Nanosheet Photocatalyst

$Bi_4Ti_3O_{12}$ nanosheets were synthesized using a sol-gel hydrothermal technique reported in our previous work (see Supplementary Materials) [9]. $Bi_4Ti_3O_{12-x}$ catalysts were prepared as follows: 4 g of the as-prepared $Bi_4Ti_3O_{12}$ nanosheets and 1 g $NaBH_4$ were ground together for 45 min. Then, the mixture was put into a quartz tube placed in a tubular furnace; the temperature of the furnace was controlled by a heating device. The mixture was calcined at 350 °C for 20~100 min or 300~400 °C for 60 min under a nitrogen atmosphere with a ramping up rate of 10 °C/min. After the completion of this solid-state chemical reduction process, the powders were allowed to cool down to ambient temperature. Then, the final product was filtered, rinsed with deionized water and ethanol, and then dried at 80 °C. The samples were marked as $Bi_4Ti_3O_{12-x}$ (T, t), where T is the temperature and t is the time of the solid-state chemical reduction procedure.

2.2. Characterization

The crystalline structures of the samples were examined by an X-ray diffractometer (XRD, D/Max-3C, Rigaku Co., Tokyo, Japan) using Cu Kα radiation (λ = 1.5418 Å). The chemical states and composition were examined using an X-ray photoelectron spectrometer (XPS, Axis uhru DCD, Manchester, UK) with a monochromatic Mg Kα X-ray source. The morphology was characterized using a scanning electron microscopy (SEM, LEO 1530 VP, Zeiss, Oberkochen, Germany) with 20 kV of accelerating voltage, and the composition was examined using energy dispersive spectroscopy (EDS, Zeiss, Oberkochen, Germany) attached to the SEM. Transmission electron microscopy (TEM) and high-resolution transmission electron microscopy (HRTEM) were performed using JEOL-2011 instrument (JEOL, Tokyo, Japan). UV-Vis spectrophotometer (lambda 35, Perkin-Elmer, Shelton, WA, USA) was used to analyze the samples' diffuse reflection spectra with $BaSO_4$ as a reference. TGAs over a temperature range of 25–800 °C were conducted during progressive heating (10 °C/min) in an air atmosphere by a thermogravimetric analyzer (SDTA 851e, Mettler Toledo, Zurich, Switzerland). The EPR analysis was performed on an Endor spectrometer JEOL ES-ED3X (JEOL, Tokyo, Japan). The specific surface areas of the powders were determined using the Brunauer-Emmett-Teller (BET) method after cooling down the samples with liquid nitrogen. PALS were obtained by ORTEC-583 fast-slow coincident system.

Photocurrent and electrochemical impedance spectroscopy (EIS) measurements were conducted at frequencies between 1×10^{-5} and 100 kHz by a CHI 660 electrochemical instrument (CH Instruments, CH Instruments, Austin, TX, USA) in 0.5 M Na_2SO_4 electrolyte solution. A three-electrode cell system was implemented with ITO (Indium tin oxide)/$Bi_4Ti_3O_{12}$ (or $Bi_4Ti_3O_{12-x}$) as the working electrode, Pt wire as the counter electrode, and standard calomel electrode (SCE) as the reference. The true potentials were calculated in reference to the results from the SCE. The visible light source was a 300 W Xe lamp with a 400 nm cut-off filter. The photocatalyst photoelectric responses "on and off" were determined at 0.0 V. Moreover, the ITO/$Bi_4Ti_3O_{12}$ (or $Bi_4Ti_3O_{12-x}$) electrodes were produced using the following recipe: First, $Bi_4Ti_3O_{12}$ (or $Bi_4Ti_3O_{12-x}$) samples (5 mg) were mixed with ethyl alcohol (0.15 mL) and 5% Nafion DE 520 solution (0.35 mL). This solution was then homogenized under ultrasound agitation for 20 min. Uniform film electrodes were prepared by casting 0.1 mL of the $Bi_4Ti_3O_{12}$ (or $Bi_4Ti_3O_{12-x}$) slurry onto pre-cleaned ITO glass (<7 ohm/square). The ITO/$Bi_4Ti_3O_{12}$ (or $Bi_4Ti_3O_{12-x}$) electrodes were finalized by sintering at 100 °C for 2 h.

2.3. Photocatalytic Activity

Photocatalytic hydrogen evolution experiments were proceeded in a methanol-water mixture and performed in the outer quartz ampules attached to the airtight gas circulation system. In order to promote the dispersion of $Bi_4Ti_3O_{12}$ or $Bi_4Ti_3O_{12-x}$ in the methanol-water mixture, the optimal dispersing process was obtained through orthogonal tests. The $Bi_4Ti_3O_{12}$ or $Bi_4Ti_3O_{12-x}$ (0.25 g) was dispersed by ultrasound agitation (15 min) in the mixture of deionized water (DI water, 200 mL) and methanol (20 mL), followed by constant stirring for 30 min. To eliminate dissolved oxygen, the solution was purged with argon for 30 min. This step is necessary to minimize the recombination reaction between H_2 and O_2 during the water splitting reaction. It also improves the purity of the hydrogen and reduces explosion risks.

Upon finishing the preparation steps, the reactor was irradiated by a 300 W Xe lamp for 4 h. A cut-off filter for visible-light irradiation was used to obtain wavelengths above 400 nm. The total amount of hydrogen evolved was measured by a gas chromatographer GC-3240/TCD (Perfect Light, Beijing, China), which used Ar carrier gas and was directly connected to a gas-circulation line. The suspensions of the photocatalyst were magnetically stirred continuously during the photocatalytic hydrogen evolution. An average of five measurements was adopted to determine the yield. To study the reusability and repeatability, the same $Bi_4Ti_3O_{12-x}$ photocatalyst was used three times, and the amount of hydrogen evolved was recorded. After each measurement, the catalyst was centrifugated, filtrated, washed by DI water, and dried at 100 °C for 3 h for the next measurement. To check the dependency of H_2 production on the additives, the same experiments were also performed with formaldehyde and formic acid instead of methanol.

3. Results and Discussion

3.1. Morphology, Structure, and UV-Vis Spectra of $Bi_4Ti_3O_{12}$ and $Bi_4Ti_3O_{12-x}$

The morphology and particle size distribution of a pristine $Bi_4Ti_3O_{12}$ were scrutinized using SEM. As shown in Figure 1a, the $Bi_4Ti_3O_{12}$ sample consists mostly of regular rectangular nanosheets with a narrow distribution of rim size and thin thickness. The average rim size of the rectangular nanosheets are about 100 and 150 nm, respectively, and thickness is about 20 nm. As shown in Figure 1b, only three elements of Bi, Ti, and O are observed with the EDS measurement. The ratio of the Bi, Ti, and O elements is 20.98:15.75:63.27, further demonstrating that the pure $Bi_4Ti_3O_{12}$ was hydrothermally synthesized successfully at 160 °C.

When the mixture of $Bi_4Ti_3O_{12}$ nanosheets and $NaBH_4$ were heated at 350 °C for 20~100 min and 300~400 °C for 60 min under nitrogen, $NaBH_4$ decomposed and produced active hydrogen. It has been reported that the reduction ability of this active hydrogen is greater than that of H_2 and other reducing agents at these temperature [23,30,31]. Hydrogen is a strong reducing agent capable of reacting fast and generating oxygen vacancies in $Bi_4Ti_3O_{12}$ at relatively low temperatures, as well as maintaining the original shape of the $Bi_4Ti_3O_{12}$ nanosheets at the same time. As shown in Figure 2a, the color of the $Bi_4Ti_3O_{12-x}$ samples is changed clearly after the solid-state chemical reduction. When the mixture of $Bi_4Ti_3O_{12}$ nanosheets and $NaBH_4$ were heated at 350 °C, it could be seen that with the increased duration of the reaction, the colors of $Bi_4Ti_3O_{12-x}$ changed from white-yellow to light blue, then to dark blue. Moreover, the colors of $Bi_4Ti_3O_{12-x}$ also became darker with the increase in temperature. When the reaction temperature rises to 400 °C, black $Bi_4Ti_3O_{12-x}$ can be synthesized in 60 min, suggesting that the solid-state chemical reduction process modifies the surface features of the $Bi_4Ti_3O_{12}$ nanosheets. According to previous reports [32,33], it is highly possible that the color change could be caused by the formation of oxygen vacancies occurring during the reduction. The different colors of the $Bi_4Ti_3O_{12-x}$ samples indicate that various reduction degrees of $Bi_4Ti_3O_{12-x}$ can be obtained by adjusting the reaction time or temperature, which is helpful to understand the formation mechanism of reductive $Bi_4Ti_3O_{12}$. Moreover, the XPS results show no residue of B and Na in $Bi_4Ti_3O_{12-x}$ (see Figure S1 in Supplementary Materials), which means that the coproducts from $NaBH_4$ during the solid-state chemical reduction method can be cleaned easily by washing with water and ethanol.

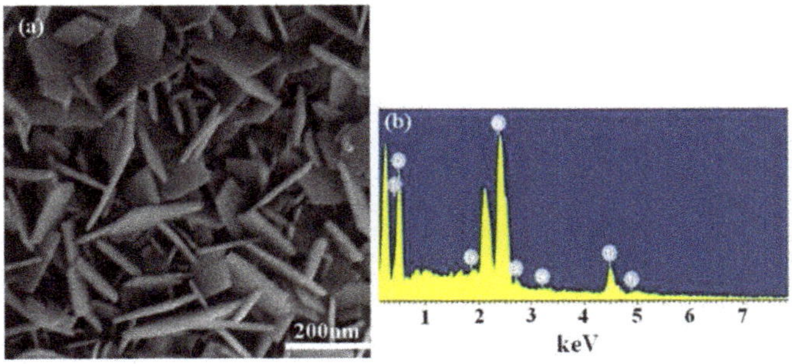

Figure 1. (**a**) SEM image; (**b**) energy dispersive spectroscopy (EDS) pattern of the pristine $Bi_4Ti_3O_{12}$ powders synthesized using a sol-gel hydrothermal technique at 160 °C for 16 h.

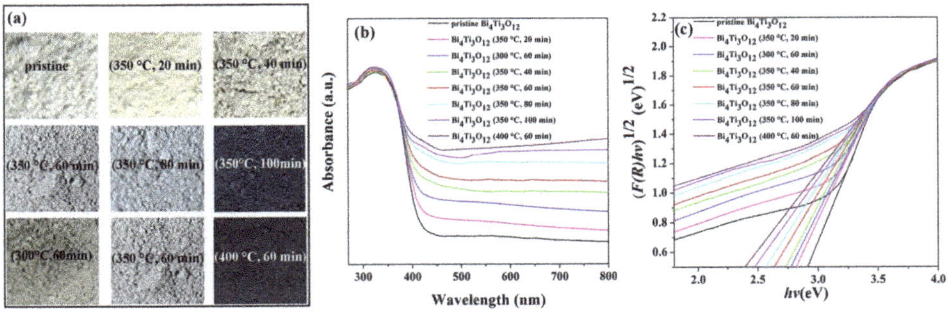

Figure 2. (**a**) photographs of pristine $Bi_4Ti_3O_{12}$ and various colored $Bi_4Ti_3O_{12-x}$; (**b**) UV-Vis absorption spectrum of the pristine $Bi_4Ti_3O_{12}$ and various $Bi_4Ti_3O_{12-x}$; (**c**) plot of the transformed Kubelka-Munk function ($F(R\infty)$) versus the photon energy ($h\upsilon$) for various $Bi_4Ti_3O_{12-x}$ and the pristine $Bi_4Ti_3O_{12}$ nanosheets.

To understand the effect of the solid-state chemical reduction treatment on the optical absorption property of the photocatalyst, the UV-Vis diffuse reflectance spectra of various $Bi_4Ti_3O_{12-x}$ and the pristine $Bi_4Ti_3O_{12}$ nanosheets were examined, as shown in Figure 2b. It can be seen that the pristine $Bi_4Ti_3O_{12}$ shows a typical spectrum with an absorption edge at about 420 nm. Compared with the pristine $Bi_4Ti_3O_{12}$ sample, the absorption edge of $Bi_4Ti_3O_{12-x}$ exhibits a clear red shift to higher wavelengths. In addition, the absorbance intensity of $Bi_4Ti_3O_{12-x}$ in the range of 400–800 nm increases with both reduction time and temperature, which agrees with the color variation in the samples. The red shift of the absorption edge and the enhanced absorbance intensity of $Bi_4Ti_3O_{12-x}$ are probably because of the different surface conditions of different samples. Surface defects, such as oxygen vacancies, generally affect the atomic structure of a photocatalyst and its surface states, which play a very important role in the overall photocatalytic activity [33,34].

The band gap can be calculated using the UV-Vis data from the following equation [8]:

$$\alpha h v = A(h v - E_g)^{n/2} \quad (1)$$

where α is an absorption coefficient, h is Planck's constant, v is light frequency, E_g is a band gap value, and A is a constant. The absorption behavior of $Bi_4Ti_3O_{12}$ demonstrates indirect transition between bands; therefore, the value of n equal to 4 is used [8,35]. The value of the band gap is estimated by extrapolating the linear part of the $(\alpha h v)^{1/2}$ versus $(h v)$ plot at $\alpha = 0$. Normally, the collected UV-Vis diffuse reflectance spectra can be converted into Kubelka-Munk function $F(R_\infty)$ based on the relationship shown in Equations (2) and (3) [36]:

$$\text{Abs} = -\log R_\infty \quad (2)$$

$$F(R_\infty) = (1 - R_\infty)^2/2R_\infty = \alpha \quad (3)$$

where Abs is absorbance, and R_∞ is reflectance. Therefore, Equation (1) can also be written as follows:

$$[F(R_\infty)\, h\nu] = A(h v - E_g)^{n/2} \quad (4)$$

In addition, $h v = hc/\lambda \approx 1241/\lambda$ (eV). Figure 2c shows the plot of the transformed Kubelka-Munk function versus the photon energy for various $Bi_4Ti_3O_{12-x}$ and the pristine $Bi_4Ti_3O_{12}$ nanosheets. The energy of the band gap values is obtained by extrapolating the linear part of $[F(R_\infty)h\nu]^{1/2}$ versus $h v$ plot at $F(R_\infty) = 0$. The band gap of the pristine $Bi_4Ti_3O_{12}$ and various $Bi_4Ti_3O_{12-x}$ samples are shown in Table 1. It is clear that the $Bi_4Ti_3O_{12-x}$ samples show a decreased band gap value when compared with the pristine $Bi_4Ti_3O_{12}$. Furthermore, it is also clear that the band gap of $Bi_4Ti_3O_{12-x}$ decreases as reaction time and temperature increase.

Table 1. Band gaps of the pristine $Bi_4Ti_3O_{12}$ and various $Bi_4Ti_3O_{12-x}$ samples.

Samples	Band Gap (eV)
$Bi_4Ti_3O_{12}$	2.91
$Bi_4Ti_3O_{12-x}$ (350 °C, 20 min)	2.83
$Bi_4Ti_3O_{12-x}$ (350 °C, 40 min)	2.74
$Bi_4Ti_3O_{12-x}$ (350 °C, 60 min)	2.63
$Bi_4Ti_3O_{12-x}$ (350 °C, 80 min)	2.57
$Bi_4Ti_3O_{12-x}$ (350 °C, 100 min)	2.48
$Bi_4Ti_3O_{12-x}$ (300 °C, 60 min)	2.77
$Bi_4Ti_3O_{12-x}$ (400 °C, 60 min)	2.39

XRD analyses were performed to characterize the changes of the crystalline structures of the $Bi_4Ti_3O_{12-x}$ samples. Figure 3 shows a comparison of XRD patterns of the pristine $Bi_4Ti_3O_{12}$ and various $Bi_4Ti_3O_{12-x}$ samples after they were treated at 350 °C for 20~100 min and at 400 °C for 60 min. No impurities can be seen for the $Bi_4Ti_3O_{12-x}$ samples, indicating that the reduction process has no effect on the crystal structure. The diffraction peaks suggest that $Bi_4Ti_3O_{12-x}$ samples have a high degree of crystalline similarity to $Bi_4Ti_3O_{12}$. However, new peaks appear in the XRD patterns of the samples treated for a longer reaction time (350 °C for 120 min) or at a higher temperature (400 °C for 80 min) (Figure S2). We could not match these new peaks to any known powder diffraction file (PDF); thus, it is possible that new phases have been formed.

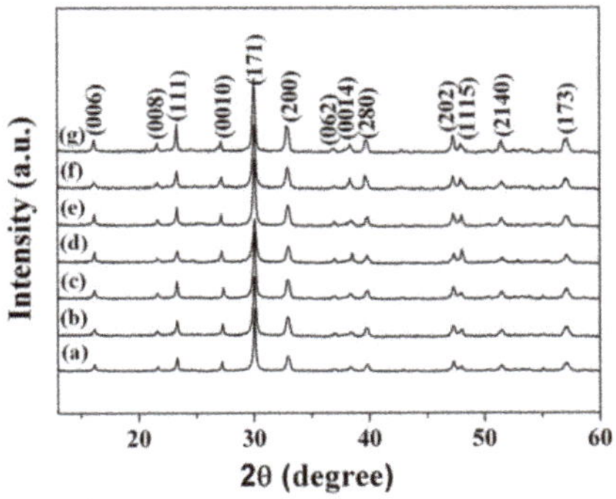

Figure 3. X-ray diffraction patterns of (a) the pristine $Bi_4Ti_3O_{12}$; (b) $Bi_4Ti_3O_{12-x}$ (350 °C, 20 min); (c) $Bi_4Ti_3O_{12-x}$ (350 °C, 40 min); (d) $Bi_4Ti_3O_{12-x}$ (350 °C, 60 min); (e) $Bi_4Ti_3O_{12-x}$ (350 °C, 80 min); (f) $Bi_4Ti_3O_{12-x}$ (350 °C, 100 min); and (g) $Bi_4Ti_3O_{12-x}$ (400 °C, 60 min).

3.2. *Photocatalytic Performance and Stability*

Photocatalytic conversion of H_2O into H_2 using $Bi_4Ti_3O_{12}$ or $Bi_4Ti_3O_{12-x}$ in a methanol-water medium was performed in a quartz cell. Methanol was used to trap holes. Figure 4a,b show the photocatalytic activity of the pristine $Bi_4Ti_3O_{12}$ and $Bi_4Ti_3O_{12-x}$ after chemical reduction treatment at 350 °C for various times and at various temperatures for 60 min for water splitting into H_2 under visible-light irradiation. It is clear that both the reduction time and temperature have significant influences on the photocatalysis ability of $Bi_4Ti_3O_{12-x}$. The pristine $Bi_4Ti_3O_{12}$ photocatalyst displays the H_2 evolution rate of around 38 $\mu mol \cdot g^{-1} \cdot h^{-1}$. After the solid-state chemical reduction treatment, the $Bi_4Ti_3O_{12-x}$ samples all show enhanced photocatalytic activity of the hydrogen evolution. Figure 4a,b show that the photocatalytic activity of $Bi_4Ti_3O_{12-x}$ improves with both reduction time and temperature increase, until a maximum activity is achieved at 350 °C for 60 min. The H_2 evolution rate over $Bi_4Ti_3O_{12-x}$ (350 °C, 60 min) reaches 129 $\mu mol \cdot g^{-1} \cdot h^{-1}$, which is 3.4 times that of the pristine $Bi_4Ti_3O_{12}$. This value is also higher than those reported previously (as presented in Table 2). Further increases in the reaction time or temperature results in a reduced hydrogen evolution rate, even though the rates are still higher than when using a pristine $Bi_4Ti_3O_{12}$ nanosheet photocatalyst.

Table 2. Comparison of H_2 evolution rate of $Bi_4Ti_3O_{12-x}$ (350 °C, 60 min) and other $Bi_4Ti_3O_{12}$ photocatalysts recently reported.

Sample	Light Source	Reactant Solution	H_2 Evolution Rate/$\mu mol \cdot g^{-1} \cdot h^{-1}$	Reference
$Bi_4Ti_3O_{12-x}$ (350 °C, 60 min)	300 W Xe Lamp (λ > 400 nm)	200 mL water + 20 mL methanol	129	This work
$Bi_4Ti_3O_{12}$	350 W high pressure Xe lamp (λ > 400 nm)	400 mL water + 20 mL methanol	36	[37]
$Bi_4Ti_{2.6}Cr_{0.4}O_{12}$	350 W high pressure Xe lamp (λ > 400 nm)	400 mL water + 20 mL methanol	58.1	[37]
$Bi_4Ti_3O_{12}$	300 W Xe Lamp (λ > 400 nm)	400 mL water + 20 mL methanol	42	[38]
$Bi_4Ti_{2.6}Cr_{0.4}O_{12}$	300 W Xe Lamp (λ > 400 nm)	400 mL water + 20 mL methanol	98	[38]
$Bi_4Ti_{2.6}Cr_{0.4}O_{12}$	300 W Xe Lamp (λ > 420 nm)	400 mL water + 30 mL methanol	117	[39]

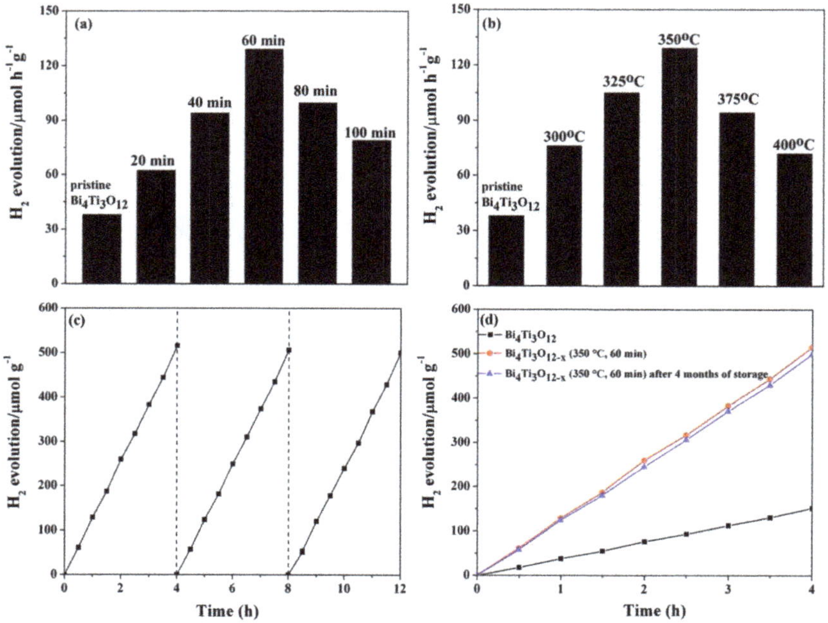

Figure 4. The hydrogen evolution rate over the pristine $Bi_4Ti_3O_{12}$ and the $Bi_4Ti_3O_{12-x}$ after chemical reduction treated (**a**) at 350 °C for various times; (**b**) at various temperature for 60 min under visible-light irradiation (λ > 400 nm); (**c**) recycling measure of hydrogen evolution with $Bi_4Ti_3O_{12-x}$ (350 °C, 60 min) under visible-light irradiation (λ > 400 nm); (**d**) visible-light photocatalytic hydrogen evolution by fresh $Bi_4Ti_3O_{12-x}$ (350 °C, 60 min) and $Bi_4Ti_3O_{12-x}$ (350 °C, 60 min) after four months of storage, compared with the pristine $Bi_4Ti_3O_{12}$.

To study the reusability and stability of the photocatalyst, cycling experiments using the optimal $Bi_4Ti_3O_{12-x}$ (350 °C, 60 min) under constant visible-light irradiation were performed. The results obtained from three consecutive experiments are shown in Figure 4c. The first run shows that around 517 $\mu mol \cdot g^{-1}$ of the total hydrogen evolved after 4 h when $Bi_4Ti_3O_{12-x}$ (350 °C, 60 min) is used. The second run of the experiment shows a 1.2% decrease in the hydrogen evolution rate comparing to the first run. The hydrogen evolution rate remains almost the same during the third run. The H_2 evolution rates for the $Bi_4Ti_3O_{12-x}$ (350 °C, 60 min) photocatalyst remain stable over the three times of cycling testing, confirming good operational stability even after introducing oxygen vacancies into the $Bi_4Ti_3O_{12}$ structure. Furthermore, in order to study the long-term stability of $Bi_4Ti_3O_{12-x}$, the photocatalytic H_2 production ability of fresh $Bi_4Ti_3O_{12-x}$ (350 °C, 60 min) and $Bi_4Ti_3O_{12-x}$ (350 °C, 60 min) after four months of storage were also examined. As shown in Figure 4d, the visible-light photocatalytic activity of the sample stored for four months is only slightly reduced compared with the fresh one, and it is still much higher than that of the pristine $Bi_4Ti_3O_{12}$. The results indicate that $Bi_4Ti_3O_{12-x}$ has good reusability and storage stability. In addition, to measure the apparent quantum efficiency (AQE), the same photocatalytic hydrogen evolution experiment was performed under 420 nm monochromatic lights irradiation, which were obtained by using band-pass filters for 1 h. The AQE was then calculated by the following Equation (5) [40]:

$$\text{AQE} = \frac{N_e}{N_p} \times 100\% = \frac{2MN_A hc}{SPt\lambda} \times 100\% \quad (5)$$

where N_e is the amount of reaction electrons, N_p is the amount of incident photons, N_A is the Avogadro constant, M is the amount of H_2 molecules, h is the Planck constant, c is the speed of light, S is irradiation area, P is the average intensity of the irradiation, t is the irradiation time, and λ is the wavelength of the monochromatic light. For the AQE at 420 nm, the average intensity of the irradiation P was determined to be 40 mW/cm^2, and the irradiation area S was 37.5 cm^2. The calculated AQE for $Bi_4Ti_3O_{12-x}$ (350 °C, 60 min) is 1.37% under irradiation at 420 nm, which is the highest among all the samples.

In the present work, methanol acts as a sacrificial agent, which is consumed during the formation of H_2. The photocatalytic mechanisms of methanol-assisted hydrogen evolution are as follows: the electron-hole pairs are produced when a $Bi_4Ti_3O_{12}$ or $Bi_4Ti_3O_{12-x}$ photocatalyst is irradiated with visible light (Equation (6)). The photo-generated carriers either recombine in the bulk or participate in the oxidation-reduction process on the surface of the photocatalyst. In a methanol-water mixture system, methanol can capture photo-generated holes and experience hole oxidation to form formaldehyde (Equation (7)), which reduces the recombination of the electron-hole pairs. Meanwhile, two protons are released during methanol oxidation, which react with the generated electrons to produce H_2 gas (Equation (8)). When accumulated to a certain degree, formaldehyde is further oxidized into formic acid and releases hydrogen gas (Equations (9) and (10)). The formic acid eventually dissociates into CO_2 and two protons (Equation (11)); then, the protons react with the photo-generated electrons and produce hydrogen gas (Equation (12)). Equation (13) can be used to represent the overall reaction. The aforementioned photocatalytic hydrogen evolution reactions are summarized below [41,42]:

$$Bi_4Ti_3O_{12} \xrightarrow{hv} h^+ + e^- \tag{6}$$

$$CH_3OH + 2h^+ \rightarrow HCHO + 2[H]^+ \tag{7}$$

$$2[H]^+ + 2e^- \rightarrow H_2(g) \tag{8}$$

$$HCHO + 2h^+ + H_2O \rightarrow HCOOH + 2[H]^+ \tag{9}$$

$$2[H]^+ + 2e^- \rightarrow H_2(g) \tag{10}$$

$$HCOOH + 2h^+ \rightarrow CO_2(g) + 2[H]^+ \tag{11}$$

$$2[H]^+ + 2e^- \rightarrow H_2(g) \tag{12}$$

Overall:

$$CH_3OH + H_2O \xrightarrow{hv, Bi_4Ti_3O_{12}} CO_2(g) + 3H_2(g) \tag{13}$$

In addition, photocatalytic conversion of H_2O into H_2 using the pristine $Bi_4Ti_3O_{12}$ and the optimal $Bi_4Ti_3O_{12-x}$ (350 °C, 60 min) in pure water were also performed in a quartz cell. The results show that when no methanol is used as a sacrificial reagent, $Bi_4Ti_3O_{12-x}$ (350 °C, 60 min) shows a very low photocatalytic H_2 evolution rate of 18 μmol·g^{-1}·h^{-1} under visible-light irradiation, and the pristine $Bi_4Ti_3O_{12}$ exhibits no H_2 evolution at all. In a pure water system, water can capture photo-generated holes produced from $Bi_4Ti_3O_{12}$ (Equation (6)) and experience hole oxidation to form oxygen gas (Equation (14)). Meanwhile, two protons can be released during water oxidation, which react with the photo-generated electrons to generate hydrogen gas (Equation (15)). Equation (16) can be used to represent the overall reaction:

$$H_2O + 2h^+ \rightarrow \frac{1}{2}O_2 + 2[H]^+ \tag{14}$$

$$2[H]^+ + 2e^- \rightarrow H_2(g) \tag{15}$$

$$H_2O \xrightarrow{hv, Bi_4Ti_3O_{12}} \frac{1}{2}O_2(g) + H_2(g) \tag{16}$$

However, to trigger this reaction, the energy of the absorbed photon must be at least 1.23 eV, which is much higher than the decomposition energy for methanol (0.7 eV, Equation (13)) [43]. It has been demonstrated that without the addition of a sacrificial agent, the water acts as an inefficient electron acceptor and donor [42–44]. As a result, the oxygen radicals and protons tend to recombine to form water, leading to limited hydrogen gas production [44].

The effect of the concentration of the $Bi_4Ti_3O_{12-x}$ (350 °C, 60 min) photocatalyst on H_2 production was investigated. Photocatalytic hydrogen evolution experiments were proceeded in a methanol (20 mL)-water (200 mL) mixture. The concentration of the photocatalyst ranged from 0.455 g/L to 1.591 g/L. As shown in Figure 5a, the H_2 evolution rate increases with the increase of the concentration until a maximum rate is achieved with 1.136 g/L $Bi_4Ti_3O_{12-x}$ (350 °C, 60 min). Further increase of the concentration results in a reduced hydrogen evolution rate. This reduction may be caused by the unsuited light scattering effect or the light shadowing due to the high turbidity of the solution that reduces the penetration depth of the visible light [45]. These effects reduce the effective incident light, thus significantly reducing the number of photo-induced electron-hole pairs necessary for the maintenance of the reaction. Therefore, the concentration of 1.136 g/L catalyst (i.e., 0.25 g $Bi_4Ti_3O_{12-x}$ (350 °C, 60 min)) is found to be the optimal concentration for H_2 generation in the present work.

In addition, the influence of additives, such as formaldehyde and formic acid, on H_2 production was further investigated. These two kinds of additives are considered mainly because they are byproducts/intermediates of methanol conversion (see Equations (7) and (9)) and also considered to be industrial wastes or model pollutants [46]. A mixture of water-formaldehyde (200/20, v/v) or water-formic acid (200/20, v/v) was used in the experiment. Figure 5b shows how different aqueous mixtures (with water-formaldehyde, water-formic acid, and water-methanol) affect visible-light photocatalytic H_2 production ($\lambda > 400$ nm) when 0.25 g of the $Bi_4Ti_3O_{12-x}$ (350 °C, 60 min) photocatalyst is used. It can be seen that the H_2 evolution rate for formic acid reaches 218 $\mu mol \cdot g^{-1} \cdot h^{-1}$, which is much higher than that for methanol and formaldehyde. It is assumed that this phenomenon is due to the low dissociation energy (-95.8 $kJ \cdot mol^{-1}$) of formic acid that is much smaller than that of methanol (64.1 $kJ \cdot mol^{-1}$) and formaldehyde (47.8 $kJ \cdot mol^{-1}$) [42,47]. Therefore, the –COOH group of formic acid can dissociate spontaneously [42], which results in a large hydrogen evolution rate. The results show that the $Bi_4Ti_3O_{12-x}$ photocatalyst can be used to decompose a variety of pollutants (such as formaldehyde and formic acid) for hydrogen production, and it also suggests a significant way to produce hydrogen by using formic acid as an additive.

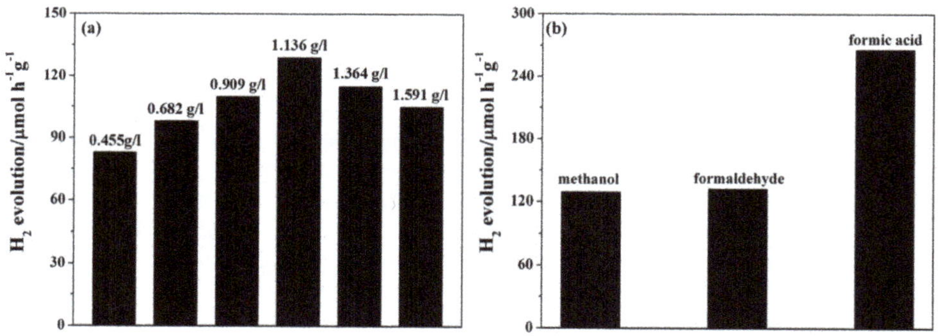

Figure 5. (**a**) Effect of the concentration of the photocatalyst on hydrogen production over the $Bi_4Ti_3O_{12-x}$ (350 °C, 60 min) nanosheets under visible-light irradiation ($\lambda > 400$ nm); and (**b**) Effect of various wastes as additives on hydrogen production over the $Bi_4Ti_3O_{12-x}$ (350 °C, 60 min) nanosheets under visible-light irradiation ($\lambda > 400$ nm).

3.3. *Surface Oxygen Vacancy Formation*

As discussed above, it is highly probable that the color change of $Bi_4Ti_3O_{12-x}$ (Figure 2a) could be caused by the formation of oxygen vacancies occurring during the chemical reduction process.

To study the presence of oxygen vacancies, room temperature EPR was performed on the pristine $Bi_4Ti_3O_{12}$ nanosheets and various $Bi_4Ti_3O_{12-x}$ samples. It is known that EPR is a highly sensitive and immediate way to characterize oxygen defects [48,49]. As shown in Figure 6a,b, the intensity of the EPR signal at g factor = 2.001 for $Bi_4Ti_3O_{12-x}$ are all higher than for the pristine $Bi_4Ti_3O_{12}$ nanosheets. Typically, a peak at 2.001~2.004 is attributed to natural surface oxygen vacancies as reported in the literature [50,51]. We attribute the enhancement of the EPR signal for $Bi_4Ti_3O_{12-x}$ at g = 2.001 to the electron-trapped center located around the site of the oxygen vacancies [52]. In addition, it can be seen from Figure 6a,b that the signal intensity at g ~ 2.001 increases with the reduction time and temperature, demonstrating that the number of oxygen vacancies in $Bi_4Ti_3O_{12-x}$ increases with the reduction reaction time and temperature.

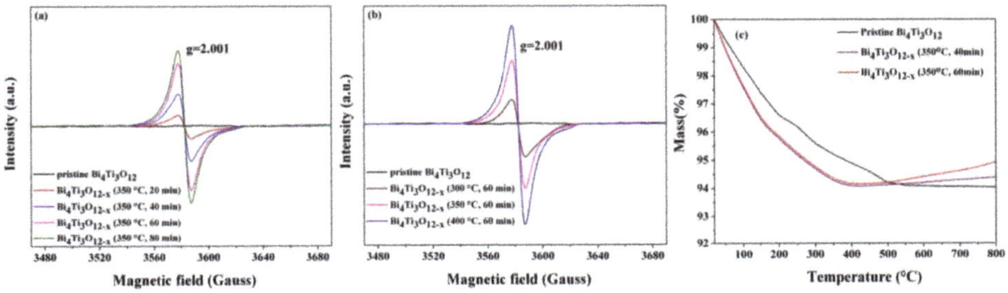

Figure 6. (a) electron paramagnetic resonance (EPR) spectra of pristine $Bi_4Ti_3O_{12}$ and $Bi_4Ti_3O_{12-x}$ after chemical reduction treatment (**a**) at 350 °C for different times; (**b**) at different temperatures for 60 min; and (c) thermogravimetric analysis (TGA) curves of the pristine $Bi_4Ti_3O_{12}$, $Bi_4Ti_3O_{12-x}$ (350 °C, 40 min), and $Bi_4Ti_3O_{12-x}$ (350 °C, 60 min).

The existence of oxygen vacancies was also proven by TGA testing in the air atmosphere. As shown in Figure 6c, the mass of the pristine $Bi_4Ti_3O_{12}$ decreases as the temperature increases because of the desorption of hydroxyl groups physically adsorbed on the surface [53]. When the temperature is above 510 °C, the mass of the pristine $Bi_4Ti_3O_{12}$ remains constant. While, as for $Bi_4Ti_3O_{12-x}$ (350 °C, 60 min), one can notice that when the temperature is below 405 °C, the variation trend of the TGA curve is the same as that of the pristine $Bi_4Ti_3O_{12}$. However, when the temperature exceeds 405 °C, there is a clear difference between the mass loss of the pristine $Bi_4Ti_3O_{12}$ and $Bi_4Ti_3O_{12-x}$ (350 °C, 60 min). A slight increase in the mass of $Bi_4Ti_3O_{12-x}$ (350 °C, 60 min) is observed, which is finished at 800 °C. The same phenomenon was also observed by Li et al. [54] and Yang et al. [55]. It was previously reported by Li et al. that an obvious mass difference between black TiO_{2-x} and white TiO_2 existed during TGA testing. A mass gain for black TiO_{2-x} was assigned to the oxidation of oxygen vacancies on the surface [54]. Yang et al. also reported an obvious difference in weight loss between TiO_2-SO (TiO_2 with surface oxygen vacancies) and conventional TiO_2 when the temperature exceeded 440 °C during TGA testing [55]. It was deduced that when the TiO_2-SO sample was heated in air, its surface oxygen vacancies can be compensated by the external oxygen, resulting in the mass increase for TiO_2-SO. Therefore, it is postulated that the slight increase in the mass of $Bi_4Ti_3O_{12-x}$ (350 °C, 60 min) can be explained by the formation of oxygen vacancies on the surface of $Bi_4Ti_3O_{12-x}$ (350 °C, 60 min) in the present work. When $Bi_4Ti_3O_{12-x}$ (350 °C, 60 min) with oxygen vacancies is heated in air, its unsaturated surface will be compensated by the external oxygen, leading to a mass increase. In addition, as shown in Figure 6c, the variation trend of the TGA curve of $Bi_4Ti_3O_{12-x}$ (350 °C, 40 min) is the same as that of $Bi_4Ti_3O_{12-x}$ (350 °C, 60 min). However, the mass increase of the former is smaller than that of the latter, which may be due to the smaller number of surface oxygen vacancies of $Bi_4Ti_3O_{12-x}$ (350 °C, 40 min).

XPS can provide useful information on the chemical states of elements and surface defects [56]. Figure 7 shows the high-resolution Bi 4f and O1s spectra of the pristine $Bi_4Ti_3O_{12}$ and the optimal $Bi_4Ti_3O_{12-x}$ (350 °C, 60 min). As shown in Figure 7a, the Bi 4f spectrum of the pristine $Bi_4Ti_3O_{12}$ sample exhibits two main peaks at 159.9 eV (Bi $4f_{7/2}$) and 165.3 eV (Bi $4f_{5/2}$) ascribed to Bi^{3+}, which are in accordance with the reported values of Bi_2O_3 powders [57,58]. Figure 7b reveals the fitted O1s spectra, where the peaks correspond to the lattice oxygen (O_L, 530.1 eV) and chemisorbed oxygen species (O_C, 532.4 eV) on the pristine $Bi_4Ti_3O_{12}$ sample, respectively. The oxygen vacancies (O_V) peak, which should appear at 531.5 eV is not observed in this spectrum, further revealing the stoichiometric properties of the pristine $Bi_4Ti_3O_{12}$ sample [59,60]. Figure 7c,d show the high-resolution XPS spectra of the Bi 4f and O1s core levels for the optimal $Bi_4Ti_3O_{12-x}$ (350 °C, 60 min). The Bi 4f spectrum shows two main peaks centered at 158.8 and 164.3 eV, which are identified as the Bi $4f_{7/2}$ and Bi $4f_{5/2}$, respectively. However, $4f_{7/2}$ and $4f_{5/2}$ peaks of the metallic Bi are located at 156.8 and 162.2 eV [57]. The chemical shift of the Bi 4f doublet relative to the metallic Bi is about 2.1 eV, which is smaller than the reported value of 3.1 eV between Bi_2O_3 and the metallic Bi [61]. This result indicates that the valence state of bismuth in the optimal $Bi_4Ti_3O_{12-x}$ (350 °C, 60 min) should be (+3 − x) owing to an increased concentration of oxygen defects in the vicinity of Bi ions, which are probably in the Bi_2O_2 layer [61]. The XPS spectrum of O1s of the optimal $Bi_4Ti_3O_{12-x}$ (350 °C, 60 min) is shown in Figure 7d. The O1s XPS spectrum is broad and unsymmetrical, indicating more than one chemical state for oxygen in the optimal $Bi_4Ti_3O_{12-x}$ (350 °C, 60 min) sample. Gaussian divided features at 530.1 eV, 531.5 eV, and 532.4 eV are credited to the lattice oxygen, oxygen vacancies, and surface chemisorbed oxygen, respectively [59,60]. The O_V peak appearing at 531.5 eV indicates that oxygen vacancies are generated in the optimal $Bi_4Ti_3O_{12-x}$ (350 °C, 60 min) during the solid-state chemical reduction process.

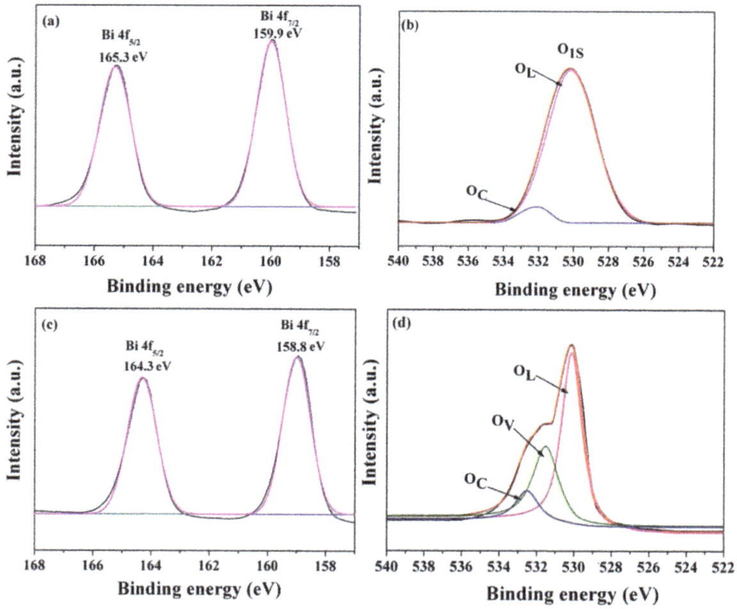

Figure 7. High-resolution X-ray photoelectron spectrometer (XPS) spectra: (**a**) Bi 4f and (**b**) O1s of the pristine $Bi_4Ti_3O_{12}$; (**c**) Bi 4f and (**d**) O1s of $Bi_4Ti_3O_{12-x}$ (350 °C, 60 min).

In addition, the change in morphology of the pristine $Bi_4Ti_3O_{12}$ and $Bi_4Ti_3O_{12-x}$ were also scrutinized to prove the formation of oxygen vacancies. Figure 8a,c show representative TEM images of both the pristine $Bi_4Ti_3O_{12}$ and the optimal $Bi_4Ti_3O_{12-x}$ (350 °C, 60 min) photocatalyst, respectively.

It is seen that both samples consist of rectangular nanosheets with sides around ~100 and ~150 nm. The particle size of $Bi_4Ti_3O_{12-x}$ (350 °C, 60 min) has not changed after the solid-state chemical reduction process. HRTEM micrographs offer a more complete view of the microstructures of the samples. As shown in Figure 8b, the pristine $Bi_4Ti_3O_{12}$ nanocrystals display a highly crystalline composition, as well as perfect lattice structures throughout the entire particles. The measured spacings are equal to 0.271 nm and to 0.273 nm, which are in agreement with the (020) and (200) planes of $Bi_4Ti_3O_{12}$, respectively [62]. However, after the solid-state reduction reaction process at 350 °C for 60 min, a disordered layer is observed on the surface of the $Bi_4Ti_3O_{12-x}$ (350 °C, 60 min) nanosheet (Figure 8d). Compared with the pristine $Bi_4Ti_3O_{12}$ nanocrystals, the lattice features shown in the HRTEM image of $Bi_4Ti_3O_{12-x}$ (350 °C, 60 min) became highly blurred. The surface structure of the $Bi_4Ti_3O_{12-x}$ (350 °C, 60 min) nanosheet is imperfect, which may have been damaged by the reduction reaction induced oxygen vacancies [33]. In summary, the EPR, XPS, TGA, and TEM results confirm the existence of oxygen vacancies on the $Bi_4Ti_3O_{12-x}$ nanosheets, which can be attributed to the reduction of the active hydrogen produced by the decomposition of $NaBH_4$.

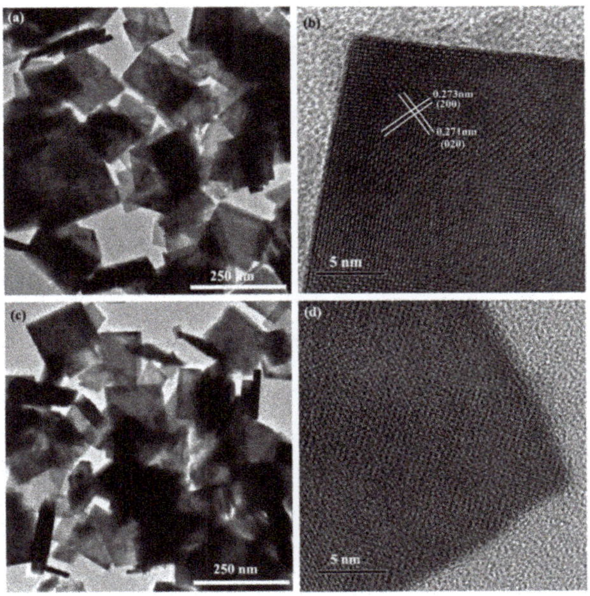

Figure 8. Transmission electron microscopy (TEM) images of (**a**) the pristine $Bi_4Ti_3O_{12}$ and (**c**) $Bi_4Ti_3O_{12-x}$ (350 °C, 60 min); HRTEM images of (**b**) the pristine $Bi_4Ti_3O_{12}$ and (**d**) $Bi_4Ti_3O_{12-x}$ (350 °C, 60 min).

The concentration and species of the oxygen vacancies in the $Bi_4Ti_3O_{12-x}$ nanosheets were studied using the positron annihilation life technique [55,63–66]. Lifetime components (τ_1, τ_2, and τ_3), as well as corresponding intensities (I_1, I_2, and I_3) for the pristine $Bi_4Ti_3O_{12}$ and the $Bi_4Ti_3O_{12-x}$ samples are shown in Table 3. The longest component (τ_3) is typically ascribed to the annihilation of the orthopositronium atom in the material voids [63], and the shortest one (τ_1) is typically due to the annihilation of the positron in the small defects in the bulk, such as the bulk oxygen vacancies [55,64]. Another component (τ_2) arises from positrons trapped by larger-sized defects on the surface of the materials, such as surface oxygen vacancies [55,65]. The relative intensity (I_1/I_2) reflects the ratio of the corresponding defects [55,63] and in the present work, reflects the relative concentration ratio of bulk and surface oxygen vacancies [55]. As shown in Table 3, when the $Bi_4Ti_3O_{12-x}$ samples were exposed to solid-state chemical reduction treatment at 350 °C, the I_1/I_2 ratio decreased with the increasing chemical reduction reaction time and reached a minimum at 60 min. With further increases in the reaction time and reaction temperature, the I_1/I_2 ratio increased instead. The results of the positron annihilation analysis indicate that the concentrations and types of oxygen vacancies can be controlled by adjusting the chemical reaction time and temperature. This means that due to the presence of active hydrogen produced from the decomposition of $NaBH_4$, $Bi_4Ti_3O_{12-x}$ nanosheets with different reduction degrees can be obtained by tuning the reduction reaction conditions.

Table 3. Positron lifetime and relative intensities of the pristine $Bi_4Ti_3O_{12}$ and various $Bi_4Ti_3O_{12-x}$ samples.

Sample	τ_1 (ps)	τ_2 (ps)	τ_3 (ns)	I_1 (%)	I_2 (%)	I_3 (%)	I_1/I_2
$Bi_4Ti_3O_{12}$	193	376	2.33	50.24	47.78	1.98	1.05
$Bi_4Ti_3O_{12-x}$ (350 °C, 20 min)	196	387	2.47	46.26	51.97	1.77	0.89
$Bi_4Ti_3O_{12-x}$ (350 °C, 40 min)	199	389	2.49	38.72	59.64	1.64	0.65
$Bi_4Ti_3O_{12-x}$ (350 °C, 60 min)	205	393	2.77	23.87	74.24	1.89	0.32
$Bi_4Ti_3O_{12-x}$ (350 °C, 80 min)	209	396	2.92	36.48	61.57	1.95	0.59
$Bi_4Ti_3O_{12-x}$ (350 °C, 100 min)	214	402	3.05	41.73	56.42	1.85	0.74
$Bi_4Ti_3O_{12-x}$ (400 °C, 60 min)	216	405	3.11	44.86	53.33	1.81	0.84

3.4. *Mechanism of Enhanced Photocatalytic Activity of $Bi_4Ti_3O_{12-x}$*

The photocatalyst's BET specific surface area was measured to examine a correlation between the surface area and the photocatalytic activity. This is because the large surface areas could influence the number of available active sites [67] and affect the interfacial charge transfer quantum efficiency [68]. As outlined in Table 4, the BET specific surface areas for the $Bi_4Ti_3O_{12-x}$ samples are almost indistinguishable from the pristine $Bi_4Ti_3O_{12}$. In addition, as described earlier, the crystal phase structures are not changed (Figure 3). These results strongly suggest that it is not the surface area or structural features that lead to the large divergence in photocatalysis ability. Therefore, this implies that the photocatalytic kinetics of the $Bi_4Ti_3O_{12-x}$ samples are mainly enhanced by other factors.

Table 4. Brunauer-Emmett-Teller (BET) specific surface areas of the pristine $Bi_4Ti_3O_{12}$ and various $Bi_4Ti_3O_{12-x}$ samples.

Samples	BET Specific Surface Area (m^2/g)
$Bi_4Ti_3O_{12}$	6.45
$Bi_4Ti_3O_{12-x}$ (350 °C, 20 min)	6.39
$Bi_4Ti_3O_{12-x}$ (350 °C, 40 min)	6.35
$Bi_4Ti_3O_{12-x}$ (350 °C, 60 min)	6.32
$Bi_4Ti_3O_{12-x}$ (350 °C, 80 min)	6.46
$Bi_4Ti_3O_{12-x}$ (350 °C, 100 min)	6.48
$Bi_4Ti_3O_{12-x}$ (300 °C, 60 min)	6.38
$Bi_4Ti_3O_{12-x}$ (400 °C, 60 min)	6.51

As is well-known, generation and disassociation of photo-generated electron-hole pairs are crucial for a semiconductor photocatalyst. The efficiency of this disassociation is central to the enhancement of photocatalytic activity. EIS was used to fully probe this efficiency. Figure 9a shows the EIS of the pristine $Bi_4Ti_3O_{12}$ and various $Bi_4Ti_3O_{12-x}$ electrodes. Each sample diagram contains a semi-circular section, reflecting the process of the charge transfer, as well as a linear section with a 45° slope corresponding to the diffusion-controlled step [69]. The value for the electron-transfer resistance (R_{ct}) is obtained by calculating the diameter of the semi-circle, and this acts as a proxy for the system's charge transfer effectiveness. In other words, a smaller R_{ct} value means a higher charge transfer efficiency of the system [70]. The order in the R_{ct} value is the pristine $Bi_4Ti_3O_{12}$ > $Bi_4Ti_3O_{12-x}$ (350 °C, 40 min) > $Bi_4Ti_3O_{12-x}$ (350 °C, 80 min) > $Bi_4Ti_3O_{12-x}$ (350 °C, 60 min), coinciding with the increased activity order of the photocatalysts. The R_{ct} of the $Bi_4Ti_3O_{12-x}$ (350 °C, 60 min) electrode is the smallest among all the catalysts. Therefore, the photo-generated electron-hole pairs are most easily separated and transferred to the surface in the $Bi_4Ti_3O_{12-x}$ (350 °C, 60 min) sample, thus leading to the highest photocatalytic activity of all the catalysts. The photocurrent analysis was also conducted to confirm the hindering efficiency of $Bi_4Ti_3O_{12-x}$ during the recombination of electron-hole pairs. Figure 9b details the photocurrent responses of the pristine $Bi_4Ti_3O_{12}$, $Bi_4Ti_3O_{12-x}$ (350 °C, 40 min), $Bi_4Ti_3O_{12-x}$ (350 °C, 60 min), and $Bi_4Ti_3O_{12-x}$ (350 °C, 80 min) after their deposition on ITO electrodes under visible light. The results show that the responses are prompt, uniform, and reproducible with the light irradiation switched on and off. Under visible light, the photocurrent density of the $Bi_4Ti_3O_{12-x}$ (350 °C, 60 min) electrode is the highest among the samples. The enhanced photocurrent indicates the amplification of the photo-induced carrier transport rate, as well as the dwindling photo-generated electron-hole pair recombination rate [71]. The results of the photocurrent investigation are in agreement with the changes in catalytic activity for the pristine $Bi_4Ti_3O_{12-x}$ and $Bi_4Ti_3O_{12-x}$ (Figure 4). Therefore, we believe that the improved charge separation and transportation are the major reasons for the enhanced photocatalytic activity of $Bi_4Ti_3O_{12-x}$.

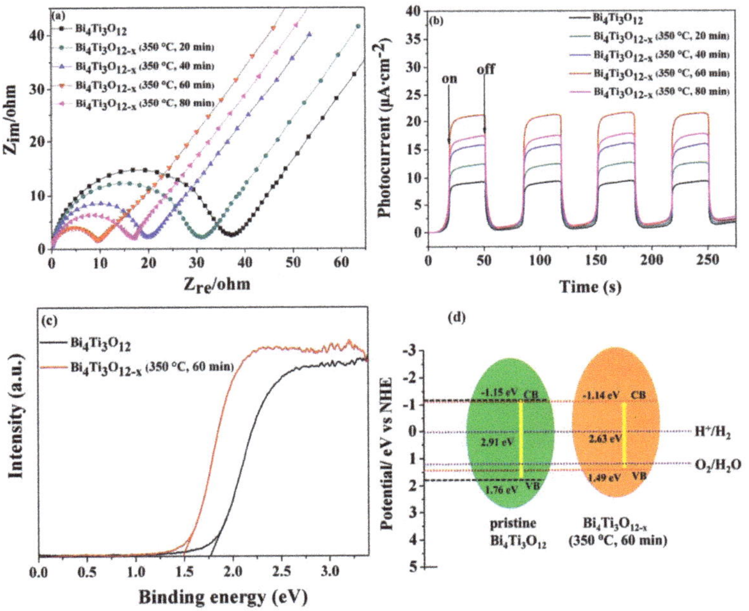

Figure 9. (**a**) The electrochemical impedance spectroscopy (EIS) Nyquist plots of the pristine $Bi_4Ti_3O_{12}$ and various $Bi_4Ti_3O_{12-x}$ samples after the buildup on the ITO electrodes with visible-light ($\lambda > 400$ nm) irradiation; (**b**) Photocurrents of the pristine $Bi_4Ti_3O_{12}$ and various $Bi_4Ti_3O_{12-x}$ samples after the buildup on the ITO electrodes under visible-light irradiation ($\lambda > 400$ nm); (**c**) valence band XPS spectra of the pristine $Bi_4Ti_3O_{12}$ and $Bi_4Ti_3O_{12-x}$ (350 °C, 60 min); (**d**) the probable band energy diagram of the pristine $Bi_4Ti_3O_{12}$ and $Bi_4Ti_3O_{12-x}$ (350 °C, 60 min).

The position of the valence band (VB) on top of the pristine $Bi_4Ti_3O_{12}$ and $Bi_4Ti_3O_{12-x}$ (350 °C, 60 min) was determined by VB XPS spectra (see Figure 9c). The top of the valence band (E_{VB}) of the pristine $Bi_4Ti_3O_{12}$ and $Bi_4Ti_3O_{12-x}$ (350 °C, 60 min) vs. the normal hydrogen electrode (NHE) are estimated to be 1.76 and 1.49 eV, respectively. Moreover, the band gaps of the pristine $Bi_4Ti_3O_{12}$ and $Bi_4Ti_3O_{12-x}$ (350 °C, 60 min) are 2.91 and 2.63 eV, respectively (Table 1). Therefore, using the formula $E_{CB} = E_{VB} - Eg$ [16], the bottom of conduction band (E_{CB}) is −1.15 and −1.14 eV for the pristine $Bi_4Ti_3O_{12}$ and $Bi_4Ti_3O_{12-x}$ (350 °C, 60 min), respectively. According to the values of E_{VB}

and E_{CB}, a suggested band energy diagram is illustrated in Figure 9d. It can be observed that the conduction band energy of $Bi_4Ti_3O_{12-x}$ (350 °C, 60 min) is almost the same as that of the pristine $Bi_4Ti_3O_{12}$. However, compared to the pristine $Bi_4Ti_3O_{12}$, the VBM of $Bi_4Ti_3O_{12-x}$ (350 °C, 60 min) rises considerably, leading to a narrowing band gap of $Bi_4Ti_3O_{12-x}$ (350 °C, 60 min). We attribute this rise in band gap to the formation of new energy states near the VB top because of the presence of oxygen vacancy in the $Bi_4Ti_3O_{12-x}$ (350 °C, 60 min) sample [72].

Based on the discussion above, the reasons for the better photocatalytic activity of the $Bi_4Ti_3O_{12-x}$ nanosheets could be explained from the point of view of surface defects. The surface oxygen vacancies are located on the top of the VBM or below the conduction band minimum (CBM) [24,48,49,72] and are considered as shallow defects. Zhu et al. discovered that the surface oxygen defect states were formed on the top of the VBM for ZnO and $BiPO_4$ [48,49]. Huang et al. demonstrated that a high number of oxygen vacancies created an impurity energy level near the valence band and caused a decrease in the band gap [24]. Zhao et al. reported that under poor oxygen conditions, the oxygen vacancy states at the top of the valence band decreased the band gap of $LiTi_2(PO_4)_3$ significantly [73]. The rise of the VBM and the reduction of the band of anatase TiO_2 have been observed by scanning tunneling microscopy [74]. In the present work, the EPR, TGA, and TEM confirm that oxygen vacancies are formed in $Bi_4Ti_3O_{12-x}$ after the chemical reduction treatment. Figure 10 shows the schematic diagram of the charge separation and photocatalytic reaction for the $Bi_4Ti_3O_{12-x}$ photocatalyst under visible-light irradiation. Many shallow surface oxygen vacancy states should be above the valence band and partially overlap with the valence band of $Bi_4Ti_3O_{12-x}$, which can cause rise of the VBM to VBM´. Hence, the VBM of $Bi_4Ti_3O_{12-x}$ (350 °C, 60 min) is higher than that of the pristine $Bi_4Ti_3O_{12}$, which is proven by the valence band XPS spectra (Figure 9c). Correspondingly, the rise of the VBM can further expand the valence band, which can increase the transport rate of photo-generated carriers, resulting in the improved separation efficiency of the photo-generated electron-hole pairs and leading to an obvious improvement of the photocatalytic activities of $Bi_4Ti_3O_{12-x}$. In addition, due to the VBM' rise, the band gap of $Bi_4Ti_3O_{12-x}$ narrows, thus expanding the photoresponse range of $Bi_4Ti_3O_{12-x}$ (from under 420 nm for the pristine $Bi_4Ti_3O_{12}$ to above 460 nm for $Bi_4Ti_3O_{12-x}$ (350 °C, 60 min)).

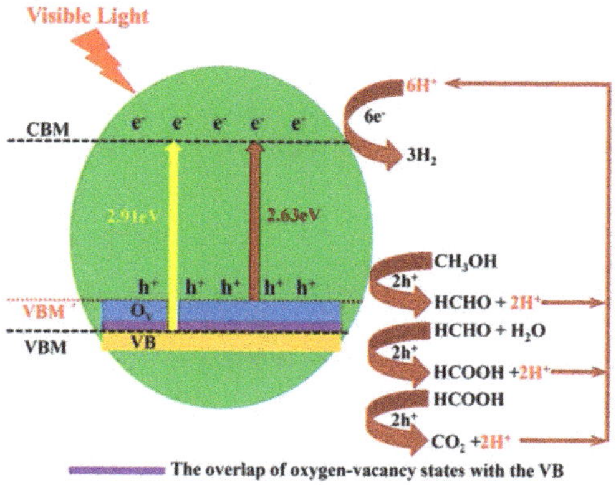

Figure 10. Schematic diagram illustrating the mechanism of the charge separation and photocatalytic reaction for the $Bi_4Ti_3O_{12-x}$ photocatalyst under visible-light irradiation. VB: valence band; CBM: conduction band minimum; and VBM: valence band maximum.

Furthermore, as shown in Figure 4, both the reduction time and temperature have significant influence on the photocatalysis ability of the $Bi_4Ti_3O_{12-x}$ photocatalysts. As discussed above, the positron annihilation analysis (Table 3) indicated that $Bi_4Ti_3O_{12-x}$ nanosheets with different concentrations and types of oxygen vacancies can be obtained by tuning the conditions of the solid-state chemical reduction process. The relative intensity (I_1/I_2) reflects the relative concentration ratios of bulk and surface oxygen vacancies [55]. When $Bi_4Ti_3O_{12-x}$ is exposed to chemical reduction treatment at a low temperature or for a short time, the I_1/I_2 ratio continues to decrease with the increasing reaction time and temperature (Table 3), thus increasing the number of surface oxygen vacancies. These surface oxygen vacancies are located above the VB and even partially overlap with it. At the same time, the photocatalytic activity improves gradually with the duration of the treatment and temperature, until a maximum is achieved at 350 °C for 60 min. With additional increases in temperature, as well as prolonged (or longer) reaction time, the I_1/I_2 ratio increases instead (Table 3), which means that the bulk oxygen vacancies continue to increase in $Bi_4Ti_3O_{12-x}$. The bulk oxygen vacancies' defect levels form easily in the forbidden band and provide a position for the recombination of the electron-hole pairs, thus reducing the photocatalytic activity [63]. Therefore, the best activity of the $Bi_4Ti_3O_{12-x}$ photocatalyst can be achieved when the minimum bulk oxygen vacancies exist simultaneously with large numbers of surface oxygen vacancies. Thus, not only the number but also the types of oxygen vacancies induced with varying chemical reduction temperature and duration are essential for a catalyst with high photoactivity.

4. Conclusions

In summary, a facile, economic solid-state chemical reduction method has been proposed to fabricate the $Bi_4Ti_3O_{12-x}$ photocatalyst with abundant oxygen vacancies. The concentration and types of oxygen vacancies could be adjusted by changing the reduction reaction time and temperature. The $Bi_4Ti_3O_{12-x}$ catalyst showed significantly improved photoactivity during visible-light driven hydrogen evolution from water compared to the pristine $Bi_4Ti_3O_{12}$. The hydrogen production rate reaches up to 129 $\mu mol \cdot g^{-1} \cdot h^{-1}$ under visible-light irradiation for the optimal $Bi_4Ti_3O_{12-x}$ photocatalyst (reduction treated at 350 °C for 60 min), which is about 3.4 times that of the pristine $Bi_4Ti_3O_{12}$. It is proposed that energy levels corresponding to the surface oxygen vacancies should be above and partially overlap with the $Bi_4Ti_3O_{12-x}$ valence band. This can raise the top of the valence band maximum. The improved photoactivity of the photocatalyst is the result of the enhanced separation ability of photo-generated electron-hole pairs that originates from the valence band expansion by the surface oxygen vacancy states. The extended photoresponse is due to the decrease in the band gap caused by the rise of the top of the valence band maximum.

Supplementary Materials: The following are available online at http://www.mdpi.com/2079-4991/8/4/261/s1.

Acknowledgments: We appreciate the financial support of the National Natural Science Foundation of China (No. 50702022), the Natural Science Foundation of Guangdong Province (No. 2014A030313245), and the State Key Laboratory of Pulp and Paper Engineering (No. 201624).

Author Contributions: Zhiwu Chen conceived of and designed the experiments; Yizeng Zhang and Zhiwu Chen performed the experiments; Yizeng Zhang, Zhiwu Chen, and Zhenya Lu analyzed the data; and Zhiwu Chen wrote the paper with input from all authors.

Conflicts of Interest: The authors declare no conflict of interest.

References

1. Wang, W.; Tadé, M.O.; Shao, Z.P. Research progress of perovskite materials in photocatalysis-and photovoltaics-related energy conversion and environmental treatment. *Chem. Soc. Rev.* **2015**, *44*, 5371–5408. [CrossRef] [PubMed]
2. Moniz, S.J.A.; Shevlin, S.A.; Martin, D.J.; Guo, Z.X.; Tang, J.W. Visible-light driven heterojunction photocatalysts for water splitting–a critical review. *Energy Environ. Sci.* **2015**, *8*, 731–759. [CrossRef]
3. Liu, L.; Chen, X. Titanium dioxide nanomaterials: Self-structural modifications. *Chem. Rev.* **2014**, *114*, 9890–9918. [CrossRef] [PubMed]
4. Wang, J.; Xia, Y.; Dong, Y.; Chen, R.S.; Xiang, L.; Komarneni, S. Defect-rich ZnO nanosheets of high surface area as an efficient visible-light photocatalyst. *Appl. Catal. B* **2016**, *192*, 8–16. [CrossRef]

5. Yu, J.G.; Yu, Y.F.; Zhou, P.; Xiao, W.; Cheng, B. Morphology-dependent photocatalytic H_2-production activity of CdS. *Appl. Catal. B* **2014**, *156*, 184–191. [CrossRef]
6. Kisch, H. Semiconductor photocatalysis-mechanistic and synthetic aspects. *Angew. Chem. Int. Ed.* **2013**, *52*, 812–847. [CrossRef] [PubMed]
7. Tong, H.; Ouyang, S.; Bi, Y.; Umezawa, N.; Oshikiri, M.; Ye, J. Nano-photocatalytic materials: Possibilities and challenges. *Adv. Mater.* **2012**, *24*, 229–251. [CrossRef] [PubMed]
8. Hou, D.F.; Luo, W.; Huang, Y.H.; Yu, J.C.; Hu, X.L. Synthesis of porous $Bi_4Ti_3O_{12}$ nanofibers by electrospinning and their enhanced visible-light-driven photocatalytic properties. *Nanoscale* **2013**, *5*, 2028–2035. [CrossRef] [PubMed]
9. Chen, Z.W.; Jiang, H.; Jin, W.L.; Shi, C.K. Enhanced photocatalytic performance over $Bi_4Ti_3O_{12}$ nanosheets with controllable size and exposed {001} facets for Rhodamine B degradation. *Appl. Catal. B* **2016**, *180*, 698–706. [CrossRef]
10. Park, B.H.; Kang, B.S.; Bu, S.D.; Noh, T.W.; Lee, J.; Jo, W. Lanthanum-substituted bismuth titanate for use in non-volatile memories. *Nature* **1999**, *401*, 682–684. [CrossRef]
11. Liu, Y.; Zhang, M.; Li, L.; Zhang, X. One-dimensional visible-light-driven bifunctional photocatalysts based on $Bi_4Ti_3O_{12}$ nanofiber frameworks and Bi_2XO_6 (X = Mo, W) nanosheets. *Appl. Catal. B* **2014**, *160–161*, 757–766. [CrossRef]
12. Kudo, A.; Hijii, S. H_2 or O_2 Evolution from aqueous solutions on layered oxide photocatalysts consisting of Bi^{3+} with 6s2 configuration and d0 transition metal ions. *Chem. Lett.* **1999**, *28*, 1103–1104. [CrossRef]
13. Yao, W.; Xu, X.; Wang, H.; Zhou, J.T.; Yang, X.; Zhang, Y.; Shang, S.; Huang, B. Photocatalytic property of perovskite bismuth titanate. *Appl. Catal. B* **2004**, *52*, 109–116. [CrossRef]
14. Zhao, W.; Jia, Z.; Lei, E.; Wang, L.; Li, Z.; Dai, Y. Photocatalytic degradation efficacy of $Bi_4Ti_3O_{12}$ micro-scale platelets over methylene blue under visible light. *J. Phys. Chem. Solids* **2013**, *74*, 1604–1607. [CrossRef]
15. Zhang, H.; Lu, M.; Liu, S.; Wang, L.; Xiu, Z.; Zhou, Y.; Qiu, Z.; Zhang, A.; Ma, Q. Preparation and photocatalytic property of perovskite $Bi_4Ti_3O_{12}$ films. *Mater. Chem. Phys.* **2009**, *114*, 716–721. [CrossRef]
16. Hou, D.F.; Hu, X.; Hu, P.; Zhang, W.; Zhang, M.; Huang, Y.H. $Bi_4Ti_3O_{12}$ nanofibers-BiOI nanosheets p-n junction: Facile synthesis and enhanced visible-light photocatalytic activity. *Nanoscale* **2013**, *5*, 9764–9772. [CrossRef] [PubMed]
17. Shi, H.; Tan, H.; Zhu, W.B.; Sun, Z.; Ma, Y.; Wang, E. Electrospun Cr-doped $Bi_4Ti_3O_{12}/Bi_2Ti_2O_7$ heterostructure fibers with enhanced visible-light photocatalytic properties. *J. Mater. Chem. A* **2015**, *3*, 6586–6591. [CrossRef]
18. Zhang, M.; Liu, Y.; Li, L.; Gao, H.; Zhang, X. BiOCl nanosheet/$Bi_4Ti_3O_{12}$ nanofiber heterostructures with enhanced photocatalytic activity. *Catal. Commun.* **2015**, *58*, 122–126. [CrossRef]
19. Zheng, C.X.; Yang, H.; Cui, Z.M.; Zhang, H.M.; Wang, X.X. A novel $Bi_4Ti_3O_{12}/Ag_3PO_4$ heterojunction photocatalyst with enhanced photocatalytic performance. *Nanoscale Res. Lett.* **2017**, *12*, 608. [CrossRef] [PubMed]
20. Guo, Y.; Li, J.H.; Gao, Z.Q.; Zhu, X.; Liu, Y.; Wei, Z.B.; Zhao, W.; Sun, C. A simple and effective method for fabricating novel p-n heterojunction photocatalyst g-$C_3N_4/Bi_4Ti_3O_{12}$ and its photocatalytic performances. *Appl. Catal. B* **2016**, *192*, 57–71. [CrossRef]
21. Kim, H.G.; Hwang, D.; Lee, J. An undoped, single-phase oxide photocatalyst working under visible light. *J. Am. Chem. Soc.* **2004**, *126*, 8912–8913. [CrossRef] [PubMed]
22. Zhang, Y.H.; Dai, R.Y.; Hu, S.R. Study of the role of oxygen vacancies as active sites in reduced graphene oxide-modified TiO_2. *Phys. Chem. Chem. Phys.* **2017**, *19*, 7307–7315. [CrossRef] [PubMed]
23. Zuo, F.; Wang, L.; Wu, T.; Zhang, Z.Y.; Borchardt, D.; Feng, P.Y. Self-doped Ti^{3+} enhanced photocatalyst for hydrogen production under visible light. *J. Am. Chem. Soc.* **2010**, *132*, 11856–11857. [CrossRef] [PubMed]
24. Wang, J.P.; Wang, Z.Y.; Huang, B.B.; Ma, Y.D.; Liu, Y.Y.; Qin, X.Y.; Zhang, X.Y.; Dai, Y. Oxygen vacancy induced band-gap narrowing and enhanced visible light photocatalytic activity of ZnO. *ACS Appl. Mater. Interfaces* **2012**, *4*, 4024–4030. [CrossRef] [PubMed]
25. Lv, Y.H.; Yao, W.Q.; Ma, X.G.; Pan, C.S.; Zong, R.L.; Zhu, Y.F. Surface oxygen vacancy induced visible activity and enhanced UV activity of ZnO_{1-x} photocatalyst. *Catal. Sci. Technol.* **2013**, *3*, 3136–3146. [CrossRef]
26. Ling, Y.C.; Wang, G.M.; Reddy, J.; Wang, C.C.; Zhang, J.Z.; Li, Y. The influence of oxygen content on the thermal activation of hematite nanowires. *Angew. Chem. Int. Ed.* **2012**, *51*, 4074–4079. [CrossRef] [PubMed]

27. Thompson, T.L.; Yates, J.T. Surface science studies of the photoactivation of TiO_2-new photochemical processes. *Chem. Rev.* **2006**, *106*, 4428–4453.
28. Chen, X.; Liu, L.; Yu, P.Y.; Mao, S.S. Increasing solar absorption for photocatalysis with black hydrogenated titanium dioxide nanocrystals. *Science* **2011**, *331*, 746–750. [CrossRef] [PubMed]
29. Wang, Z.; Yang, C.; Lin, T.; Yin, H.; Chen, P.; Wan, D.; Xu, F.; Huang, F.; Lin, J.; Xie, X.; et al. Visible-light photocatalytic, solar thermal and photoelectrochemical properties of aluminium- reduced black titania. *Energy Environ. Sci.* **2013**, *6*, 3007–3014. [CrossRef]
30. Wang, G.; Wang, H.; Ling, Y.; Tang, Y.; Yang, X.; Fitzmorris, R.C.; Wang, C.; Zhang, J.Z.; Li, Y. Hydrogen-treated TiO_2 nanowire arrays for photoelectrochemical water splitting. *Nano Lett.* **2011**, *11*, 3026–3033. [CrossRef] [PubMed]
31. Wang, Z.; Yang, C.; Lin, T.; Yin, H.; Chen, P.; Wan, D.; Xu, F.; Huang, F.; Lin, J.; Xie, X.; et al. H-Doped black titania with very high solar absorption and excellent photocatalysis enhanced by localized surface plasmon resonance. *Adv. Funct. Mater.* **2013**, *23*, 5444–5450. [CrossRef]
32. Cushing, S.K.; Meng, F.; Zhang, J.Y.; Ding, B.F.; Chen, C.K.; Chen, C.J.; Liu, R.S.; Bristow, A.D.; Bright, J.; Zheng, P.; et al. Effects of defects on photocatalytic activity of hydrogen-treated titanium oxide nanobelts. *ACS Catal.* **2017**, *7*, 1742–1748. [CrossRef]
33. Lv, Y.H.; Yao, W.Q.; Zong, R.L.; Zhu, Y.F. Fabrication of wide-range-visible photocatalyst Bi_2WO_{6-x} nanoplates via surface oxygen vacancies. *Sci. Rep.* **2016**, *6*, 19347. [CrossRef] [PubMed]
34. Wu, Q.; Krol, R.V.D. Selective photoreduction of nitric oxide to nitrogen by nanostructured TiO_2 photocatalysts: Role of oxygen vacancies and iron dopant. *J. Am. Chem. Soc.* **2012**, *134*, 9369–9375. [CrossRef] [PubMed]
35. Wei, W.; Dai, Y.; Huang, B. Density functional characterization of the electronic structure and optical properties of N-doped, La-doped, and N/La-codoped $SrTiO_3$. *J. Phys. Chem. C* **2009**, *113*, 5658–5663. [CrossRef]
36. Mou, P.; Pal, U.; Jiménez, J.M.G.Y.; Pérezrodríguez, F. Effects of crystallization and dopant concentration on the emission behavior of TiO_2: Eu nanophosphors. *Nanoscale Res. Lett.* **2012**, *7*, 1.
37. Zhang, H.J.; Chen, G.; Li, X. Synthesis and visible light photocatalysis water splitting property of chromium-doped $Bi_4Ti_3O_{12}$. *Solid State Ionics* **2009**, *180*, 1599–1603. [CrossRef]
38. Hou, J.G.; Cao, R.; Wang, Z.; Jiao, S.Q.; Zhu, H.M. Chromium-doped bismuth titanate nanosheets as enhanced visible-light photocatalysts with a high percentage of reactive {110} facets. *J. Mater. Chem.* **2011**, *21*, 7296–7301. [CrossRef]
39. Chen, Z.W.; Jiang, X.Y.; Zhu, C.B.; Shi, C.K. Chromium-modified $Bi_4Ti_3O_{12}$ photocatalyst: Application forhydrogen evolution and pollutant degradation. *Appl. Catal. B* **2016**, *199*, 241–251. [CrossRef]
40. Yang, P.J.; Ou, H.H.; Fang, Y.X.; Wang, X.C. A facile steam reforming strategy to delaminate layered carbon nitride semiconductors for photoredox catalysis. *Angew. Chem. Int. Ed.* **2017**, *56*, 3992–3996. [CrossRef] [PubMed]
41. Chen, X.B.; Shen, S.H.; Guo, L.J.; Mao, S.S. Semiconductor-based photocatalytic hydrogen generation. *Chem. Rev.* **2010**, *110*, 6503–6570. [CrossRef] [PubMed]
42. Allured, B.; DelaCruz, S.; Darling, T.; Huda, M.N.; Subramanian, V. Enhancing the visible light absorbance of $Bi_2Ti_2O_7$ through Fe-substitution and its effects on photocatalytic hydrogen evolution. *Appl. Catal. B* **2014**, *144*, 261–268. [CrossRef]
43. Choi, H.J.; Kang, M. Hydrogen production from methanol/water decomposition in a liquid photosystem using the anatase structure of Cu loaded TiO_2. *Int. J. Hydrogen Energy* **2007**, *32*, 3841–3848. [CrossRef]
44. Yang, X.; Salzmann, C.; Shi, H.; Malcolm, L.; Green, H.; Xiao, T. The role of photoinduced defects in TiO_2 and its effects on hydrogen evolution from aqueous methanol solution. *J. Phys. Chem. C* **2008**, *112*, 10784–10789. [CrossRef] [PubMed]
45. Bhosale, R.; Pujari, S.; Muley, G.; Pagare, B.; Gambhire, A. Visible-light-activated nanocomposite photocatalyst of Cr_2O_3/SnO_2. *J. Nanostruct. Chem.* **2013**, *3*, 46. [CrossRef]
46. Chen, J.; Ollis, D.F.; Rulkens, W.H.; Bruning, H. Photocatalyzed oxidation of alcohols and organochlorides in the presence of native TiO_2 and metallized TiO_2 suspensions. Part (II): Photocatalytic mechanisms. *Water Res.* **1999**, *33*, 669–676. [CrossRef]

47. Gupta, S.; De Leon, L.; Subramanian, V. Mn-modified $Bi_2Ti_2O_7$ photocatalysts: Bandgap engineered multifunctional photocatalysts for hydrogen generation. *Phys. Chem. Chem. Phys.* **2014**, *16*, 12719–12727. [CrossRef] [PubMed]
48. Lv, Y.H.; Pan, C.S.; Ma, X.G.; Zong, R.L.; Bai, X.J.; Zhu, Y.F. Production of visible activity and UV performance enhancement of ZnO photocatalyst via vacuum deoxidation. *Appl. Catal. B* **2013**, *138–139*, 26–32. [CrossRef]
49. Lv, Y.H.; Liu, Y.F.; Zhu, Y.Y.; Zhu, Y.F. Surface oxygen vacancy induced photocatalytic performance enhancement of a $BiPO_4$ nanorod. *J. Mater. Chem. A* **2014**, *2*, 1174–1182. [CrossRef]
50. Ischenko, V.; Polarz, S.; Grote, D.; Stavarache, V.; Fink, K. Zinc oxide nanoparticles with defects. *Adv. Funct. Mater.* **2005**, *15*, 1945–1954. [CrossRef]
51. Lv, Y.H.; Zhu, Y.Y.; Zhu, Y.F. Enhanced photocatalytic performance for the $BiPO_{4-x}$ nanorod induced by surface oxygen vacancy. *J. Phys. Chem. C* **2013**, *117*, 18520–18528. [CrossRef]
52. Zhang, L.W.; Wang, L.; Zhu, Y.F. Synthesis and performance of $BaAl_2O_4$ with a wide spectral range of optical absorption. *Adv. Funct. Mater.* **2007**, *173*, 781–790. [CrossRef]
53. Zhuang, W.; Li, L.; Zhu, J.; An, R.; Lu, L.; Lu, X.; Wu, X.; Ying, H. Facile synthesis of mesoporous MoS_2-TiO_2 nanofibers for ultrastable lithium ion battery anodes. *Chemelectrochem* **2015**, *2*, 374–381. [CrossRef]
54. Li, L.; Shi, K.; Tu, R.; Qian, Q.; Li, D.; Yang, Z.; Lu, X. Black TiO_2(B)/anatase bicrystalline TiO_{2-x} nanofibers with enhanced photocatalytic performance. *Chin. J. Catal.* **2015**, *36*, 1943–1948. [CrossRef]
55. Li, J.L.; Zhang, M.; Guan, Z.J.; Li, Q.Y.; He, C.Q.; Yang, J.J. Synergistic effect of surface and bulk single-electron-trapped oxygen vacancy of TiO_2 in the photocatalytic reduction of CO_2. *Appl. Catal. B* **2017**, *206*, 300–307. [CrossRef]
56. Haider, Z.; Kang, Y.S. Facile preparation of hierarchical TiO2 nano structures: Growth mechanism and enhanced photocatalytic H_2 production from water splitting using methanol as a sacrificial reagent. *ACS Appl. Mater. Interfaces* **2014**, *6*, 10342–10352. [CrossRef] [PubMed]
57. Moulder, J.F.; Stickle, W.F.; Sobol, P.E.; Bomben, K.D. *Handbook of X-ray Photoelectron Spectroscopy*; Physical Electronics Inc.: Chanhassen, MN, USA, 1992.
58. Chu, M.W.; Ganne, M.; Caldes, M.T.; Brohan, L. X-ray photoelectron spectroscopy and high resolution electron microscopy studies of Aurivillius compounds: $Bi_{4-x}La_xTi_3O_{12}$(x = 0, 0.5, 0.75, 1.0, 1.5, and 2.0). *J. Appl. Phys.* **2002**, *91*, 3178–3187. [CrossRef]
59. Leelavathi, A.; Madras, G.; Ravishankar, N. Origin of enhanced photocatalytic activity and photoconduction in high aspect ratio ZnO nanorods. *Phys. Chem. Chem. Phys.* **2013**, *15*, 10795–10802. [CrossRef] [PubMed]
60. Han, X.G.; He, H.Z.; Kuang, Q.; Zhou, X.; Zhang, X.H.; Xu, T.; Xie, Z.X.; Zheng, L.S. Controlling morphologies and tuning the related properties of nano/microstructured ZnO crystallites. *J. Phys. Chem. C* **2009**, *113*, 584–589. [CrossRef]
61. Jovalekic, C.; Pavlovic, M.; Osmokrovic, P.; Atanasoska, L. X-ray photoelectron spectroscopy study of $Bi_4Ti_3O_{12}$ ferroelectric ceramics. *Appl. Phys. Lett.* **1998**, *72*, 1051–1053. [CrossRef]
62. He, H.; Yin, J.; Li, Y.; Zhang, Y.; Qiu, H.; Xu, J.; Xu, T.; Wang, C. Size controllable synthesis of single-crystal ferroelectric $Bi_4Ti_3O_{12}$ nanosheet dominated with {001} facets toward enhanced visible-light-driven photocatalytic activities. *Appl. Catal. B* **2014**, *156–157*, 35–43. [CrossRef]
63. Kong, M.; Li, Y.; Chen, X.; Tian, T.; Fang, P.; Zheng, F.; Zhao, X. Tuning the relative concentration ratio of bulk defects to surface defects in TiO_2 nanocrystals leads to high photocatalytic efficiency. *J. Am. Chem. Soc.* **2011**, *133*, 16414–16417. [CrossRef] [PubMed]
64. Dutta, S.; Chattopadhyay, S.; Jana, D.; Banerjee, A.; Manik, S.; Pradhan, S.K.; Sutradhar, M.; Sarkar, A. Annealing effect on nano-ZnO powder studied from positron lifetime and optical absorption spectroscopy. *J. Appl. Phys.* **2006**, *100*, 114328. [CrossRef]
65. Liu, X.; Zhou, K.; Wang, L.; Wang, B.; Li, Y. Oxygen vacancy clusters promoting reducibility and activity of ceria nanorods. *J. Am. Chem. Soc.* **2009**, *131*, 3140–3141. [CrossRef] [PubMed]
66. Guan, M.; Xiao, C.; Zhang, J.; Fan, S.; An, R.; Cheng, Q.; Xie, J.; Zhou, M.; Ye, B.; Xie, Y. Vacancy associates promoting solar-driven photocatalytic activity of ultrathin bismuth oxychloride nanosheets. *J. Am. Chem. Soc.* **2013**, *135*, 10411–10417. [CrossRef] [PubMed]
67. Hoffmann, M.R.; Martin, S.T.; Choi, W.; Bahnemann, D.W. Environmental applications of semiconductor photocatalysis. *Chem. Rev.* **1995**, *95*, 69–96. [CrossRef]

68. Becker, W.G.; Truong, M.M.; Ai, C.C.; Hamel, N.N. Interfacial factors that affect the photoefficiency of semiconductor-sensitized oxidations in nonaqueous media. *ACS J. Phys. Chem.* **1989**, *93*, 4882–4886. [CrossRef]
69. Pei, R.; Cheng, Z.; Wang, E.; Yang, X. Amplification of antigen-antibody interactions based on biotin labeled protein-streptavidin network complex using impedance spectroscopy. *Biosens. Bioelectron.* **2001**, *16*, 355–361. [CrossRef]
70. Wang, Y.; Wang, F.; He, J. Controlled fabrication and photocatalytic properties of a three-dimensional ZnO nanowire/reduced graphene oxide/CdS heterostructure on carbon cloth. *Nanoscale* **2013**, *5*, 1129–11297. [CrossRef] [PubMed]
71. Bai, X.; Zong, R.; Li, C.; Liu, D.; Liu, Y.; Zhu, Y.F. Enhancement of visible photocatalytic activity via Ag@C_3N_4 core–shell plasmonic composite. *Appl. Catal. B* **2014**, *147*, 82–91. [CrossRef]
72. Sinhamahapatra, A.; Jeon, J.P.; Kang, J.; Han, B.; Yu, J.S. Oxygen-deficient zirconia (ZrO_{2-x}): A new material for solar light absorption. *Sci. Rep.* **2016**, *6*, 27218. [CrossRef] [PubMed]
73. Chen, L.J.; Zhao, Y.J.; Luo, J.Y.; Xia, Y.Y. Oxygen vacancy in $LiTiPO_5$ and $LiTi_2(PO_4)_3$: A first-principles study. *Phys. Lett. A* **2011**, *375*, 934–938. [CrossRef]
74. Dette, C.; Perez-Osorio, M.A.; Kley, C.S.; Punke, P.; Patrick, C.E.; Jacobson, P.; Giustino, F.; Jung, S.J.; Kern, K. TiO_2 anatase with a bandgap in the visible region. *Nano Lett.* **2014**, *14*, 6533–6538. [CrossRef] [PubMed]

© 2018 by the authors. Licensee MDPI, Basel, Switzerland. This article is an open access article distributed under the terms and conditions of the Creative Commons Attribution (CC BY) license (http://creativecommons.org/licenses/by/4.0/).

nanomaterials

MDPI

Article

Effect of Organic Substrates on the Photocatalytic Reduction of Cr(VI) by Porous Hollow Ga_2O_3 Nanoparticles

Jin Liu [1], Huihui Gan [2], Hongzhang Wu [1], Xinlei Zhang [1], Jun Zhang [1], Lili Li [1,3,*] and Zhenling Wang [1,*]

1 Henan Key Laboratory of Rare Earth Functional Materials, International Joint Research Laboratory for Biomedical Nanomaterials of Henan, The Key Laboratory of Rare Earth Functional Materials and Applications, Zhoukou Normal University, Zhoukou 466001, China; jinliu@zknu.edu.cn (J.L.); hongzhang_wu@zknu.edu.cn (H.W.); zhangxinlei@zknu.edu.cn (X.Z.); 18238260132@163.com (J.Z.)

2 School of Civil Engineering and Architecture, Ningbo Institute of Technology, Zhejiang University, Ningbo 315100, China; hhgan@nit.zju.edu.cn

3 School of Life Science and Agriculture, Zhoukou Normal University, Zhoukou 466001, China

* Correspondence: lilili@zknu.edu.cn (L.L.); zlwang2007@hotmail.com (Z.W.); Tel.: +86-394-8178518 (Z.W.)

Received: 6 March 2018; Accepted: 19 April 2018; Published: 22 April 2018

Abstract: Porous hollow Ga_2O_3 nanoparticles were successfully synthesized by a hydrolysis method followed by calcination. The prepared samples were characterized by field emission scanning electron microscope, transmission electron microscope, thermogravimetry and differential scanning calorimetry, UV-vis diffuse reflectance spectra and Raman spectrum. The porous structure of Ga_2O_3 nanoparticles can enhance the light harvesting efficiency, and provide lots of channels for the diffusion of Cr(VI) and Cr(III). Photocatalytic reduction of Cr(VI), with different initial pH and degradation of several organic substrates by porous hollow Ga_2O_3 nanoparticles in single system and binary system, were investigated in detail. The reduction rate of Cr(VI) in the binary pollutant system is markedly faster than that in the single Cr(VI) system, because Cr(VI) mainly acts as photogenerated electron acceptor. In addition, the type and concentration of organic substrates have an important role in the photocatalytic reduction of Cr(VI).

Keywords: Ga_2O_3; porous; Cr(VI); organic pollutants

---

1. Introduction

Heavy metal ions from wastewater have become the primary threat to the human environment with the development of industrial civilization [1–4]. Hexavalent chromium (Cr(VI)) is a typical heavy metal contaminant with high solubility and toxicity, which originates from various industrial processes such as electroplating, leather tanning, and paint manufacture [5]. A common method of treating Cr(VI) in wastewater is to convert it into low toxic Cr(III), which can be precipitated as $Cr(OH)_3$ in neutral or alkaline solutions, and removed as a solid waste [6]. Recently, photocatalytic reduction of Cr(VI) to Cr(III) has been recognized as an efficient and economical form of technology [7–11]. Briefly, photocatalytic reduction of Cr(VI) is based on the photogenerated electrons in the conduction band of semiconductor when it is irradiated by UV/visible light having energy greater than the band gap energy of the semiconductor. In addition, organic and inorganic pollutants usually co-exist in industrial wastewater and natural aqueous environment, and no doubt the presence of organic pollutants in wastewater will greatly increase the difficulty of photocatalytic reduction of Cr(VI) [12–14].

During the photocatalytic process, the photocatalyst is the key factor, and it is necessary to design and fabricate efficient and stable photocatalysts. Ga_2O_3 is one of most popular photocatalysts used in the photocatalytic degradation of organic pollutants and reduction of CO_2 [15–18], owing to its high activity and environmental friendliness. Its activity can be further enhanced through a proper synthetic strategy to obtain nanostructured materials, as the morphology, size and pore structure of materials can significantly influence their properties and applications [19–21]. As one promising field of research, porous hollow nanostructures have been investigated for a long time. Compared with bulk materials, porous hollow materials have higher porosity, larger specific surface areas, and lots of active chemical sites, which could enhance light harvesting efficiency and provide lots of channels for the diffusion of pollutants, while also improving photocatalytic activity efficiently [22–24].

In this study, porous hollow Ga_2O_3 nanoparticles were prepared via a hydrolysis method followed by calcination. The effect of parameters including pH and concentration of metronidazole on the reduction rate of Cr(VI) by the porous hollow Ga_2O_3 nanoparticles was also studied. Meanwhile, the photocatalytic reduction of Cr(VI) was also systematically investigated in the absence and presence of organic substrates. To the best of our knowledge, this is the first report on the simultaneous treatment of organics and Cr(VI) using porous hollow Ga_2O_3.

2. Materials and Methods

2.1. Materials

Ga_2O_3 (99.999%), $NaCO_3$ (AR, 99.8%), and NaOH (AR, 96%) were purchased from Shanghai Aladdin Bio-Chem Technology Co., Ltd. (Shanghai, China) and HCl (AR, 36–38%) were purchased from Sinopharm Chemical Reagent Co., Ltd. (Shanghai, China) Deionized water was used throughout the experiments.

2.2. Synthesis of Porous Hollow Ga_2O_3 Nanoparticles

Porous hollow Ga_2O_3 nanoparticles were synthesized via a thermal transformation of GaOOH precursor based on our previous study [25]. The $NaGaO_2$ powders were prepared by heating a stoichiometric mixture of Na_2CO_3 and Ga_2O_3 at 850 °C for 12 h. The GaOOH precursor was prepared by a hydrolysis reaction of $NaGaO_2$ colloidal solution. The $NaGaO_2$ powders (1.0 g) was dispersed in deionized water (100 mL) to obtain a colloidal solution with ultrasonic oscillations. Then, 5 mol/L HCl solution was added to the $NaGaO_2$ colloidal solution with magnetic stirring; the final pH value was kept at 9.0. The obtained white suspension was treated thermally at 80 °C for 12 h. The obtained GaOOH precursor was separated by centrifugation and dried at 70 °C for 10 h. The Ga_2O_3 was prepared by calcining GaOOH precursor with a programmed temperature (400 °C, 5 h and 700 °C, 1.5 h, 1 °C/min).

2.3. Characterization

The Raman spectrum was recorded using a Raman spectrometer (RM2000) (Renishaw, Gloucestershire, UK). Field emission scanning electron microscope (FESEM) images were obtained using a MERLIN scanning microscope at an accelerating voltage of 10 kV (ZEISS, Oberkochen, German). Scanning transmission electron microscopy (STEM), transmission electron microscope (TEM) and high-resolution transmission electron microscopy (HRTEM) images were obtained using a JEOL-2010 transmission electron microscope (JEOL Ltd., Kyoto, Japan) at an accelerating voltage of 200 kV. TEM is equipped with an energy-dispersive X-ray spectroscopy (EDS) analysis system. The quantitation method for Ga and O elements is Cliff Lorimer thin ratio section. Thermogravimetry and differential scanning calorimetry (TG-DSC) analysis was performed on a STA 6000 (Perkin Elmer, Waltham, MA, USA) instrument at a heating rate of 10 °C/min. UV-vis diffuse reflectance spectra (UV-vis DRS) were obtained by a UV-2600 UV-vis spectrophotometer (Shimadzu Corporation, Kyoto, Japan).

2.4. Photocatalytic Experiments

Photoreduction of Cr(VI) ($K_2Cr_2O_7$) and photocatalytic degradation of rhodamine B (RhB), acid red 1 (AR1), methyl orange (MO) and metronidazole (MNZ) as well as their binary mixtures, were adopted to evaluate the photocatalytic activity of the as-synthesized Ga_2O_3 sample. The concentration of Cr(VI) and organics is the same in single and binary pollutants (Table 1). Typically, 20 mg of the Ga_2O_3 sample was added into a 50 mL Cr(VI) aqueous solution. The initial pH of the Cr(VI) solution was adjusted to 2–9 by adding HCl or NaOH. Prior to irradiation, the suspensions were magnetically stirred for 30 min to establish the adsorption-desorption equilibrium. The irradiation was performed with a 30 W UV light lamp (λ = 253.7 nm). At a given time interval, about 3 mL suspension was taken for further analysis after centrifugation. The concentration of organic pollutants, including RhB, AR1, MO and MNZ, were analyzed by UV-vis spectroscopy at 554, 505, 464 and 320 nm, respectively. Meanwhile, the concentration of Cr(VI) was analyzed by a 1,5-diphenylcarbazide spectrophotometric method with a spectrophotometer at 540 nm (GB 7466-87, Standards of China). The characteristic absorbance peaks of organic pollutants (rhodamine B, acid red 1, methyl orange, metronidazole) are different; their absorbances are different when the concentrations are same. In order to quickly measure the absorbance of organic pollutants by UV-vis spectroscopy, we created the proper concentrations.

Table 1. The concentration of Cr(VI) and organics in single and binary pollutant system.

Name	Single Pollutant System, mg/L	Binary Pollutant System, mg/L			
		RhB + Cr(VI)	AR1 + Cr(VI)	MO + Cr(VI)	MNZ + Cr(VI)
Cr(VI)	2.5	2.5	2.5	2.5	2.5
RhB	5	5	×	×	×
AR1	20	×	20	×	×
MO	10	×	×	10	×
MNZ	20	×	×	×	20

3. Results and Discussion

3.1. Composition and Morphology

Previous X-ray diffraction (XRD) results show that the phase composition of the GaOOH precursor and its calcined product are α-GaOOH (JCPDS No. 06-0180) and β-Ga_2O_3 (JCPDS No. 41-1103), respectively [25]. The average crystallite size of β-Ga_2O_3 sample is about 27.3 nm by the Scherrer equation. [26] The composition of Ga_2O_3 samples was further investigated by Raman spectra, owing to the greater sensitivity of Raman spectroscopy to the outer region of the solid samples than XRD [27]. The Raman spectra of Ga_2O_3 samples from the GaOOH precursor—calcined at 600 °C and 700 °C—are shown in Figure 1. The characteristic Raman bands of α-Ga_2O_3 and β-Ga_2O_3 are shown in Figure 1a,b, respectively; this is consistent with the reported results [28,29]. This result indicates that the α-Ga_2O_3 gradually transforms into β-Ga_2O_3 with the increase of calcination temperature; pure phase β-Ga_2O_3 is finally obtained at 700 °C.

The composition of porous hollow Ga_2O_3 nanoparticles was also analyzed by EDS elemental mapping images and EDS spectrum. As shown in Figure 2, it is clearly seen that Ga_2O_3 nanoparticles possess a porous structure, and that Ga and O elements are distributed homogenously in the Ga_2O_3 sample and their atomic ratio was close to 2:3, which further indicates that the synthesized sample is pure Ga_2O_3.

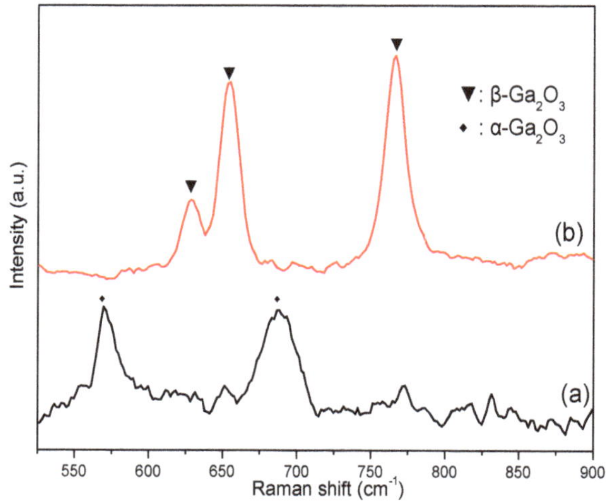

Figure 1. Raman spectra of Ga_2O_3 from the GaOOH precursor calcined at (**a**) 600 °C and (**b**) 700 °C.

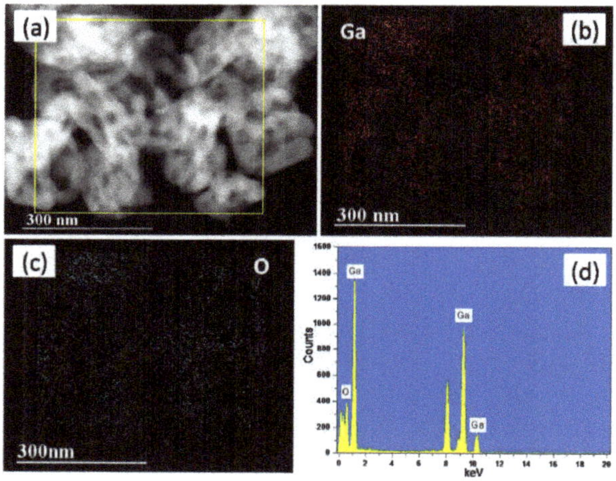

Figure 2. (**a**) Scanning transmission electron microscopy (STEM) image, (**b**,**c**) energy-dispersive X-ray spectroscopy (EDS) elemental mapping images and (**d**) EDS spectrum of Ga_2O_3.

The morphology and microstructure of GaOOH precursor and its calcined product Ga_2O_3 were investigated by TEM and SEM. As shown in Figure 3a,b, the GaOOH precursor presents monodisperse nanoplate-like structure. Compared with GaOOH (Figure 3a,b), Ga_2O_3 nanoparticles (Figure 3c) present porous hollow structures. Moreover, the clearly resolved lattice fringes with d spacing of 0.23 nm (distance between two arrow heads in Figure 3d) correspond to the ($\bar{3}11$) lattice planes of monoclinic β-Ga_2O_3, which is in good agreement with the XRD result. Figure 4 also shows that the Ga_2O_3 nanoparticles possess hollow structures [25]. The porous hollow structure of Ga_2O_3 nanoparticles is mainly ascribed to the thermal dehydration of the GaOOH precursor. These porous structure can enhance the light harvesting efficiency and provide lots of channels for the diffusion of Cr(VI) and Cr(III), resulting in the improvement of photocatalytic efficiency [30]. The size distribution of Ga_2O_3 in Figure 5 was evaluated from the SEM image (Figure 4) by measuring the diamenter of about 100 nanoparticles. It is clearly seen that the size of most Ga_2O_3 nanoparticles is 160–230 nm.

Figure 3. Transmission electron microscopy (TEM) and high-resolution transmission electron microscopy (HRTEM) images of (**a**,**b**) GaOOH and (**c**,**d**) Ga_2O_3.

Figure 4. Scanning electron microscope (SEM) image of Ga_2O_3 nanoparticles.

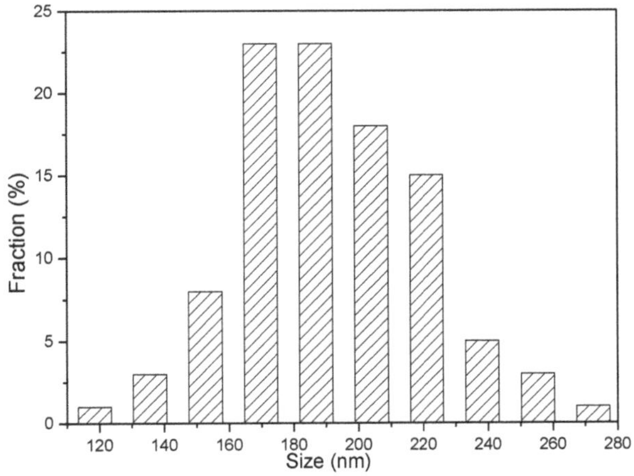

Figure 5. The size distribution of Ga_2O_3 nanoparticles.

3.2. Hermogravimetry and Differential Scanning Calorimetry TG-DSC Analysis

To understand the thermal conversion of the GaOOH precursor to Ga_2O_3, a TG-DSC measurement was performed. The TG-DSC measurement, performed from 40 to 900 °C for the GaOOH precursor, is shown in Figure 6. The major exothermic peak at about 394 °C was probably caused by the phase transformation of the sample from GaOOH to Ga_2O_3, as evidenced by a weight loss of 12% in the range of 40–400 °C in TG curve. A weight loss of 3% in the range of 400–600 °C, which is demarcated by weak endothermic peak in DSC curve, indicates the conversion of α-Ga_2O_3 to β-Ga_2O_3 above 600 °C [31]. The result is consistent with the Raman band of the GaOOH precursor calcined at 600 °C (Figure 1). The XRD, Raman spectra, and TG-DSC results indicate that the pure phase β-Ga_2O_3 can be obtained at 700 °C, and that α-Ga_2O_3 gradually transforms into β-Ga_2O_3 in the temperature range of 600–700 °C.

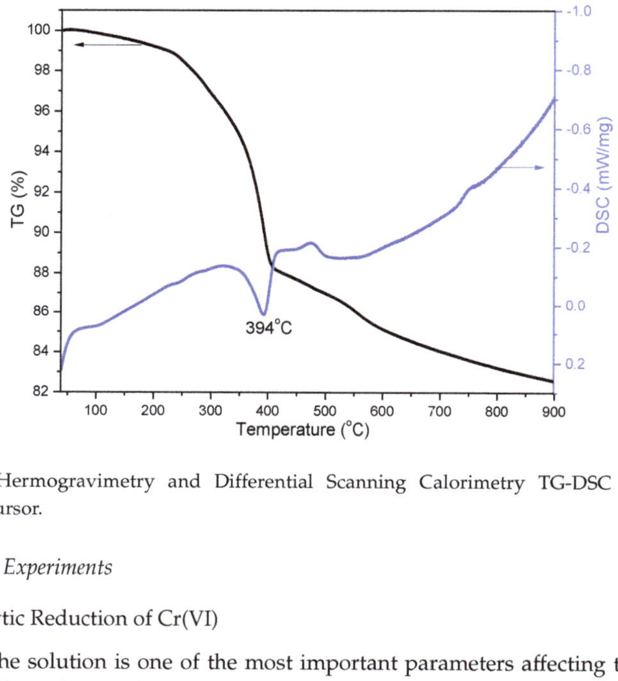

Figure 6. Hermogravimetry and Differential Scanning Calorimetry TG-DSC curve of the GaOOH precursor.

3.3. Photocatalytic Experiments

3.3.1. Photocatalytic Reduction of Cr(VI)

The pH of the solution is one of the most important parameters affecting the photocatalytic reduction of Cr(VI) on photocatalysts. The temporal concentration variation of Cr(VI) reduction by the porous hollow Ga_2O_3 nanoparticles at a pH range from 2.0 to 9.0 is shown in Figure 7. Obviously, the reduction of Cr(VI) is increased rapidly by decreasing the pH when the initial pH is in the range of 3–9. In general, the predominant form of Cr(VI) is $Cr_2O_7^{2-}$ at a pH range of 2–6, while the major form was CrO_4^{2-} at pH > 7 [32,33]. The photocatalytic reduction of Cr(VI) to Cr(III) consumes H^+ in an acidic solution (Equation (1)), and produces OH^- in an alkaline solution (Equation (2)). At a low pH, the Ga_2O_3 nanoparticles are highly protonated and have a strong affinity toward the anion $Cr_2O_7^{2-}$, and thus enhance the photocatalytic reduction of Cr(VI). However, the photocatalytic reduction of Cr(VI) is decreased when the initial pH is kept at 2–2.5, which may be attributed the dissolution of Ga_2O_3 nanoparticles. At a higher pH, the surface charge of the Ga_2O_3 nanoparticles will be less positively charged, or even negatively charged, which tends to electrostatically repel the anionic Cr(VI), and adsorb the cationic Cr(III) [34,35]. The electrostatical repulsion makes it more difficult for the anionic Cr(VI) to obtain the photogenerated electrons. Meanwhile, $Cr(OH)_3$ precipitate will be formed at pH > 6, and occupies the active sites of Ga_2O_3 nanoparticles, leading to the decrease in the photocatalytic reduction of Cr(VI). Therefore, it is concluded that the photocatalytic reduction of Cr(VI) to Cr(III) is highly efficient at a suitable acidic condition, and is restrained at an alkaline condition.

$$Cr_2O_7^{2-} + 14H^+ + 6e^- \rightarrow 2Cr^{3+} + 7H_2O \quad (1)$$

$$CrO_4{}^{2-} + 4H_2O + 3e^- \rightarrow Cr(OH)_3 + 5OH^- \quad (2)$$

3.3.2. Photocatalytic Degradation of Organic Pollutants

Besides the photocatalytic reduction of Cr(VI), typical organic pollutants such as RhB, AR1, MO and MNZ were also used to evaluate the photocatalytic activity of porous hollow Ga_2O_3 nanoparticles. As shown in Figure 8, these four pollutants can be effectively degraded by the Ga_2O_3 nanoparticles in 60 min when they are in the single pollutant system (Table 1). The results indicate that the porous hollow Ga_2O_3 is a promising photocatalyst in water treatment.

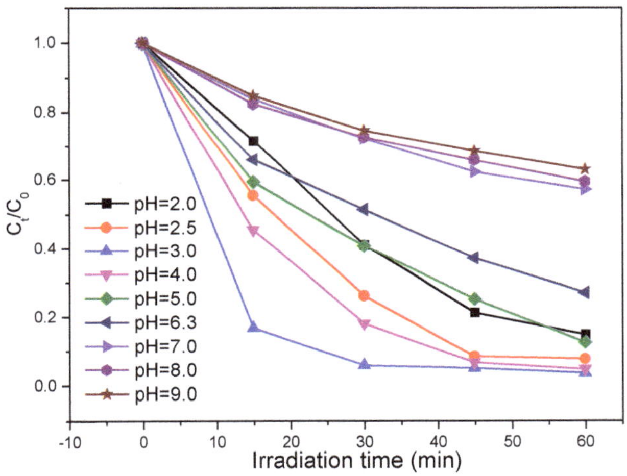

Figure 7. Photocatalytic reduction of Cr(VI) by Ga_2O_3 at different initial pH.

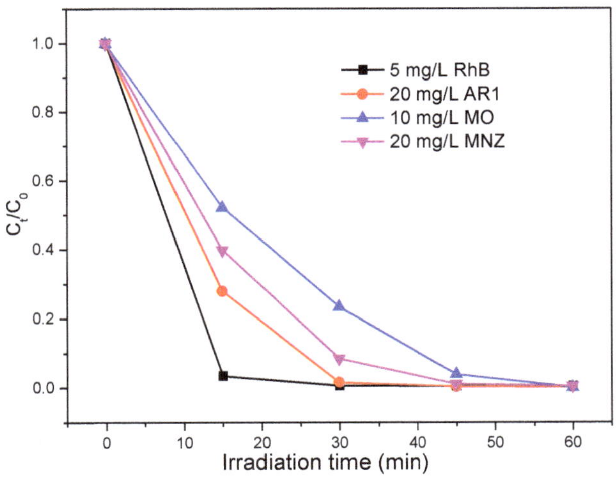

Figure 8. Photocatalytic degradation of single pollutant by Ga_2O_3.

3.3.3. Simultaneous Treatment of Cr(VI) and Organic Pollutants

Cr(VI) is often discharged together with hazardous organics from industrial wastewater. To further study the photocatalytic activity of porous hollow Ga_2O_3 nanoparticles, several binary pollutants were simulated using organics as the additional substrates. As shown in Figure 9, the reduction rate of Cr(VI) in the binary pollutants is markedly faster than that of the single Cr(VI). Cr(VI) mainly acts as photogenerated electron acceptor. However, the degradation rate of the organic pollutants in the binary pollutant system is lower than that of the corresponding single pollutant system. Based on our previous study, the photogenerated electrons play an important role in the degradation of organic pollutants by Ga_2O_3. The photogenerated electrons in the conduction band of Ga_2O_3 are assumed by Cr(VI), reducing the degradation rate of organic pollutants. The photocatalytic stability of the prepared Ga_2O_3 for the treatment of pollutants has been investigated by the recycling experiments. However, the photocatalytic reduction rate of Cr(VI) is 67% after two cycling runs. In order to activate the recycled Ga_2O_3 photocatalyst, an ultrasound treatment is used; a photocatalytic reduction rate of Cr(VI) is able to maintain 81%, under the same conditions. However, the photocatalytic reduction rate of Cr(VI) is only 44%, even after four cycling runs with the ultrasound treatment. The photocatalytic degradation rate of organic pollutants (RhB, AR1, MO, MNZ) is stable, even after five cycling runs.

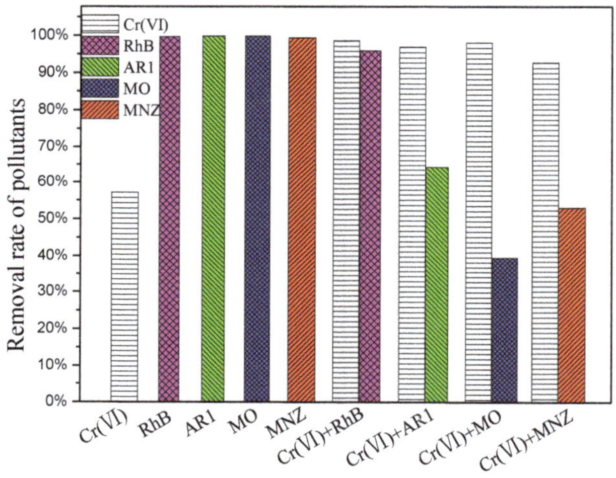

Figure 9. Photocatalytic treatment of single and binary pollutant system by Ga_2O_3.

3.3.4. Effect of Substrate Concentration on Photocatalytic Reduction of Cr(VI)

To further assess the effect of organic pollutants as substrates on the photocatalytic reduction of Cr(VI), several different concentrations of MNZ in the Cr(VI)/MNZ binary pollutants were investigated. As shown in Figure 10, only 57% of Cr(VI) is photocatalytically reduced in the absence of MNZ after 60 min, and the reduction of Cr(VI) is increased striking when MNZ is added into the system. By increasing the concentration of MNZ to 10 mg/L, 94% of Cr(VI) is reduced. The reason may be that the presence of MNZ can consume the photogenerated holes in photocatalyst, and more photogenerated electrons are captured by Cr(VI), improving the photocatalytic reduction of Cr(VI). The reduction rate of Cr(VI) is not markedly changed when the concentration of MNZ is further increased.

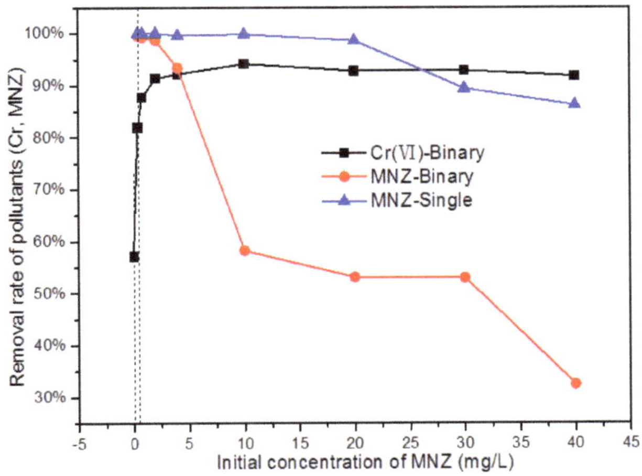

Figure 10. Photocatalytic treatment of Cr(VI) and MNZ in single and binary pollutant system.

4. Conclusions

In summary, porous hollow Ga_2O_3 nanoparticles were successfully synthesized by a hydrolysis method followed by calcination. It was demonstrated that the Ga_2O_3 photocatalyst is effective for the treatment of Cr(VI) and organic pollutants—as well as a mixture of them. The photocatalytic removal rate of Cr(VI) is highest when the initial pH of Cr(VI) is 3.0. The presence of organic pollutants in the reaction system improves the photocatalytic reduction of Cr(VI) by acting as a holes scavenger, leading to better charge carrier separation. The results broaden the range of approaches for the treatment of practical wastewater.

Acknowledgments: This work was supported by the National Natural Science Foundation of China (21477167, 51509220, 21701203), Key Scientific and Technological Project of Henan Province (182102210603, 172102210612), Zhejiang Provincial Natural Science Foundation of China (LQ14E090003) and Ningbo Science and Technology Plan Projects (2014C50007).

Author Contributions: Lili Li and Zhenling Wang conceived and designed the experiments; Jin Liu and Jun Zhang performed the experiments; Lili Li and Jin Liu analyzed the data; Huihui Gan, Hongzhang Wu and Xinlei Zhang contributed reagents/materials/analysis tools; all authors contributed to the writing of the paper.

Conflicts of Interest: The authors declare no conflict of interest.

References

1. Jiang, W.J.; Cai, Q.; Xu, W.; Yang, M.W.; Cai, Y.; Dionysiou, D.D.; O'Shea, K.E. Cr(VI) adsorption and reduction by humic acid coated on magnetite. *Environ. Sci. Technol.* **2014**, *48*, 8078–8085. [CrossRef] [PubMed]
2. Naimi-Joubani, M.; Shirzad-Siboni, M.; Yang, J.-K.; Gholami, M.; Farzadkia, M. Photocatalytic reduction of hexavalent chromium with illuminated ZnO/TiO_2 composite. *J. Ind. Eng. Chem.* **2015**, *22*, 317–323. [CrossRef]
3. Yuan, X.; Zhou, C.; Jing, Q.; Tang, Q.; Mu, Y.; Du, A.-K. Facile Synthesis of g-C_3N_4 Nanosheets/ZnO Nanocomposites with Enhanced Photocatalytic Activity in Reduction of Aqueous Chromium(VI) under Visible Light. *Nanomaterials* **2016**, *6*, 173. [CrossRef] [PubMed]
4. Luo, S.; Qin, F.; Ming, Y.A.; Zhao, H.P.; Liu, Y.L.; Chen, R. Fabrication uniform hollow Bi_2S_3 nanospheres via Kirkendall effect for photocatalytic reduction of Cr(VI) in electroplating industry wastewater. *J. Hazard. Mater.* **2017**, *340*, 253–262. [CrossRef] [PubMed]

5. Abdel Moniem, S.M.; Ali, M.E.M.; Gad-Allah, T.A.; Khalil, A.S.G.; Ulbricht, M.; El-Shahat, M.F.; Ashmawy, A.M.; Ibrahim, H.S. Detoxification of hexavalent chromium in wastewater containing organic substances using simonkolleite-TiO_2 photocatalyst. *Process Saf. Environ. Prot.* **2015**, *95*, 247–254. [CrossRef]
6. Zhang, Y.C.; Li, J.; Zhang, M.; Dionysiou, D.D. Size-tunable hydrothermal synthesis of SnS_2 nanocrystals with high performance in visible light-driven photocatalytic reduction of aqueous Cr(VI). *Environ. Sci. Technol.* **2011**, *45*, 9324–9331. [CrossRef] [PubMed]
7. Xu, X.Y.; Zhou, X.S.; Zhang, L.L.; Xu, L.M.; Ma, L.; Luo, J.; Li, M.J.; Zeng, L.H. Photoredox degradation of different water pollutants (MO, RhB, MB, and Cr(VI)) using Fe-N-S-tri-doped TiO_2 nanophotocatalyst prepared by novel chemical method. *Mater. Res. Bull.* **2015**, *70*, 106–113. [CrossRef]
8. Meng, X.D.; Zhang, G.K.; Li, N. $Bi_{24}Ga_2O_{39}$ for visible light photocatalytic reduction of Cr(VI): Controlled synthesis, facet-dependent activity and DFT study. *Chem. Eng. J.* **2017**, *314*, 249–256. [CrossRef]
9. Wan, Z.; Zhang, G.K.; Wu, X.Y.; Yin, S. Novel visible-light-driven Z-scheme $Bi_{12}GeO_{20}/g\text{-}C_3N_4$ photocatalyst: Oxygen-induced pathway of organic pollutants degradation and proton assisted electron transfer mechanism of Cr(VI) reduction. *Appl. Catal. B* **2017**, *207*, 17–26. [CrossRef]
10. Wang, P.; Ji, W.D.; Li, M.M.; Zhang, G.K.; Wang, J.L. $Bi_{25}VO_{40}$ microcube with step surface for visible light photocatalytic reduction of Cr(VI): Enhanced activity and ultrasound assisted regeneration. *Ultrason. Sonochem.* **2017**, *38*, 289–297. [CrossRef] [PubMed]
11. Vera, M.L.; Traid, H.D.; Henrikson, E.R.; Ares, A.E.; Litter, M.I. Heterogeneous photocatalytic Cr(VI) reduction with short and long nanotubular TiO_2 coatings prepared by anodic oxidation. *Mater. Res. Bull.* **2018**, *97*, 150–157. [CrossRef]
12. Sun, B.; Reddy, E.P.; Smirniotis, P.G. Visible light Cr(VI) reduction and organic chemical oxidation by TiO_2 photocatalysis. *Environ. Sci. Technol.* **2005**, *39*, 6251–6259. [CrossRef] [PubMed]
13. Huang, L.H.; Chan, Q.Z.; Wu, X.J.; Wang, H.J.; Liu, Y.L. The simultaneous photocatalytic degradation of phenol and reduction of Cr(VI) by TiO_2/CNTs. *J. Ind. Eng. Chem.* **2012**, *18*, 574–580. [CrossRef]
14. Liu, F.H.; Yu, J.; Tu, G.Y.; Qu, L.; Xiao, J.C.; Liu, Y.D.; Wang, L.Z.; Lei, J.Y.; Zhang, J.L. Carbon nitride coupled Ti-SBA15 catalyst for visible-light-driven photocatalytic reduction of Cr(VI) and the synergistic oxidation of phenol. *Appl. Catal. B* **2017**, *201*, 1–11. [CrossRef]
15. Li, X.F.; Zhen, X.Z.; Meng, S.G.; Xian, J.J.; Shao, Y.; Fu, X.Z.; Li, D.Z. Structuring β-Ga_2O_3 photonic crystal photocatalyst for efficient degradation of organic pollutants. *Environ. Sci. Technol.* **2013**, *47*, 9911–9917. [CrossRef] [PubMed]
16. Shao, T.; Zhang, P.Y.; Jin, L.; Li, Z.M. Photocatalytic decomposition of perfluorooctanoic acid in pure water and sewage water by nanostructured gallium oxide. *Appl. Catal. B* **2013**, *142–143*, 654–661. [CrossRef]
17. Girija, K.; Thirumalairajan, S.; Mastelaro, V.R.; Mangalaraj, D. Photocatalytic degradation of organic pollutants by shape selective synthesis of β-Ga_2O_3 microspheres constituted by nanospheres for environmental remediation. *J. Mater. Chem. A* **2015**, *3*, 2617–2627. [CrossRef]
18. Pan, Y.X.; Sun, Z.Q.; Cong, H.P.; Men, Y.L.; Xin, S.; Song, J.; Yu, S.H. Photocatalytic CO_2 reduction highly enhanced by oxygen vacancies on Pt-nanoparticle-dispersed gallium oxide. *Nano Res.* **2016**, *9*, 1689–1700. [CrossRef]
19. Chen, S.L.; Li, D.; Liu, Y.X.; Huang, W.X. Morphology-dependent defect structures and photocatalytic performance of hydrogenated anatase TiO_2 nanocrystals. *J. Catal.* **2016**, *341*, 126–135. [CrossRef]
20. Liu, J.; Lu, W.; Tian, B.S.; Hu, B.; Jin, L.; Shi, Y.R.; Li, L.; Wang, Z.L. Shape-controlled synthesis and facet-dependent performance of single-crystal $Bi_{25}GaO_{39}$ photocatalysts. *CrystEngComm* **2016**, *18*, 7715–7721. [CrossRef]
21. Liu, J.; Lu, W.; Zhong, Q.; Jin, X.D.; Wei, L.Y.; Wu, H.Z.; Zhang, X.L.; Li, L.L.; Wang, Z.L. Single-crystal $Bi_2Ga_4O_9$ nanoplates with exposed {110} facets for photocatalytic degradation of Acid Red 1. *Mol. Catal.* **2017**, *433*, 354–362. [CrossRef]
22. Parlett, C.M.A.; Wilson, K.; Lee, A.F. Hierarchical porous materials: Catalytic applications. *Chem. Soc. Rev.* **2013**, *42*, 3876–3893. [CrossRef] [PubMed]
23. Deng, Y.C.; Tang, L.; Zeng, G.M.; Zhu, Z.J.; Yan, M.; Zhou, Y.Y.; Wang, J.J.; Liu, Y.N.; Wang, J.J. Insight into highly efficient simultaneous photocatalytic removal of Cr(VI) and 2,4-diclorophenol under visible light irradiation by phosphorus doped porous ultrathin g-C_3N_4 nanosheets from aqueous media: Performance and reaction mechanism. *Appl. Catal. B* **2017**, *203*, 343–354. [CrossRef]

24. Zhang, M.; Xu, J.; Zong, R.L.; Zhu, Y.F. Enhancement of visible light photocatalytic activities via porous structure of g-C_3N_4. *Appl. Catal. B* **2014**, *147*, 229–235. [CrossRef]
25. Liu, J.; Lu, W.; Zhong, Q.; Wu, H.Z.; Li, Y.L.; Li, L.L.; Wang, Z.L. Effect of pH on the microstructure of β-Ga_2O_3 and its enhanced photocatalytic activity for antibiotic degradation. *J. Colloid Interface Sci.* **2018**, *519*, 255–262. [CrossRef] [PubMed]
26. Liu, J.; Lu, W.; Wu, H.Z.; Jin, L.; Hu, B.; Li, L.L.; Wang, Z.L. In situ synthesis of rice-like $ZnGa_2O_4$ for the photocatalytic removal of organic and inorganic pollutants. *Mater. Sci. Semicond. Process.* **2016**, *56*, 251–259. [CrossRef]
27. Wang, X.; Xu, Q.; Li, M.R.; Shen, S.; Wang, X.L.; Wang, Y.C.; Feng, Z.C.; Shi, J.Y.; Han, H.X.; Li, C. Photocatalytic overall water splitting promoted by an α-β phase junction on Ga_2O_3. *Angew. Chem. Int. Ed.* **2012**, *51*, 13089–13092. [CrossRef] [PubMed]
28. Machon, D.; McMillan, P.F.; Xu, B.; Dong, J.J. High-pressure study of the β-to-α transition in Ga_2O_3. *Phys. Rev. B* **2006**, *73*, 094125. [CrossRef]
29. Hou, Y.D.; Wu, L.; Wang, X.C.; Ding, Z.X.; Li, Z.H.; Fu, X.Z. Photocatalytic performance of α-, β-, and γ-Ga_2O_3 for the destruction of volatile aromatic pollutants in air. *J. Catal.* **2007**, *250*, 12–18. [CrossRef]
30. Guo, S.; Yuan, N.; Zhang, G.K.; Yu, J.C. Graphene modified iron sludge derived from homogeneous Fenton process as an efficient heterogeneous Fenton catalyst for degradation of organic pollutants. *Microporous Mesoporous Mater.* **2017**, *238*, 62–68. [CrossRef]
31. Girija, K.; Thirumalairajan, S.; Patra, A.K.; Mangalaraj, D.; Ponpandian, N.; Viswanathan, C. Enhanced photocatalytic performance of novel self-assembled floral β-Ga_2O_3 nanorods. *Curr. Appl. Phys.* **2013**, *13*, 652–658. [CrossRef]
32. Barrera-Díaz, C.E.; Lugo-Lugo, V.; Bilyeu, B. A review of chemical, electrochemical and biological methods for aqueous Cr(VI) reduction. *J. Hazard. Mater.* **2012**, *223–224*, 1–12. [CrossRef] [PubMed]
33. Alidokht, L.; Khataee, A.R.; Reyhanitabar, A.; Oustan, S. Reductive removal of Cr(VI) by starch-stabilized Fe^0 nanoparticles in aqueous solution. *Desalination* **2011**, *270*, 105–110. [CrossRef]
34. Liang, R.W.; Shen, L.J.; Jing, F.F.; Wu, W.M.; Qin, N.; Lin, R.; Wu, L. NH_2-mediated indium metal-organic framework as a novel visible-light-driven photocatalyst for reduction of the aqueous Cr(VI). *Appl. Catal. B* **2015**, *162*, 245–251. [CrossRef]
35. Wang, H.; Yuan, X.Z.; Wu, Y.; Zeng, G.M.; Chen, X.H.; Leng, L.J.; Wu, Z.B.; Jiang, L.B.; Li, H. Facile synthesis of amino-functionalized titanium metal-organic frameworks and their superior visible-light photocatalytic activity for Cr(VI) reduction. *J. Hazard. Mater.* **2015**, *286*, 187–194. [CrossRef] [PubMed]

© 2018 by the authors. Licensee MDPI, Basel, Switzerland. This article is an open access article distributed under the terms and conditions of the Creative Commons Attribution (CC BY) license (http://creativecommons.org/licenses/by/4.0/).

nanomaterials

MDPI

Article

Enhanced Photocatalytic Activity of {110}-Faceted TiO_2 Rutile Nanorods in the Photodegradation of Hazardous Pharmaceuticals

Tran Thi Thuong Huyen [1,2,*], Tran Thi Kim Chi [1], Nguyen Duc Dung [3], Hendrik Kosslick [2] and Nguyen Quang Liem [1]

1 Institute of Materials Science (IMS), Vietnam Academy of Science and Technology (VAST), 18 Hoang Quoc Viet, Cau Giay, Hanoi 100000, Vietnam; chittk@ims.vast.ac.vn (T.T.K.C.); liemnq@ims.vast.ac.vn (N.Q.L.)

2 Institute of Chemistry, Department of Inorganic Chemistry, University of Rostock, Albert-Einstein-Str. 3a, Rostock 18051, Germany; hendrik.kosslick@uni-rostock.de

3 Advanced Institute of Science and Technology (AIST), Hanoi University of Science and Technology, 1 Dai Co Viet, Hanoi 100000, Vietnam; dung.nguyenduc@hust.edu.vn

* Correspondence: huyenttt@ims.vast.ac.vn; Tel.: +84-243-756-4129 or +84-906-072-519

Received: 3 April 2018; Accepted: 20 April 2018; Published: 25 April 2018

Abstract: Rutile TiO_2 with highly active facets has attracted much attention owing to its enhanced activity during the photocatalytic degradation of pollutants such as pharmaceuticals in wastewater. However, it is difficult to obtain by controlling the synthetic conditions. This paper reports a simple hydrothermal synthesis of rutile TiO_2 nanorods with highly exposed {110} facets. The obtained rutile was characterized by X-ray diffraction (XRD), scanning electron microscopy (SEM), high-resolution transmission electron microscopy (HR-TEM), and Raman spectroscopy. The main contribution to the photocatalytic activity comes from rutile nanorods with highly dominant active {110} facets, which were studied in the photodegradation of reactive cinnamic acid and more recalcitrant ibuprofen. The contribution of active species was also investigated. The present work further confirmed the hydrothermal synthesis route for controlling the preparation of highly crystalline and active rutile nanocrystals.

Keywords: rutile; active facets; pharmaceutical; degradation; mineralization; active species

1. Introduction

Heterogeneous photocatalysis on semiconductor photocatalysts has attracted considerable interest due to its applicability in the treatment of hazardous organic pollutants [1–5]. Among the ultraviolet light driven photocatalysts, TiO_2 has received much attention. The use of a TiO_2 photocatalyst and a wide band gap (~3.2 eV) [6] offers several advantages including its low cost, chemical stability, high oxidizing ability, safety, and reusability [7–9]. TiO_2 occurs in three main phases: anatase, brookite, and rutile, with anatase being the most commonly used in photocatalytic applications [10–14]. Recently, some photocatalytic studies on rutile TiO_2 have been published. Kalaivani's group reported the enhanced photocatalytic decolourization efficiency of methylene blue over the bio-nanocomposite inulin-TiO_2 rutile under ultraviolet (UV) irradiation [15]. In this case, the inulin-TiO_2 rutile was obtained by embedding rutile nanoparticles into a novel biopolymer-inulin. The approach allowed the diminishing of the agglomeration of rutile nanoparticles, providing a larger surface area, thus improving the activity of rutile. In another report, Nair's group found that the high photocatalytic activity of three-dimensional rutile micro-flowers in the decolourization of Rhodamine B under UV light was due to the large surface area contributed by the highly dense spiky nanostructures [16]. In terms of enhanced activity, rutile nanocrystals with exposed active facets are relatively unknown due to its difficulty in synthesis. Therefore, acquiring a large percentage of active facets by controlling the synthetic conditions is highly desirable. According to the nature of rutile rods, the crystal growth is indeed a result of the competitive growth of {111} facets and {110} facets [17,18]. The reported rutile with exposed {111} facets that had both the advantages of large specific surface area and exposed high active facets were active in the decolourization of methylene blue under UV irradiation [19]. In our work, the large-sized rutile nanocrystals with highly exposed {110} facets were successfully prepared by hydrothermal synthesis and exhibited an enhanced activity in the photodegradation of reactive cinnamic acid (CA) and recalcitrant pharmaceutical ibuprofen (IBP). The photocatalytic performances were investigated under low power solarium lamps, high organic loading, and a small amount of the photocatalyst. Such testing conditions have been rarely reported in wastewater treatment. In addition, trapping experiments were carried out to analyze the contribution of active species to the photodegradation of organic compounds.

2. Materials and Methods

2.1. Materials

All chemicals were of analytical grade and used without further purification: titanium (IV) *i*-propoxide (TTIP, Merck, Kenilworth, NJ, USA, 98%), hydrochloric acid (HCl, Chemsolute, Th. Geyer, Berlin, Germany, 35–38%), cinnamic acid ($C_9H_8O_2$, Reachim, Moscow, Russia, 99%), ibuprofen sodium salt ($C_{13}H_{17}O_2Na$, Sigma-Aldrich, St. Louis, MA, USA, 98%), ethylenediaminetetraacetic acid ($C_{10}H_{16}N_2O_8$, Sigma-Aldrich, 99%), *i*-propanol (C_3H_8O, Sigma-Aldrich, >99%), tert-butanol ($C_4H_{10}O$, Sigma-Aldrich, 99%), benzoquinone ($C_6H_4O_2$, Sigma-Aldrich, ≥98%), and titania P25 (TiO_2, Evonik, Essen, Germany, 99.5%).

2.2. Synthesis of TiO_2 Rutile

The rutile was prepared based on a previously reported procedure in [20] with improved synthesis parameters. The procedure consisted of two main steps:

(i) *Sol-gel synthesis of an amorphous titania precursor*: 20 mL of titanium (IV) *i*-propoxide was dissolved in 105 mL of *i*-propanol. The solution was kept at 0 °C under vigorous stirring. To this colorless solution, a stock solution containing 105 mL of *i*-propanol and 1 mL of distilled water prepared at room temperature (RT) was slowly dropped over a period of 5 h. The suspension gradually changed into a white/milky color. This was further stirred at RT for 24 h. Once the reaction was completed, the white product was removed from the suspension by centrifugation and the obtained clear colorless solution was again diluted with 1000 mL of distilled water and further stirred at RT for 24 h. The obtained white amorphous titania was separated by centrifugation and washed with distilled water and ethanol, then dried under vacuum at 60 °C. The final white powder was used as the precursor for the hydrothermal treatment step.

(ii) *Hydrothermal treatment of amorphous titania precursor*: 1.0 g of amorphous titania was placed into a 120 mL Teflon cup and then an appropriate amount of concentrated aqueous 4.0 M hydrochloric acid was added and stirred at RT for 30 min. Next, the Teflon cup was transferred into a stainless steel-lined autoclave, which was placed into an oven and heated at 200 °C for 7 h. Thereafter, the autoclave was allowed to cool down to RT. The precipitate was decanted from the reaction mixture, washed thoroughly with distilled water and ethanol, and finally dried at 60 °C overnight in an oven. The final product was ground in a porcelain mortar with a pistil to obtain fine powders.

2.3. Characterization

The morphology and microscopic structure of the samples were characterized by scanning electron microscopy (SEM) (FE-SEM S-4800, Hitachi, Tokyo, Japan) operating at 5 kV, and transmission electron microscopy (TEM) and high-resolution transmission electron microscopy (HR-TEM) (JEM 2100, JEOL, Tokyo, Japan) operating at 200 kV.

The crystal structures and phases of the samples were measured using an X-ray diffractometer (STADI-P, STOE, Darmstadt, Germany) with monochromatic Cu Kα radiation (λ = 1.5406 Å). Raman spectra were recorded using a LabRAM HR 800 Raman microscope system (Horiba Jobin YVON, Kyoto, Japan) equipped with a high stability BX40 microscope (Focus 1 μm). A blue laser (473 nm, 20 mW air-cooled solid-state laser) was used as an excitation source.

Brunauer-Emmett-Teller (BET) surface areas (S_{BET}) were determined using the adsorption data in the relative pressure (p/p_0) range of 0.05–0.35. The measurements were performed at 77 K on a Thermo Sorptomatic 1990 nitrogen adsorption apparatus (Thermo Fisher Scientific, Waltham, MA, USA).

The weight loss of the samples was evaluated from the thermogravimetric curve analyzed on a TGA Labsys 1600 DSC instrument (Setaram, Caluire, France) under argon gas at a heating rate of 10 K/min where 0.1 cm^3 alumina (Al_2O_3) crucibles were used.

2.4. Photocatalysis

The photocatalytic performance of the rutile samples was evaluated in the photocatalytic degradation reactions of CA and IBP under ultraviolet-visible (UV-Vis) irradiation using batch-conditions. In each experiment, a glass beaker containing 10 mg of the photocatalyst and 250 mL of an aqueous 10 ppm organic solution was used. The reaction mixture was magnetically stirred in the dark at RT for 30 min to reach the adsorption-desorption equilibrium. Four UV-Vis solarium lamps with a total power of 60 W were used as the light source. These lamps simulate the UV part of sunlight (by light energy distribution and intensity) and emit a continuous spectrum range of about 370–400 nm. The distance between the applied lamps and the surface of the pollutant solution was 15 cm. Parallel tests were performed by placing four batches into a closed aluminum box. Four magnetic stirrers were arranged below. After certain time intervals (0 min, 15 min, 30 min, 1 h, 2 h, 3 h, 4 h, and 5 h), 5 mL aliquots were taken from the reaction mixture, and 5 mL aliquots were taken from the reaction mixture with a syringe and separated from the catalyst by a 0.45 μm polytetrafluoroethylene (PTFE) syringe filter. The abatement of IBP and CA was determined by the change in the absorbance at 221 and 273 nm, respectively, as follows [21]:

$$\text{abatement (\%)} = (A_0 - A_t)/A_0 \times 100, \quad (1)$$

where A_0 and A_t are the initial absorbance and the absorbance after various time intervals of UV-Vis irradiation (t), respectively. All data were measured at RT using a Lambda 19 UV/Vis spectrometer (Perkin Elmer, Waltham, MA, USA).

The trapping experiments using different scavengers (1.46 mg of ethylenediaminetetraacetic acid (EDTA) as scavenger for holes, 0.1 mL of tert-butanol (t-BuOH) as scavenger for hydroxyl radicals, and 2.7 mg of benzoquinone (BQ) as scavenger for superoxide anion radicals) were performed in a similar manner to the above photocatalytic degradation reaction of IBP and CA except that the mentioned scavengers were added to the reaction.

3. Results

3.1. Characterization

A combined study of the XRD pattern and Raman scattering spectrum was performed to confirm the formation of the pure rutile phase. The XRD pattern (Figure 1a) revealed that the obtained hydrothermal product was present in the rutile form with a high intense diffraction peak at 27.4° (2θ) corresponding to the {110} facets and others with high intensities located at 36.1°, 41.2°, and 54.3° representing the {101}, {111}, and {211} facets, respectively [16,19]. The results were consistent with the theoretical diffraction pattern from the JCPDS database (No. 96-900-7532). The average crystallite size D_{XRD} of the rutile was calculated using the Scherrer equation from the width of the most intense rutile reflection (2θ = 27.4°) [22] and determined to be about 90 nm. Representative Raman spectrum

(Figure 1b) showed that three bands appeared at 234, 447, and 608 cm^{-1}, which are characteristic of the TiO_2 rutile phase [23,24].

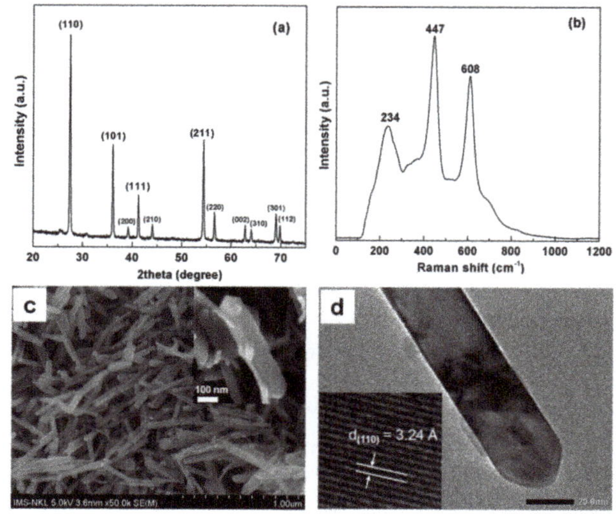

Figure 1. (**a**) XRD pattern; (**b**) Raman spectrum; and (**c**) SEM image of the as-synthesized rutile nanorods. The inset of (**c**) is the corresponding high-resolution SEM image; (**d**) Low resolution TEM image of rutile with exposed {110} facets. Inset: HR-TEM image taken from the body of the rutile nanorods.

The SEM image of the rutile nanocrystals is shown in Figure 1c. Under the present hydrothermal treatment conditions, nearly uniformed rutile nanorods were formed with the size of about 50–100 nm in width and about 300–500 nm in length. Each nanorod consisted of four lateral smooth facets and two pyramidal ends (inset of Figure 1c), which were confirmed by the TEM image shown in Figure 1d. The inset of Figure 1d revealed that the main exposed facets of the obtained rutile were {110} facets corresponding to a spacing value of 3.24 Å.

Based on the weight loss and BET surface [25], the densities of the adsorbed water (physisorbed water) and surface hydroxyl groups (chemisorbed OH groups) were calculated in the temperature range of RT–250 °C and 250–700 °C, respectively [26]. The amount of adsorbed water was estimated as follows: (weight loss × Avogadro's number)/(molecular mass of water × S_{BET}), while surface hydroxyl groups were calculated as follows: (weight loss × Avogadro's number)/(molecular mass of OH group × S_{BET}). The rutile was covered with about twice the amount of OH groups than with anatase and titania P25 (Table 1). This property correlated with the surface hydroxylation, which might explain the difference in the photocatalytic performances between rutile and anatase [27].

Table 1. Densities of the adsorbed water and surface hydroxyl groups calculated based on the weight loss and S_{BET} for anatase, rutile, and titania P25.

Sample	Weight Loss (%)		S_{BET} [1)] (m^2/g)	Density of Adsorbed Water (Molecules/nm^2)	Density of Surface OH Groups (Molecules/nm^2)
	RT–250 °C	250–700 °C			
Rutile	1.01	0.67	12	28	25
Titania P25	2.65	0.79	46	19	8
Anatase [2)]	4.86	3.03	132	12	10

[1),2)] published in Ref. [26].

3.2. *Photocatalysis*

Photocatalyic Activity

The photocatalytic performance of rutile was investigated in the degradation of CA. For comparison, an identical assay was conducted using a commercial TiO_2 (P25, Evonik, Essen, Germany). Figure 2a shows the CA abatement curves determined by the change in the CA absorbance. The CA adsorption on the surfaces of the rutile and P25 was similar (about 1.7%). Through a comparison of the specific surface area of rutile (12 m^2/g) and TiO_2 P25 (46 m^2/g) determined by the BET method using the relative pressure range of 0.05–0.35 in the present nitrogen adsorption-desorption measurement (Table 1), the loading of CA molecules on the rutile surface was higher than that on the P25 surface. Obviously, the quite large planar facets of the rutile crystals allow for the alignment of an increased amount of CA molecules adsorbed on its surface. Such high loadings can be realized with aligned adsorbed molecule multilayers (Langmuir Blodgett-type) [28]. Therefore, the photocatalytic degradation of CA over low surface area rutile is not limited by mass transfer or adsorption.

After the first hour of exposure to UV irradiation, the rutile had slightly higher photocatalytic activity than the P25 (about 91% and 84% of CA was degraded with rutile and P25, respectively). Based on the Langmuir-Hinshelwood first-order kinetic model, the apparent first-order rate constant was derived from the slope of the linear transform $\ln(C/C_0) = f(t)$, where C_0 is the initial concentration of CA and C_t is the concentration of CA at various irradiation times (0 min, 5 min, 15 min, 30 min, and 60 min). Rutile exhibited a fast reaction rate of 0.04214, which was higher than that of the P25 (i.e., 0.03027) as seen in the insert of Figure 2a. A similar high CA abatement (98–99%) was achieved at the end of the reaction (duration 5 h) showing the unexpectedly high photocatalytic activity of the as-synthesized rutile even with its large crystal size and low surface area. It should be noted that the TiO_2 P25 photocatalyst, being a mixture of anatase (80%) and rutile (20%), consisted of spherical aggregated nanoparticles with a diameter of about 25 nm (Figure 2b). In a reported comparison with spherical anatase TiO_2 nanoparticles prepared by a similar hydrothermal procedure [26], rutile was more active for the CA photodegradation. Anatase had a much smaller particle size (10 nm) and higher BET surface area (132 m^2/g). These findings showed that the shape, particle size, and surface area were not limiting factors when explaining the high activity of rutile in the CA photodegradation.

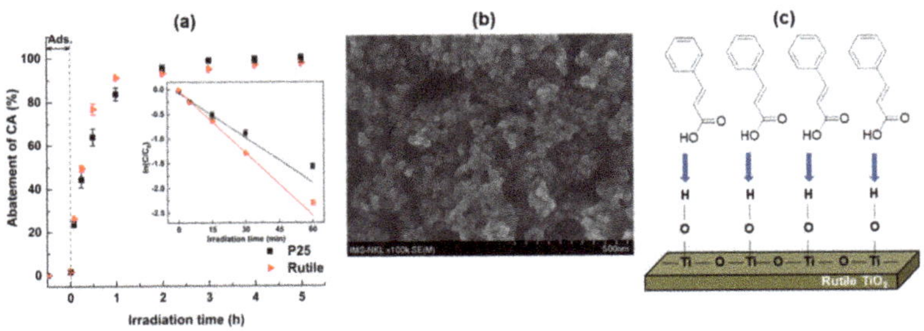

Figure 2. (**a**) Photocatalytic abatement of cinnamic acid (CA) in the photocatalytic performances of TiO_2 P25 and rutile, inset: corresponding plots for linear fitting followed by the apparent first-order reaction model. Reaction conditions: RT, 10 ppm CA, 250 mL reaction solution, 10 mg catalyst loading; (**b**) SEM image of TiO_2 P25; (**c**) Langmuir-Blodgett assembly of CA molecules on the planar crystal rutile facets.

To further understand the photocatalytic behavior of rutile, the degradation of IBP, known as a recalcitrant compound [29], was studied. Figure 3 shows the abatement and mineralization curves determined by the change in the absorbance of CA, IBP, and in the total organic carbon (TOC) removal, respectively. Under UV irradiation, the CA and IBP molecules were degraded immediately, especially with cinnamic acid (Figure 3a). In the photodegradation of IBP, the aromatic ring opening occurred gradually with irradiation time. After 5 h, about 80% of IBP and 98% of CA were degraded. The obtained abatement showed that rutile exhibited different photocatalytic activities depending on the reactivity or recalcitrance of the organic compounds. Namely, the reactive olefinic (–C=C–) double bond presented in the side chain of the CA molecules was immediately attacked at the onset of photocatalytic reaction [30]. As a result, the CA was nearly completely degraded after 5 h of reaction, while the abatement of IBP might take a longer time. This behavior was more pronounced in the mineralization. Figure 3b shows the markedly lower degree of mineralization when compared to the rapid degradation rate. This might be due to the formation of reaction intermediates and/or by-products such as hydroxylation products, ring-opened products, etc. About 98% (UV) and 69% (TOC) were achieved in the CA degradation after 5 h of the photocatalytic reaction, whereas about 80% (UV) and 28% (TOC) were obtained in the IBP degradation. These results indicated that rutile behaves according to the chemical reactivity of the organic compounds. In this work, the recalcitrance of the studied organic compounds increased as follows: CA < IBP.

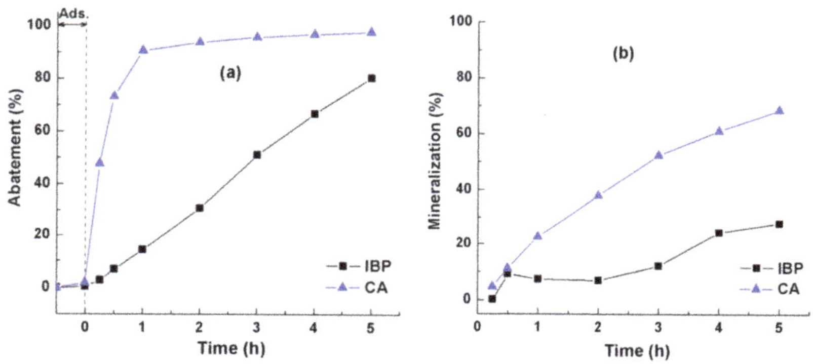

Figure 3. (**a**) Abatement and (**b**) mineralization of (CA) cinnamic acid and (IBP) ibuprofen in the photocatalytic performance of rutile. Reaction conditions: RT, 10 ppm pollutant, 250 mL reaction solution, 10 mg rutile loading.

The difference in the activity of rutile for the photodegradation of CA and IBP can be clarified in terms of the contribution of active species which are considered as practically involved in the photocatalytic reactions but have not been reported so far [31–34]. The role of active species was investigated by the impact of adding different scavengers (*t*-BuOH for $^{\bullet}$OH scavenger, EDTA for holes scavenger, and BQ for $O_2^{\bullet-}$ scavenger) on the photodegradation of IBP over rutile.

Figure 4a shows that the addition of *t*-BuOH remarkably reduced the photocatalytic IBP abatement from the original 80% to 30%, indicating the strong contribution of $^{\bullet}$OH radicals in the treatment with rutile. In contrast, the addition of a holes (h^+) scavenger had an unexpectedly increase in the abatement of IBP, e.g., from 30% (without addition of EDTA) to 50% (addition of EDTA) after 2 h of reaction. This means that these holes were not directly involved in the oxidation of IBP and can migrate to the rutile surface and react with surface OH groups and/or water molecules surrounding the rutile particles that ultimately lead to the formation of $^{\bullet}$OH radicals [5,34,35]. The increase in the abatement of IBP after the addition of holes also implied that EDTA injected more electrons into the valence band of rutile, thereby improving charge carrier separation by the excitation of electrons to the conduction band, thus causing an enhancement in the formation of $O_2^{\bullet-}$ radical anions via the reduction of molecular oxygen by electrons. As a result, the photodegradation of IBP was completely inhibited after the addition of the $O_2^{\bullet-}$ scavenger.

In contrast with the photocatalytic performance of rutile, the holes contributed remarkably to the cleavage of the aromatic ring for the photodegradation of IBP treated with anatase, especially at the initial stage of reaction (Figure 4b). Obviously, the impact of $^{\bullet}OH$ radicals was minor. These results indicated that the lower activity of rutile when compared to anatase in the experiment without the addition of scavengers was due to the absence of oxidative holes. Interestingly, the $O_2^{\bullet-}$ anions had a strong impact during the course of treatment with both photocatalysts (rutile and anatase).

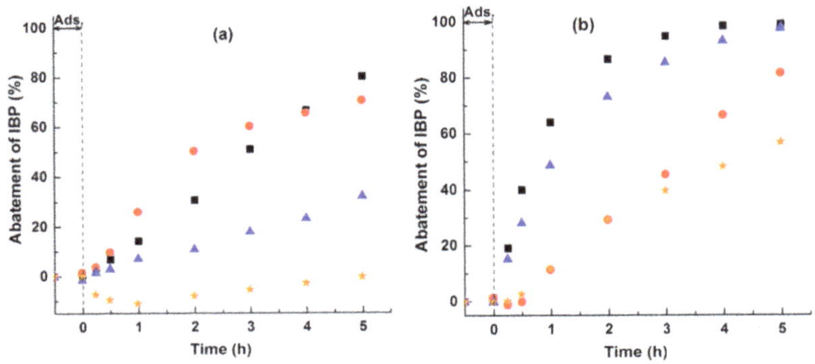

Figure 4. Impact of different scavengers on the photocatalytic abatement of IBP over (**a**) rutile and (**b**) anatase. Reaction conditions: RT, 10 ppm IBP, 250 mL aqueous reaction solution, 10 mg catalyst loading. No scavenger: (black square). Scavengers: (red circle) EDTA, (blue triangle) *t*-BuOH, and (orange star) BQ.

By means of the scavenger experiments, we found that the holes were not directly involved in the photodegradation of IBP over rutile, but that they produced a relatively high number of $^{\bullet}OH$ radicals that acted as "door openers" via the oxidation of surface OH groups which were calculated and shown in Table 1. Murai et al. also suggested that holes were trapped at the titania surface in the adsorbed OH groups yielding the reactive $^{\bullet}OH$ radicals [36]. These findings explained the lower activity of rutile for the photodegradation of recalcitrant ibuprofen when compared to that of reactive cinnamic acid. Cinnamic acid does not require strong active species like holes but reacts with $^{\bullet}OH$ radicals to attack its olefinic double bond, therefore yielding a higher photodegradation efficiency.

In terms of the contribution of the active facets, the difference in the activity of rutile was further studied. Three different rutile samples were used for this purpose including R1 (directly synthesized by the hydrothermal method), R2 (separated from the brookite/rutile mixture by peptization with water), and R3 (obtained by the short-time grinding of R1). The morphology, crystal structures, and phases of the rutile samples are shown in Figure 5. Their XRD patterns indicated that the pure rutile phase was formed and well crystallized. Using the SEM data, the order of particle size was as follows: R1 > R2 > R3.

Figure 6 shows the photocatalytic abatement of IBP over different rutile samples. The difference in the activity between these samples was clearly observed. Only about 20% of IBP over R3 was degraded after 5 h under UV irradiation, while a 2–4 times higher IBP abatement was achieved with R2 (55%) and R1 (80%), respectively. A decrease in the photocatalytic activity followed in the order: R1 > R2 > R3, which was not dependent on the order of particle size: R1 > R2 > R3. The largest rutile nanorods (R1) with the highest activity indicated the minor role of particle size. The photocatalytic activity of rutile nanorods was clearly independent of particle size for the photodegradation of cinnamic acid (Table 2). In general, the particle size reduction accompanied by an increase in the specific surface area led to an enhancement in the activity of the photocatalysts [35,37].

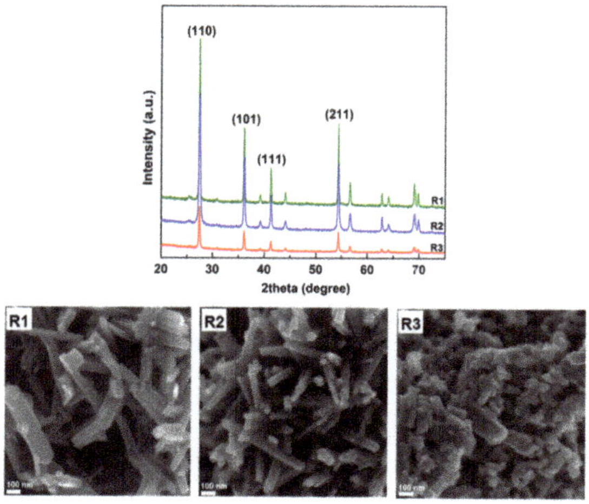

Figure 5. (**Top**) XRD and (**bottom**) SEM images of different rutile samples. R1, R2, and R3 represent the rutile synthesized by the hydrothermal method, rutile separated from the brookite/rutile mixture by peptization with water, and the rutile obtained by short-time grinding of R1, respectively.

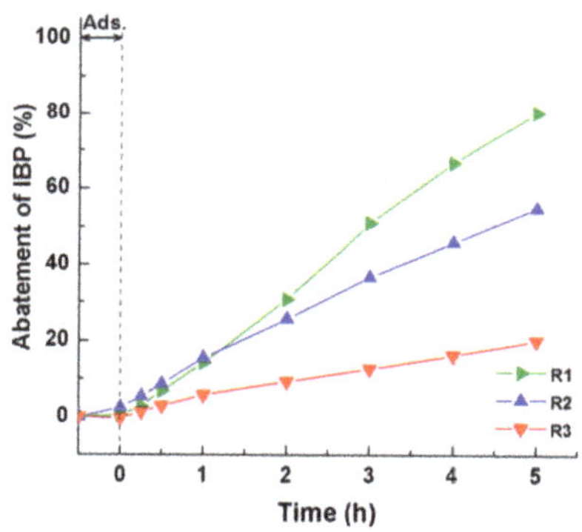

Figure 6. Photocatalytic abatement of IBP over different rutile samples (R1, R2, and R3 represent the rutile synthesized by the hydrothermal method, rutile separated from the brookite/rutile mixture by peptization with water, and the rutile obtained by the short-time grinding of R1, respectively). Reaction conditions: RT, 10 ppm IBP, 250 mL aqueous reaction solution, 10 mg rutile loading.

Table 2. Photocatalytic abatement of CA (%) over different rutile samples.

Time (h) \ Photocatalysts	R1 [a]	R2 [b]	R3 [c]
1	90	84	87
2	93	87	89
3	95	91	90
4	96	94	93
5	97	96	94

[a] Rutile synthesized by the hydrothermal method. [b] Rutile separated from the brookite/rutile mixture by peptization with water. [c] Rutile obtained by the short-time grinding of R1.

Usually, rutile TiO_2 crystals show a rod morphology with the competitive growth of {111} facets and {110} facets [38]. It has also been established that the {110} facets include some point defect types, typically bridging oxygen vacancies, which are strongly related to the surface reactivity. Additionally, these facets have a tendency to trap holes and electrons, which greatly yields the improvement in the separation efficiency of electron-hole pairs, and thus the improvement in activity [39–43]. In our case, we found that the rutile TiO_2 rods with exposed crystal facets that have been predominantly ascribed to the {110} side and {111} edge (sample R1) and {011} edge (sample R2) (Figure 7) had much higher photocatalytic activity than those with a mixture of exposed facets (sample R3 exposing {011}, {020}, and {0$\bar{1}$1} facets). Samples R1 and R2 were both hydrothermal products under different conditions given as 200 °C, HCl 4 M and 175 °C, HCl 3 M, respectively. Under a similar synthesis procedure, these rutile samples were both formed in the rod-shaped nanocrystals with the main active facets of {110}. In contrast, the grinding of R1 may have destroyed or decreased the number of this facet leading to its absence in the R3 sample, which was less active than samples R1 and R2. This finding confirmed the important role of the {110} facets to explain the high photocatalytic activity of the rutile nanorods for the photodegradation of recalcitrant ibuprofen even with a large particle size and low surface area.

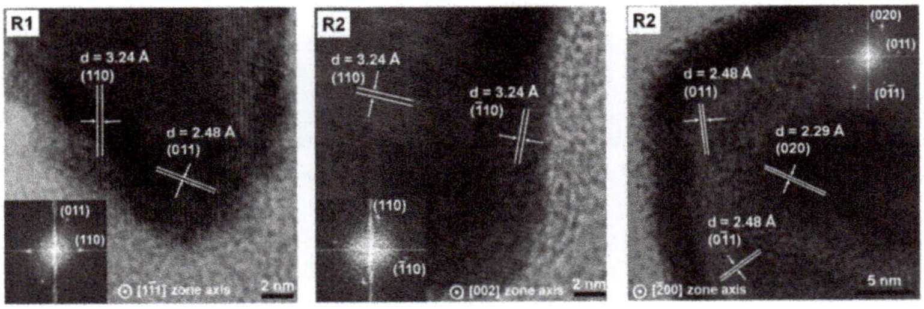

Figure 7. HR-TEM of different rutile samples (R1, R2, and R3 represent the rutile synthesized by the hydrothermal method, rutile separated from the brookite/rutile mixture by peptization with water, and the rutile obtained by the short-time grinding of R1, respectively). Inset: corresponding FFT patterns.

4. Conclusions

In conclusion, highly crystalline rutile TiO_2 nanorods were obtained by hydrothermal synthesis at 200 °C using hydrochloric acid as an acidic agent. Rutile had an unexpectedly high photocatalytic activity in the photodegradation of reactive cinnamic acid even with a large crystal size and low specific surface area. The activity of rutile was correlated to the dominant {110} facets in the photodegradation of recalcitrant ibuprofen. Larger proportion of surface exposed {110} active facets has been formed, therefore, higher photocatalytic activity has been achieved. Rutile behaves according to the chemical reactivity of the organic compounds. The lack of oxidative holes caused a lower IBP aromatic ring opening efficiency, whereas the less strong $^{\bullet}OH$ radicals preferentially formed and easily attacked the cinnamic acid molecules.

Author Contributions: T.T.T.H. and H.K. provided guidance and advice; T.T.T.H., T.T.K.C., and N.D.D. performed the experiments and measurements; T.T.T.H., H.K., and N.Q.L. provided the discussion; T.T.T.H. wrote the paper. All authors contributed to discussion and reviewed the manuscript.

Acknowledgments: This work was supported by the European Union in the frame of the FP7 program, PCATDES project (Grant number 309846); the National Key Laboratory project (Grant number CSTĐ01.17); the National Key Laboratory for Electronic Materials and Devices, Institute of Materials Science, Vietnam Academy of Science and Technology.

Conflicts of Interest: The authors declare no conflicts of interest.

References

1. Ibhadon, A.; Fitzpatrick, P. Heterogeneous Photocatalysis: Recent Advances and Applications. *Catalysts* **2013**, *3*, 189–218. [CrossRef]

2. Litter, M. Heterogeneous photocatalysis Transition metal ions in photocatalytic systems. *Appl. Catal. B Environ.* **1999**, *23*, 89–114. [CrossRef]
3. Kanakaraju, D.; Glass, B.D.; Oelgemöller, M. Titanium dioxide photocatalysis for pharmaceutical wastewater treatment. *Environ. Chem. Lett.* **2014**, *12*, 27–47. [CrossRef]
4. Hashimoto, K.; Irie, H.; Fujishima, A. Photocatalysis: A Historical Overview and Future Prospects. *Jpn. J. Appl. Phys.* **2005**, *44*, 8269–8285. [CrossRef]
5. Ribeiro, A.R.; Nunes, O.C.; Pereira, M.F.R.; Silva, A.M.T. An overview on the advanced oxidation processes applied for the treatment of water pollutants defined in the recently launched Directive 2013/39/EU. *Environ. Int.* **2015**, *75*, 33–51. [CrossRef] [PubMed]
6. Hoffmann, M.R.; Martin, S.T.; Choi, W.; Bahnemann, D.W. Environmental Applications of Semiconductor Photocatalysis. *Chem. Rev.* **1995**, *95*, 69–96. [CrossRef]
7. Herrmann, J. Heterogeneous photocatalysis: Fundamentals and applications to the removal of various types of aqueous pollutants. *Catal. Today* **1999**, *53*, 115–129. [CrossRef]
8. Xu, H.; Ouyang, S.; Liu, L.; Reunchan, P.; Umezawa, N.; Ye, J. Recent advances in TiO_2-based photocatalysis. *J. Mater. Chem. A* **2014**, *2*, 12642. [CrossRef]
9. Tseng, T.K.; Lin, Y.S.; Chen, Y.J.; Chu, H. A review of photocatalysts prepared by sol-gel method for VOCs removal. *Int. J. Mol. Sci.* **2010**, *11*, 2336–2361. [CrossRef] [PubMed]
10. Hanaor, D.A.H.; Sorrell, C.C. Review of the anatase to rutile phase transformation. *J. Mater. Sci.* **2011**, *46*, 855–874. [CrossRef]
11. Kaplan, R.; Erjavec, B.; Pintar, A. Enhanced photocatalytic activity of single-phase, nanocomposite and physically mixed TiO_2 polymorphs. *Appl. Catal. A* **2015**, *489*, 51–60. [CrossRef]
12. Li, Z.; Cong, S.; Xu, Y. Brookite vs. Anatase TiO_2 in the Photocatalytic Activity for Organic Degradation in Water. *ACS Catal.* **2014**, *4*, 3273–3280. [CrossRef]
13. Zhang, J.; Zhou, P.; Liu, J.; Yu, J. New understanding of the difference of photocatalytic activity among anatase, rutile and brookite TiO_2. *Phys. Chem. Chem. Phys.* **2014**, *16*, 20382–20386. [CrossRef] [PubMed]
14. Kaplan, R.; Erjavec, B.; Dražić, G.; Grdadolnik, J.; Pintar, A. Simple synthesis of anatase/rutile/brookite TiO_2 nanocomposite with superior mineralization potential for photocatalytic degradation of water pollutants. *Appl. Catal. B* **2016**, *181*, 465–474. [CrossRef]
15. Jayanthi Kalaivani, G.; Suja, S.K. TiO_2 (rutile) embedded inulin—A versatile bio-nanocomposite for photocatalytic degradation of methylene blue. *Carbohydr. Polym.* **2016**, *143*, 51–60. [CrossRef] [PubMed]
16. Nair, R.V.; Jijith, M.; Gummaluri, V.S.; Vijayan, C. A novel and efficient surfactant-free synthesis of Rutile TiO_2 microflowers with enhanced photocatalytic activity. *Opt. Mater* **2016**, *55*, 38–43. [CrossRef]
17. Truong, Q.D.; Kato, H.; Kobayashi, M.; Kakihana, M. Hierarchical structures of rutile exposing high-index facets. *J. Cryst. Growth* **2015**, *418*, 86–91. [CrossRef]
18. Zhang, Q.; Li, R.; Li, Z.; Li, A.; Wang, S.; Liang, Z.; Liao, S.; Li, C. The dependence of photocatalytic activity on the selective and nonselective deposition of noble metal cocatalysts on the facets of rutile TiO_2. *J. Catal.* **2016**, *337*, 36–44. [CrossRef]
19. Zhang, J.; Liu, P.; Lu, Z.; Xu, G.; Wang, X.; Qian, L.; Wang, H.; Zhang, E.; Xi, J.; Ji, Z. One-step synthesis of rutile nano-TiO_2 with exposed {111} facets for high photocatalytic activity. *J. Alloys Compd.* **2015**, *632*, 133–139. [CrossRef]
20. Reyes-Coronado, D.; Rodríguez-Gattorno, G.; Espinosa-Pesqueira, M.E.; Cab, C.; de Coss, R.; Oskam, G. Phase-pure TiO_2 nanoparticles: Anatase, brookite and rutile. *Nanotechnology* **2008**, *19*, 145605–145615. [CrossRef] [PubMed]
21. Wei, X.; Zhu, G.; Fang, J.; Chen, J. Synthesis, Characterization, and Photocatalysis of Well-Dispersible Phase-Pure Anatase TiO_2 Nanoparticles. *Int. J. Photoenergy* **2013**, *2013*, 726872. [CrossRef]
22. Mahshid, S.; Askari, M.; Ghamsari, M.S. Synthesis of TiO_2 nanoparticles by hydrolysis and peptization of titanium isopropoxide solution. *J. Mater. Process. Technol.* **2007**, *189*, 296–300. [CrossRef]
23. Tompsett, G.A.; Bowmaker, G.A.; Cooney, R.P.; Metson, J.B.; Rodgers, K.A.; Seakins, J.M. The Raman spectrum of brookite, TiO_2 (*Pbca*, $Z = 8$). *J. Raman Spectrosc.* **1995**, *26*, 57–62. [CrossRef]
24. Yang, J.; Mei, S.; Ferreira, J.M.F.; Norby, P.; Quaresmâ, S. Fabrication of rutile rod-like particle by hydrothermal method: An insight into HNO_3 peptization. *J. Colloid Interface Sci.* **2005**, *283*, 102–106. [CrossRef] [PubMed]

25. Kakuma, Y.; Nosaka, A.Y.; Nosaka, Y. Difference in TiO_2 photocatalytic mechanism between rutile and anatase studied by the detection of active oxygen and surface species in water. *Phys. Chem. Chem. Phys.* **2015**, *17*, 18691–18698. [CrossRef] [PubMed]
26. Tran, H.T.T.; Kosslick, H.; Ibad, M.F.; Fischer, C.; Bentrup, U.; Vuong, T.H.; Nguyen, L.Q.; Schulz, A. Photocatalytic Performance of Highly Active Brookite in the Degradation of Hazardous Organic Compounds Compared to Anatase and Rutile. *Appl. Catal. B* **2017**, *200*, 647–658. [CrossRef]
27. Bouras, P.; Stathatos, E.; Lianos, P. Pure versus metal-ion-doped nanocrystalline titania for photocatalysis. *Appl. Catal. B* **2007**, *73*, 51–59. [CrossRef]
28. Choina, J.; Bagabas, A.; Fischer, C.; Flechsig, G.-U.; Kosslick, H.; Alshammari, A.; Schulz, A. The influence of the textural properties of ZnO nanoparticles on adsorption and photocatalytic remediation of water from pharmaceuticals. *Catal. Today* **2015**, *241*, 47–54. [CrossRef]
29. Carballa, M.; Omil, F.; Alder, A.C.; Lema, J.M. Comparison between the conventional anaerobic digestion of sewage sludge and its combination with a chemical or thermal pre-treatment concerning the removal of pharmaceuticals and personal care products. *Water Sci. Technol.* **2006**, *53*, 109–117. [CrossRef] [PubMed]
30. Murakami, F.S.; Bernardi, L.S.; Pereira, R.N.; Valente, B.R. Comparative behavior studies of cinnamic acid using isothermal and nonisothermal kinetic methods. *Pharm. Chem. J.* **2009**, *43*, 716–720. [CrossRef]
31. Marques, R.R.N.; Sampaio, M.J.; Carrapiço, P.M.; Silva, C.G.; Morales-Torres, S.; Dražić, G.; Faria, J.L.; Silva, A.M.T. Photocatalytic degradation of caffeine: Developing solutions for emerging pollutants. *Catal. Today* **2013**, *209*, 108–115. [CrossRef]
32. Wang, Y.; Shi, R.; Lin, J.; Zhu, Y. Significant photocatalytic enhancement in methylene blue degradation of TiO_2 photocatalysts via graphene-like carbon in situ hybridization. *Appl. Catal. B Environ.* **2010**, *100*, 179–183. [CrossRef]
33. Liu, J.; Liu, R.; Li, H.; Kong, W.; Huang, H.; Liu, Y.; Kang, Z. Au nanoparticles in carbon nanotubes with high photocatalytic activity for hydrocarbon selective oxidation. *Dalton Trans.* **2014**, *43*, 12982–12988. [CrossRef] [PubMed]
34. Yang, M.-Q.; Zhang, Y.; Zhang, N.; Tang, Z.-R.; Xu, Y.-J. Visible-light-driven oxidation of primary C–H bonds over CdS with dual co-catalysts graphene and TiO_2. *Sci. Rep.* **2013**, *3*, 3314–3321. [CrossRef] [PubMed]
35. Zangeneh, H.; Zinatizadeh, A.A.L.; Habibi, M.; Akia, M.; Hasnain Isa, M. Photocatalytic oxidation of organic dyes and pollutants in wastewater using different modified titanium dioxides: A comparative review. *J. Ind. Eng. Chem.* **2015**, *26*, 1–36. [CrossRef]
36. Murai, M.; Tamaki, Y.; Furube, A.; Hara, K.; Katoh, R. Reaction of holes in nanocrystalline TiO_2 films evaluated by highly sensitive transient absorption spectroscopy. *Catal. Today* **2007**, *120*, 214–219. [CrossRef]
37. Beydoun, D.; Amal, R.; Low, G.; McEvoy, S. Role of Nanoparticles in Photocatalysis. *J. Nanopart. Res.* **1999**, *1*, 439–458. [CrossRef]
38. Matsunaga, K.; Tanaka, Y.; Toyoura, K.; Nakamura, A.; Ikuhara, Y.; Shibata, N. Existence of basal oxygen vacancies on the rutile TiO_2(110) surface. *Phys. Rev. B* **2014**, *90*, 195303. [CrossRef]
39. Nakamura, R.; Okamura, T.; Ohashi, N.; Imanishi, A.; Nakato, Y. Molecular mechanisms of photoinduced oxygen evolution, PL emission, and surface roughening at atomically smooth (110) and (100) n-TiO_2 (rutile) surfaces in aqueous acidic solutions. *J. Am. Chem. Soc.* **2005**, *127*, 12975–12983. [CrossRef] [PubMed]
40. Wallace, S.K.; Mckenna, K.P. Facet-Dependent Electron Trapping in TiO_2 Nanocrystals. *J. Phys. Chem. C* **2015**, *119*, 1913–1920. [CrossRef]
41. Kowalski, P.M.; Camellone, M.F.; Nair, N.N.; Meyer, B.; Marx, D. Charge localization dynamics induced by oxygen vacancies on the TiO_2(110) surface. *Phys. Rev. Lett.* **2010**, *105*, 146405–146409. [CrossRef] [PubMed]
42. Zuo, F.; Bozhilov, K.; Dillon, R.J.; Le Wang, An; Smith, P.; Zhao, X.; Bardeen, C.; Feng, P. Active facets on titanium(III)-doped TiO_2: An effective strategy to improve the visible-light photocatalytic activity. *Angew. Chem. Int. Ed.* **2012**, *51*, 6223–6226. [CrossRef] [PubMed]
43. Ohno, T.; Murakami, N. Murakami. Spatial Separation of Reaction Sites on Rutile TiO_2 Nanorod. In *Controlled Nanofabrication: Advances and Applications*; Liu, R.S., Ed.; Taylor & Francis: New York, NY, USA, 2012; Volume 2, pp. 17–41.

© 2018 by the authors. Licensee MDPI, Basel, Switzerland. This article is an open access article distributed under the terms and conditions of the Creative Commons Attribution (CC BY) license (http://creativecommons.org/licenses/by/4.0/).

nanomaterials

MDPI

Article

A Facile Approach for the Synthesis of Zn_2SnO_4/BiOBr Hybrid Nanocomposites with Improved Visible-Light Photocatalytic Performance

Tiekun Jia [1,2,*], Ming Liu [1], Dongsheng Yu [1], Fei Long [2], Shuyi Mo [2], Zhao Deng [3,*] and Weimin Wang [3]

1 School of Materials Science and Engineering, Luoyang Institute of Science and Technology, Luoyang 471023, China; liumingming4455@163.com (M.L.); dongsh_yu@163.com (D.Y.)
2 Guangxi Ministry-Province Jointly-Constructed Cultivation Base for State Key Laboratory of Processing for Non-ferrous Metal and Featured Materials, Guilin University of Technology, Guilin, 541004, China; longf@glut.edu.cn (F.L.); moshuyi@glut.edu.cn (S.M.)
3 State Key Lab of Materials Synthesis and Processing, Wuhan University of Technology, Wuhan 430070, China; wangwm@hotmail.com
* Correspondence: tiekunjia@126.com (T.J.); zhaodeng@whut.edu.cn (Z.D.); Tel.: + 86-379-6592-8196 (T.J.); + 86-27-8765-1843 (Z.D.)

Received: 25 March 2018; Accepted: 3 May 2018; Published: 9 May 2018

Abstract: In this study, a novel Zn_2SnO_4/BiOBr hybrid photocatalyst was prepared via a mild hydrothermal synthesis combined with a chemical deposition method. The morphological structure, chemical composition, crystal structure, and optical properties were comprehensively characterized by a series of measurement techniques. Morphological observation showed that fine Zn_2SnO_4 nanoparticles were anchored on the nanoplate surface of a flower-like BiOBr 3D hierarchical structure. The experimental results of UV-vis diffuse reflection spectroscopy revealed that the visible-light absorptive capacity of the Zn_2SnO_4/BiOBr hybrid photocatalyst was promoted, as compared to that of pure Zn_2SnO_4. Evidenced by electro-negativity theoretical calculation, Zn_2SnO_4 and BiOBr possessed matched band edges for accelerating photogenerated charge separation at the interface. The Zn_2SnO_4/BiOBr hybrid photocatalyst exhibited enhanced photocatalytic performance in the degradation of Rhodamine B (RhB) under visible light irradiation. According to the band energy structure and the experimental results, the enhanced photocatalytic performance was ascribed to the improved visible-light absorptive capacity and the contact interface between Zn_2SnO_4 nanoparticles and BiOBr nanoplates, being able to favor the prompt charge migration and suppress the recombination of photogenerated carriers in the hybrid system.

Keywords: nanocomposites; Zn_2SnO_4/BiOBr; visible light; photocatalytic performance

1. Introduction

Nowadays, photocatalysis is considered to be an effective and sustainable approach for resolving the disturbing issue of the energy crisis and environmental pollution. Owing to the high electrical conductivity and electron mobility [1–4], Zn_2SnO_4 (ZTO) has been confirmed the diversity in the photodegradation of organic contaminants [5–10], the evolution of hydrogen from ethanol solution [11], and the reduction of CO_2 to CH_4 [12]. For instance, Lou et al. demonstrated the photodegradation of three water-soluble organic dyes, using ZTO nanocrystals as photocatalyst under UV irradiation [5]. Moreover, the photodegradation of rhodamine B (RhB) [6], methyl blue [7,8], methyl orange [9], and direct black 38 azo-dye [10] was successfully achieved over ZTO nanostructured photocatalysts under UV irradiation. Besides these examples, as reported in the previous literature [11], the photocatalytic H_2 production from methanol solution was successfully accomplished using a ZTO nanostructured material as a photocatalyst under UV irradiation. Despite of the merits mentioned above, ZTO has a wide bandgap (~3.6 eV), leading to the fact that it can only absorb a small portion of the total irradiated natural sunlight. Thus, most of the above mentioned studies on ZTO nanostructured material addressed the enhancement of photocatalytic activity under UV light illumination, owing to its wide band gap. That is to say, the spectral response of bare ZTO nanostructured photocatalysts is still limited to UV region, and the utilization efficiency of solar energy is not desirable for the low portion of UV light in the total solar energy. From another point of view, like other single-component photocatalysts, ZTO also suffers from the fast recombination rate of the photo-generated charge carriers, which is considered to be another critical cause for hindering bare ZTO photocatalysts from acquiring excellent photoactivity. Based on the aforementioned research background, it is particularly essential to design a simple and efficient strategy for achieving visible-light-active ZTO based photocatalysts with high photocatalytic efficiency.

Coupling semiconductor photocatalysts with matched band energy levels has been proven to be an effective strategy for facilitating efficient photogenerated charge separation [12–22]. Owing to its desirable band gap (2.7 eV) and peculiar layered structure constituted by interlacing [Bi_2O_2] slabs with double bromine slabs, BiOBr is emerging as a relatively efficient and stable visible-light-activated photocatalyst. As a sensitizer, BiOBr can extend the visible light response range of wide band gap semiconductors, and separate the photogenerated carriers efficiently. Until now, a series of coupled photocatalysts containing BiOBr, such as, ZnO/BiOBr [18], Ni_2FeO_4/BiOBr [19], $BiPO_4$/BiOBr [20], TiO_2/BiOBr [21], $ZnWO_4$/BiOBr [22] have been reported to show improved photocatalytic performance for degrading organic pollutants. According to previous reports, BiOBr has matched the energy levels of the valence band (VB) and conduction band (CB) with those of ZTO, which provides a theoretical basis for the migration of photoinduced charge carriers in the photocatalytic process. Considering the above mentioned characteristics, ZTO/BiOBr hybrid structures are anticipated to possess high photocatalytic efficiency and photostability compared with pure ZTO. To the best of our knowledge, the preparation and investigation on ZTO/BiOBr hybrid structure has not been reported.

Herein, we developed a facile approach to fabricate ZTO/BiOBr nanocomposites via a hydrothermal synthesis followed by in situ chemical deposition method. According to the determined experimental procedure, a series of ZTO/BiOBr hybrid nanocomposites was obtained by loading different amounts of BiOBr nanoplates. Compared with pure ZTO, the resulting ZTO/BiOBr nanocomposites exhibited wider visible light response range and higher photocatalytic performance toward the degradation of RhB solution. The proposed mechanism of the enhanced photocatalytic performance was discussed based on the experimental results and energy band structure analysis.

2. Results and Discussion

The morphology and microstructure of pure ZTO, pure BiOBr, and 1ZTO/1BiOBr samples were investigated by SEM and TEM images. As displayed in Figure 1a, pure ZTO consisted of abundant fine nanoparticles with sizes ranging from 25–40 nm. From Figure 1b, it is clearly observed that pure BiOBr exhibited flower-like 3D hierarchical structure assembled by substantive nanoplates with thicknesses of 5–8 nm. As for the 1ZTO/1BiOBr sample (Figure 1c), it can be distinctly observed that fine ZTO nanoparticles stacked onto the nanoplate surface of flower-like BiOBr 3D hierarchical structure. Comparatively, the morphology of BiOBr was substantially modified after the introduction of ZTO nanoparticles; the size of nanoplates decreased also. Notably, the linked and intercrossed nanoplates endowed the 1ZTO/1BiOBr sample with an exceptive structure, which resulted in the formation of nano- and macro-pores. The existence of pores and the interface between ZTO and BiOBr are favorable for visible light harvesting and the transfer of photoinduced electron-hole pairs.

Figure 1. FESEM images of (**a**) pure ZTO; (**b**) pure BiOBr; and (**c**) 1ZTO/1BiOBr.

Figure 2a presents a typical TEM image of the 1ZTO/1BiOBr sample, having 3D flower-like hierarchical structure with ZTO nanoparticles growing on the surface of BiOBr nanoplate. Clear and ordered lattice fringes with an interplanar distance of 0.35 nm are indexed to be (101) planes of BiOBr from Figure 2b, indicating high crystallinity. Another lattice fringe with d value of 0.26 nm belongs to (311) plane of ZTO. The above microstructural observation demonstrated that BiOBr nanoplates accreted onto the surface of ZTO nanoparticles, and the intimate contact at the heterojunction, was accomplished along the interface of ZTO nanoparticles and BiOBr nanoplates. Figure 2c shows the typical high angle annular dark field (HAADF)-scanning transmission electron microscope (STEM) image of the 1ZTO/1BiOBr sample. The result of elemental analysis validated the existence of Zn, Sn, O, Bi and Br, revealing a uniform distribution of the above elements throughout the hybrid system as well. Additionally, according to the Brunauer-Emmett-Teller (BET) method, the specific surface area of the ZTO and 1ZTO/1BiOBr samples can be estimated to be 69.5 m^2/g and 46.4 m^2/g respectively, indicating that the BET surface area of the 1ZTO/1BiOBr sample exhibits a slight decrease compared with that of pure ZTO photocatalyst.

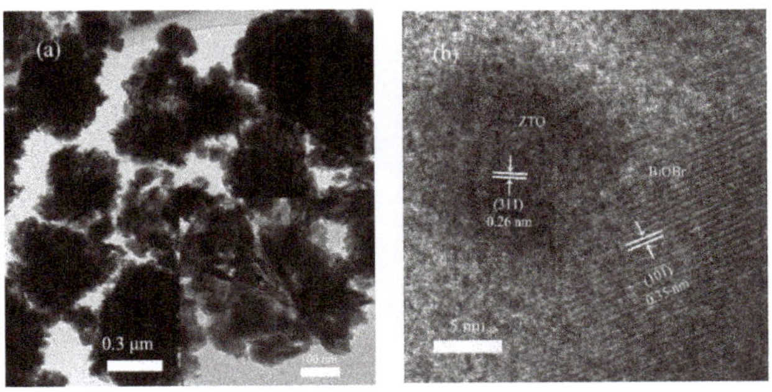

Figure 2. *Cont.*

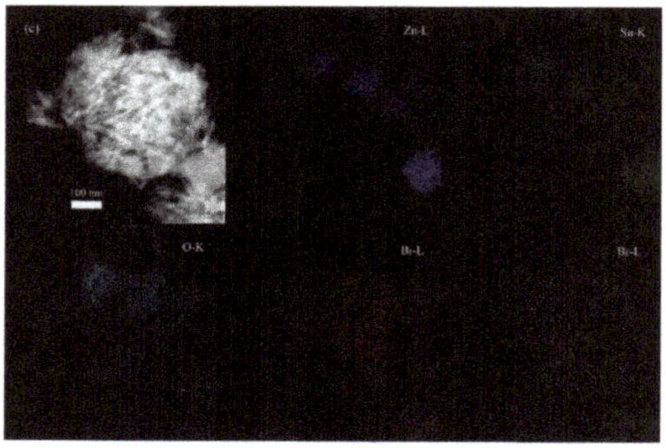

Figure 2. (**a**) TEM image and (**b**) HRTEM image of 1ZTO/1BiOBr; and (**c**) the corresponding EDS mapping of Zn, Sn, O, Bi and Br elements.

Figure 3 presents the XRD patterns of ZTO, BiOBr and ZTO/BiOBr nanocomposites with different mass ratios. All the characteristic diffraction peaks of bare ZTO can be identified to the cubic Zn_2SnO_4 (JCPDS 74-2184). Four intense diffraction peaks at 2θ angles of 17.6°, 28.9°, 34.1°, 41.4° and 60.1° are in high accordance with (111), (220), (311), (400) and (440) planes of ZTO, respectively. For bare BiOBr sample, six characteristic diffraction peaks correspond respectively to the planes of (001), (011), (012), (110), (020) and (212), which agree well with the standard card of the tetragonal phase of BiOBr (JCPDS 09-0393). Two sets of diffraction peaks resulting from ZTO and BiOBr are observable in the ZTO/BiOBr spectra, indicating the coexistence of both ZTO and BiOBr phases in the nanocomposites. Moreover, the intensities of BiOBr tend to be greater with the decrease of ZTO. The grain size of the BiOBr and ZTO crystallites was calculated by means of the Debye-scherrer formula from the broadening of the (101) and (311) XRD reflections at 2θ of 25.304° and 34.319° respectively, after $K\alpha_2$ correction. Full width at half-maximum (FWHM) values was obtained using the Highscore software program as 0.56° and 0.199°. The calculated particle sizes for the BiOBr and Zn_2SnO_4 crystallites are 14.4 nm and 41.3 nm respectively. The XRD result revealed that the obtained samples are of high crystallinty, agreeing well with that of TEM and HRTEM analysis. Additionally, no other diffraction peaks from other impurities are found in the ZTO/BiOBr nanocomposites, suggesting that the as-prepared samples are rather pure.

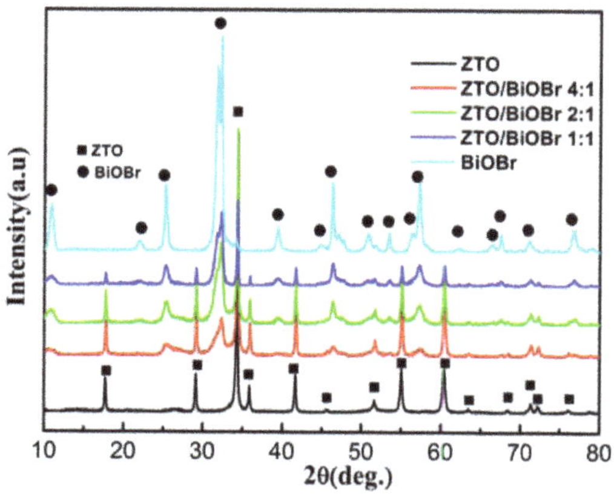

Figure 3. XRD patterns of ZTO, BiOBr, and ZTO/BiOBr nanocomposites.

Figure 4a shows the XPS survey spectrum of the 1ZTO/1BiOBr sample. From Figure 4a, we can observe that the as-obtained 1ZTO/1BiOBr sample is composed of Sn, Zn, O, Bi and Br. High resolution

XPS spectra of Zn 2p, Sn 3d, Bi 4f, Br 3d and O 1s are presented in Figure 4b–f. From Figure 4b, two typical peaks located at 1020.5 and 1043.4 eV for bare ZTO sample correspond to Zn $2p_{3/2}$ and Zn $2p_{1/2}$ respectively. When BiOBr nanoflakes were introduced into the composite system to form the heterojunction, the Zn $2p_{1/2}$ and Zn $2p_{3/2}$ peaks shifted slightly toward higher binding energies by ca. 0.2 eV. Double peaks located at 485.9 and 494.3 eV for bare ZTO sample (Figure 4c) are respectively attributed to Sn $3d_{5/2}$ and Sn $3d_{3/2}$. Similarly, the Sn $3d_{5/2}$ and Sn $3d_{3/2}$ peaks of the 1ZTO/1BiOBr sample exhibited an obvious shift toward high binding energies by ca. 0.5 eV in comparison with those of bare ZTO. From Figure 4d, we can find that the peak position of O 1s spectra for the 1ZTO/1BiOBr sample is different from the counterparts of bare ZTO and BiOBr, which is probably due to the presence of hybrid bonds of Bi-O, Zn-O and Sn-O in the nanocomposites. In terms of Bi 4f (158.9 eV for Bi $4f_{7/2}$ and 164.3 eV for Bi $4f_{5/2}$) and Br 3d (68.1 eV for Br $3d_{5/2}$ and 68.8 eV for Br $3d_{3/2}$) for the 1ZTO/1BiOBr sample (Figure 4e,f), the binding energies are a bit lower than the corresponding counterparts for pure ZTO and BiOBr. The phenomenon of the as-mentioned binding energy shift is most likely a result of the strong interaction between ZTO NPs and BiOBr nanoplates. From a theoretical point of view, the reinforcement of binding energies implies a weakened electron screening effect caused by decreased electron concentration, whereas the increase in electron concentration results in a decrease of binding energies, owing to the improved electron screening effect [23–26]. Based on the above demonstration, we think that the higher and lower shifts are respectively attributed to the decreased electron concentration of ZTO NPs and increased electron concentration of BiOBr nanoplates, due to the strong interaction between the ZTO NPs and BiOBr nanoplates, assisted by the interfacial charge carrier migration.

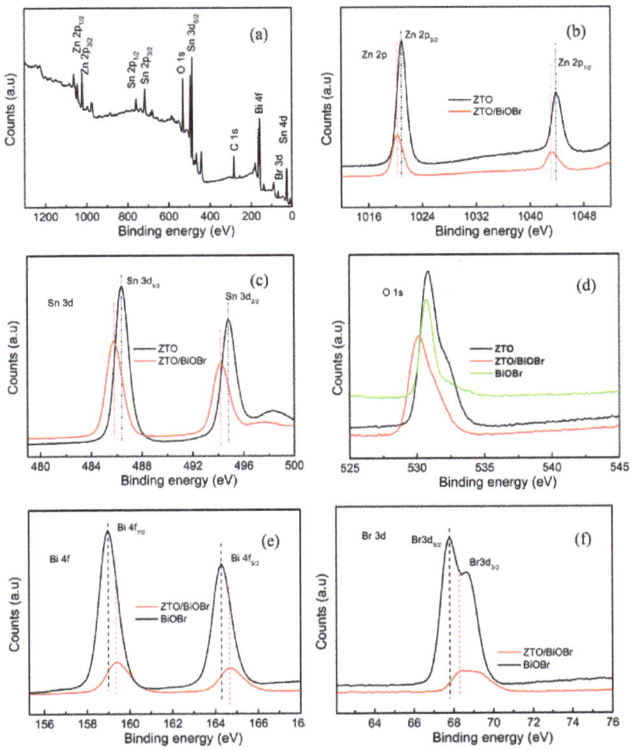

Figure 4. XPS spectra of pure ZTO, BiOBr, and 1ZTO/1BiOBr; (**a**) Survey spectrum; (**b**) Zn 2p; (**c**) Sn 3d; (**d**) O 1s; (**e**) Bi 4f; (**f**) Br 3d.

In order to characterize the optical absorption properties, the obtained diffuse reflection spectra (DRS) are presented in Figure 5. The absorption edge was approximately located at the wavelength of 350 nm for pure ZTO photocatalyst, as displayed in Figure 5a. In addition, a steep slope was found for the absorption curve of pure ZTO, indicating the feature of strong UV light absorption capacity. Comparably, BiOBr exhibited visible light absorption with wavelengths of below 450 nm. After hybridization of ZTO nanoparticles with BiOBr nanoplates, an evident redshift was seen from the absorption edge of ZTO/BiOBr nanocomposites, as compared with pure ZTO, suggesting that BiOBr has the functionality of a visible light sensitizer. Therefore, we can infer that more photon efficiency will be acquired in a ZTO/BiOBr system, resulting in the generation of more photoinduced carriers for enhancing photocatalytic performance. Moreover, ZTO/BiOBr hybrid nanocomposites possessed enhanced UV light absorption, indicating the result of an absorbance superposition of the ZTO and BiOBr components.

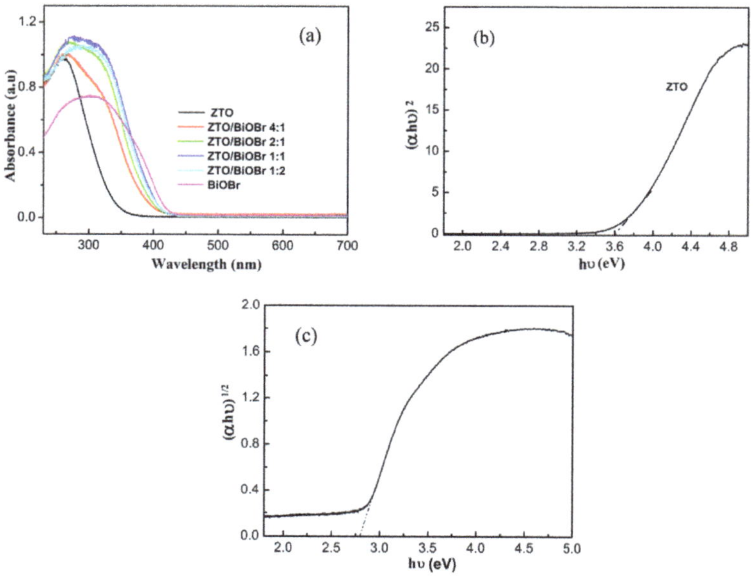

Figure 5. (**a**) UV-vis absorbance spectra of pure ZTO, BiOBr, and ZTO/BiOBr nanocomposites; (**b**) the derived plot of $(\alpha h\nu)^2$ versus hν from the absorption spectrum for pure ZTO; (**c**) the derived plot of $(\alpha h\nu)^{1/2}$ versus hν from the absorption spectrum for pure BiOBr.

The band gap energy can be determined according to the empirical equation, as follows: $(\alpha h\nu)^{2/n} = A(h\nu - E_g)$ [27]. In the empirical equation, α, hν, E_g and A correspond to the absorption coefficient, photon energy, band-gap energy, and a constant. The n values are equal to 1 for ZTO (direct semiconductor) and 4 for BiOBr (indirect semiconductor), respectively, owing to their intrinsic characteristic of electronic transition. After calculation, the E_g values of pure ZTO and BiOBr were respectively estimated to be 3.58 eV and 2.79 eV from the corresponding plots of hν against $(\alpha h\nu)^2$ and hν against $(\alpha h\nu)^{1/2}$ (Figure 5b,c). Moreover, the band edge positions of ZTO and BiOBr can be determined through the empirical formulas [28] as follows:

$$E_{CB} = X - E_e - 0.5E_g$$

$$E_{VB} = E_{CB} + E_g$$

E_{CB} and E_{VB} respectively represent the conduction band (CB) and valence band potential. X belongs to the electronegativity of ZTO and BiOBr, whose values are 7.0 and 6.18 eV respectively. E_g and E_e refer to the band gap energy and energy of free electrons on the hydrogen scale (~4.5 eV) respectively. After calculation, the E_{CB} values of ZTO and BiOBr were determined to be 0.71 and 0.28 eV, while the E_{VB} values of ZTO and BiOBr were respectively determined to be 4.3 and 3.07 eV.

Figure 6a shows the plots of absorbance with respect to irradiation time in the presence of pure ZTO. A stepwise decrease was found from the intensity of the absorption band at 554 nm, indicating that RhB solution was gradually photodegraded over ZTO photocatalyst. Additionally, a slight hypsochromic shift deviating from characteristic absorption band of RhB was also found, with prolonged illumination time during the photodegradation process, agreeing well with previous report [29]. Figure 6b displays the photodegradation activities of RhB over different photocatalysts. As seen in Figure 6b, pure ZTO and BiOBr exhibited low photocatalyitic performance, with only about 60% and 69% of the RhB photodegradation rate after 70 min of visible light illumination. After hybridizing ZTO with BiOBr, the photocatalytic performance of ZTO/BiOBr nanocomposites was obviously improved, as compared with that of single component photocatalyst. Specifically, the photodegradation rate of RhB continuously increased with an increase in BiOBr component, until the mass ratio of BiOBr reached up to 50%. The photodegradation rate of RhB over ZTO/BiOBr nanocomposites decreased steeply, relative to the maximum value (~96%), when the mass ratio of BiOBr exceeded the critical value of 50%. This phenomenon may be attributed to the cooperative roles between ZTO nanoparticles and BiOBr nanoplates exerting a remarkable influence on the charge carrier migration between two semiconductors. At a low mass ratio of BiOBr, the effect of the charge separation was not notable because of an insufficiency of BiOBr. Nevertheless, the introduction of an excessive amount of BiOBr affected the intimate contact between BiOBr nanoplates and ZTO nanoparticles, and the efficient charge carrier separation was hindered correspondingly when increasing BiOBr mass ratio beyond the optimum value (50%). Thus, an appropriate mass ratio of BiOBr is pivotal to modulate the photodegradation behavior of ZTO/BiOBr nanocomposites.

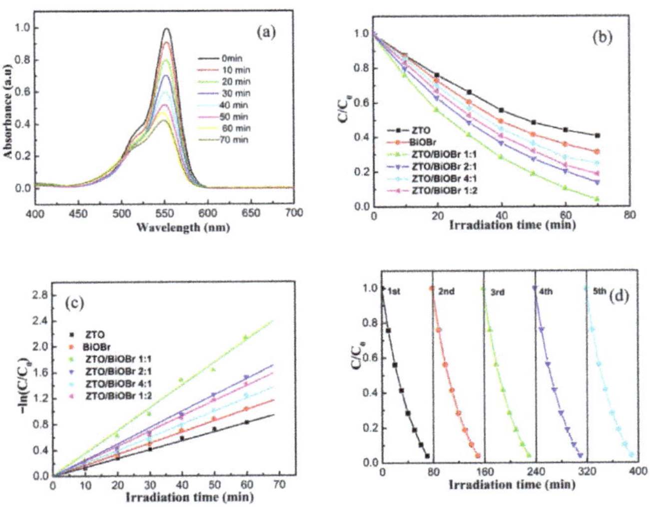

Figure 6. (**a**) Time dependent absorption spectra of RhB aqueous solution over pure ZTO; (**b**) the photodegradation performance of RhB solution over photocatalysts of pure ZTO, BiOBr, and ZTO/BiOBr nanocomposites; (**c**) the plots of $-\ln(C/C_0)$ versus irradiation time for RhB degradation over pure ZTO, BiOBr, and ZTO/BiOBr nanocomposites; (**d**) cycling performance of the photodegradation of RhB solution over 1ZTO/1BiOBr.

Considering that the photolysis of RhB process approximately obeys a pseudo-first-order kinetic equation of $-\ln(C/C_0) = kt$, in which C and C_0 refer to the RhB concentration at time t (min) and t = 0; k (min^{-1}) represents the apparent reaction rate constant, the plots of $-\ln(C/C_0)$ with respect to irradiation time are depicted in Figure 6c. The k value for RhB photodegradation over different photocatalysts followed the order of 1:1 > 2:1 > 1:2 > 4:1 > BiOBr > ZTO. Particularly, the ZTO/BiOBr nanocomposites with mass ratio of 1:1 possess the highest k value (0.037 min^{-1}), which is almost 2.7 and 2.1 times greater than that of pure ZTO and BiOBr respectively. The above result was consistent with the conclusions drawn from photocatalytic degradation plots. Taking consideration into practical applications, the stability of the 1ZTO/1BiOBr photocatalyst was investigated by cyclic photodegradation experiments under identical conditions. From Figure 6d, we can observe that there was no evident deterioration in the photcatalytic performance after five cycle runs. The photodegradation rate even exceeded 93% after five cycles, suggesting that the 1ZTO/1BiOBr photocatalyst takes on adequate stability and long life for degrading pollutants.

To definitively determine the role of the main active species, such as •OH, $•O_2^-$, and h^+, in RhB photodegradation process, different scavengers of EDTA disodium (Na_2-EDTA 4 mmol L^{-1}), tert-butyl-alcohol (*t*-BuOH 4 mmol L^{-1}), and benzoquinone (BQ 4 mmol L^{-1}) were employed as quenchers to capture h^+, •OH, and $•O_2^-$, respectively. Figure 7 displays the effect of various quenchers on the photodegradation of RhB after 70 min of visible light irradiation over a 1ZTO/1BiOBr photocatalyst. Only a bit loss of photodegradation rate was found for the 1ZTO/1BiOBr photocatalyst after the introduction of BQ or Na_2-EDTA, whereas the introduction of *t*-BuOH gave rise to a significant decrease of the photodegradation rate, demonstrating that •OH acted as the main oxidative species, and dominated the photodegradation of RhB over the ZTO/BiOBr nanocomposites under visible light irradiation.

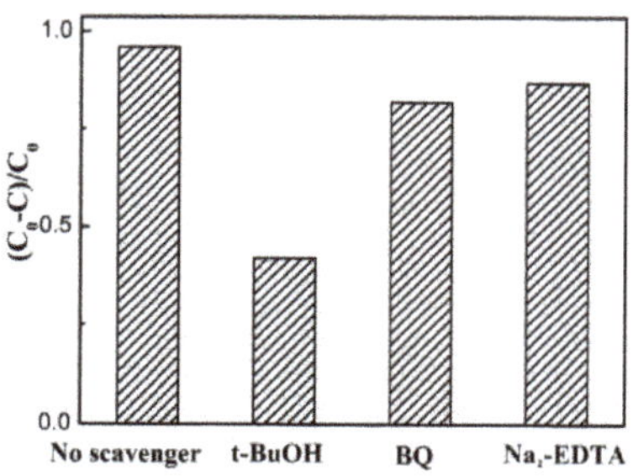

Figure 7. Effect of various scavengers on the visible light photocatalytic performance of 1ZTO/1BiOBr toward the degradation of RhB.

As previous studies have documented [30,31], photoluminescence (PL) spectra of semiconductor photocatalysts are intimately related to the migration, transfer, and the recombination rate of photo-induced e_{CB}^--h_{VB}^+ pairs. Theoretically, the PL emission originates from the recombination of charge carriers, and the intensity of PL emission peaks is approximately proportional to the recombination amount of the excited e_{CB}^--h_{VB}^+ pairs. In other words, lower intensity of the PL emission peak means that the higher the separation efficiency of photo-induced e_{CB}^--h_{VB}^+ pairs is available; accordingly, high photodegradation performance will be achieved. Based on the above, PL spectra of the fabricated photocatalysts were measured, and the results are presented in Figure 8a. The strong emission peak located about 370 nm corresponds to the band-band PL phenomenon of the photo-induced charge carriers for ZTO. Two weak emission peaks located at about 420 nm and 480 nm could be attributed to either to oxygen vacancies or to other kinds of defects, e.g., tin vacancies and O_2 interstitials [32,33]. Among these photocatalysts, pure ZTO showed the strongest PL emission intensity, suggesting a high charge carrier recombination rate. As for ZTO/BiOBr nanocomposites, the PL emission peak intensity decreased continuously, indicating that the recombination of photo-generated charge carriers was effectively restricted due to the efficient charge injection between the two components. The 1ZTO/1BiOBr photocatalyst showed the weakest recombination PL intensity compared to other photocatalysts, revealing the smallest recombination rate of photo-induced e_{CB}^{-}-h_{VB}^{+} pairs in the hybrid system. The increased separation efficiency would provide more opportunities for photogenerated carriers to form active radicals involved into the photodegradation of RhB. The transient photocurrent test was also performed to examine the separation and transfer of photo-excited holes-electrons pairs over different photocatalysts. As displayed in Figure 8b, the photocurrents followed the order of 1ZTO/1BiOBr > BiOBr > ZTO, suggesting that 1ZTO/1BiOBr photocatalyst has a more efficient separation and migration efficiency of the photoexcited carriers. Accordingly, the result of the photocurrent is in good agreement with that of PL spectra.

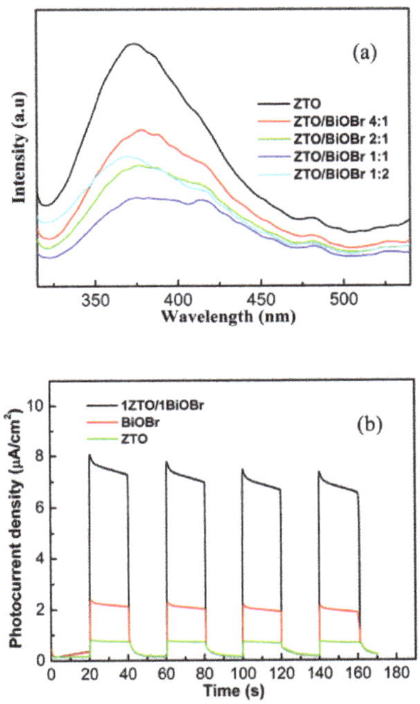

Figure 8. (**a**) PL spectra of pure ZTO, BiOBr, and ZTO/BiOBr nanocomposites; (**b**) Photocurrents of pure ZTO, BiOBr, and 1ZTO/1BiOBr electrodes under visible light irradiation (λ > 420 nm).

As demonstrated above, the 1ZTO/1BiOBr photocatalyst showed the greatest separation efficiency and highest photocatalytic performance. In order to disclose the enhancement of the photocatalytic performance of ZTO/BiOBr nanocomposites, a possible photocatalytic mechanism is proposed and illustrated in Figure 9. Under visible light illumination, BiOBr can be easily excited to produce electron-hole pairs, owing to its appropriate band gap. For ZTO, a small portion of electrons located at the VB position of a shallow energy level is able to transfer to the CB portion, due to the existence of defects, subsequently, generated electron-hole pairs are also achieved on the surface of ZTO. On the basis of the calculation result, the two components of ZTO (*n*-type) and BiOBr (*p*-type) have well-matched energy band structures. After contact, *p*-*n* junctions are formed along the interface of the two components. The arrangement of VB and CB energy band levels induced by the potentials' difference leads to the formation of an inner electric field with the direction from *n*-type ZTO to *p*-type BiOBr, promoting different Femi levels to reach an equilibrium between the two components. As depicted in Figure 9, the CB position of BiOBr is more negative than that of

ZTO, so the photogenerated electrons in the CB of BiOBr are capable of transferring to the CB of ZTO, based on the charge separation way. Simultaneously, the VB position of ZTO is more positive than that of BiOBr; thus, some available holes migrate from the surface of VB of ZTO to that of VB of BiOBr. Based on the above analysis, photogenerated electron-hole pairs are effectively separated by assisting the functionality of the inner electric field. Furthermore, the VB potential of of BiOBr (3.07 eV vs. NHE) is more positive than that of OH^-/•OH (1.99 eV) [34,35]; thus, more photogenerated holes can react with OH^- to form •OH, affecting the oxidation reaction of the RhB molecules. This result is highly consistent with the conclusion of the trapping experiment, i.e., that OH played a dominant role in the oxidation reaction process.

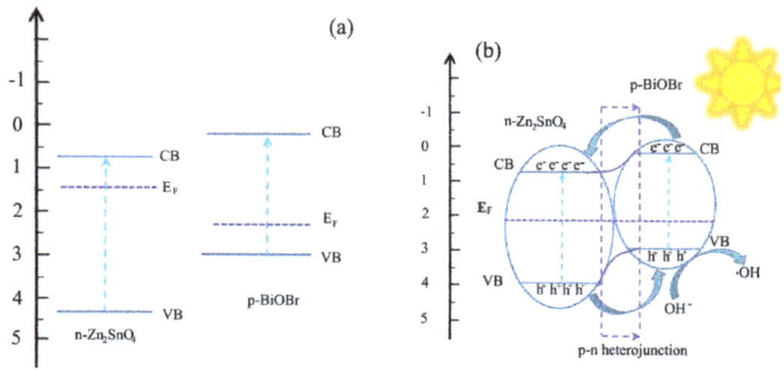

Figure 9. Schematic illustration of the energy band structures (**a**) for separate phases and (**b**) after the formation of *p-n* heterojunction of ZTO and BiOBr.

3. Experimental Section

3.1. Materials

All chemicals, including tin (IV) chloride penthydrate ($SnCl_4 \cdot 5H_2O$), zinc acetate dehydrate ($Zn(CH_3COO)_2 \cdot 2H_2O$), sodium hydroxide (NaOH), bismuth nitrate pentahydrate ($Bi(NO_3)_3 \cdot 5H_2O$), potassium bromide (BiOBr), and rhodamine B (RhB) were purchased from Sinopharm Chemical Reagent CO., Ltd., Shanghai, China, and used as received.

3.2. Synthesis Zn_2SnO_4 Nanoparticles (ZTO NPs)

ZTO NPs were prepared via a simple hydrothermal synthesis route; the detailed procedure was already described in our previous studies [8]. Briefly, after the hydrothermal treatment, a white precipitate was collected, centrifuged and washed repeatedly with deioned water and ethanol respectively, followed by drying at 60 °C for 12 h.

3.3. Synthesis ZTO/BiOBr Nanocomposites

Typically, ZTO/BiOBr nanocomposites were synthesized via a facile in situ precipitation method. 2 mmol of KBr was dissolved into 60 mL deionized water under vigorous stirring and formed a transparent solution; then a certain amount of ZTO NPs was completely dispersed into solution containing KBr under vigorous magnetic stirring, followed by ultrasonic treatment for 30 min. After that, 2 mmol $Bi(NO_3)_3 \cdot 5H_2O$ dissolved into 40 mL ethylene glycol was added dropwise into the previous suspension under constant stirring. Subsequently, the mixture was continuously stirred for another 3 h at 40 °C to make the precipitation reaction complete. After centrifuging and washing with deioned water and ethanol repeatedly, the resulting product was dried at 60 °C in an oven overnight. By varying amount of ZTO NPs, a series of ZTO/BiOBr nanocomposites with mass ratios of 4:1, 2:1, and 1:1, 1:2 (denoted as 4ZTO/1BiOBr, 2ZTO/1BiOBr, 1ZTO/1BiOBr and1ZTO/2BiOBr) was obtained via the similar synthetic process. For comparison, bare BiOBr was also prepared by the similar procedure described above without the addition of ZTO NPs.

3.4. Characterization

Morphological observation and microstructural analysis was taken on a JEM 2100F transmission electron microscope (JEOL Ltd., Tokyo, Japan) and a Hitachi S-4800 field emission scanning electron microscope (Hitachi, Tokyo, Japan). The crystalline structure identification was accomplished on a D8 Advance X-ray diffractometer (Bruker, Billerica, MA, USA) with a Cu Kα source (λ = 0.15406 nm). The XRD measurement was performed at 40 kV and 40 mA in the range of 10°–80°. The surface composition was analyzed by X-ray photoelectron spectroscopy (XPS) spectra on an ESCALAB 250Xi spectrometer (Thermo Fisher Scientific, NewYork, NY, USA) with a monochromatic Al-Kα radiation source. UV-vis diffuse reflectance spectra and room temperature photoluminescence (PL) spectra were recorded on a TU 1901 UV-vis spectrophotometer (Puxi, Beijing, China) with $BaSO_4$ as reference and a LS55 fluorescence spectrophotometer (PE, Waltham, MA, USA) with an excitation wavelength of 310 nm, respectively. The Brunauer-Emmett-Teller (BET) specific surface area of the as-prepared pure ZTO and 1ZTO/BiOBr samples was measured on a Quantachrome NOVA 2000e sorption analyzer.

3.5. Photocatalytic Experiments

The process of the photodegradation test was demonstrated in our previous studies [16,29,36,37]. In short, two daylight lamps (60 W, $\lambda \geq$ 420 nm) were designated as the visible light source for triggering the photodegration reaction of RhB. In a typical run, a thin layer of film (100 mg photocatalyst) was dispersed into 40 mL RhB aqueous solution (1.0×10^{-5} M). After a certain interval of visible light illumination, the reacted solution treated by centrifuging was analyzed by using an UV-vis spectrophotometer at its characteristic absorbance wavelength of 554 nm. The C/C_0 ratios of the RhB concentrations were employed to evaluate the degradation efficiency, in which C represented the concentration at a certain irradiation time, while C_0 was the initial concentration after adsorption-desorption equilibrium.

3.6. Photocatalytic Measurements

Transient photocurrents of the prepared photocatalysts were performed on an electrochemical workstation (CHI660E, Chenhua Instruments Co., Shanghai, China) using a standard three electrode system with a Pt wire as the counter electrode, a saturated Ag/AgCl electrode as the reference electrode, and the as-prepared photocatalysts as the working electrode, respectively. The detailed procedure of the measurement was reported in our previous study [38].

4. Conclusions

In summary, ZTO/BiOBr hybrid nanocomposites have been successfully constructed via a facile approach. The introduction of BiOBr resulted in the establishment of the contact interface of Zn_2SnO_4/BiOBr heterojunctions that endowed the photocatalyst with an improved light harvesting capacity, high efficiency of photogenerated charge separation, as well as stronger oxidation ability. Zn_2SnO_4/BiOBr hybrid photocatalysts displayed enhanced photocatalytic performance and excellent anti-photocorrosion toward the photodegradation of RhB under visible light illumination. This study will open up a new way of constructing novel hybrid photocatalysts with efficient visible-light phtocatalytic activity for use in the fields of environmental remediation and hydrogen energy production.

Author Contributions: T.J. conceived and designed the experiments; M.L. and D.Y. carried out the synthetic experiment and photo-catalytic performance of the as-prepared samples; F.L. and S.M. measured the as-prepared samples; Z.D. and W.W. provided precise instruction and analyzed the data. T.J. wrote the paper.

Funding: The authors express grateful thanks to the National Natural Science Foundation of China (Grant U1304520), Education Department of Henan Province for the fund support (18A 430020), Henan Provincial Natural Science Foundation (182300410252), and the Open Fund of Guangxi Ministry-Province Jointly-Constructed Cultivation Base for State Key Laboratory of Processing for Non-ferrous Metal and Featured Materials (No.15KF-7).

Acknowledgments: We express our thanks Xiaoqing Liu from Wuhan University of Technology for their assistance with the HRTEM characterizations.

Conflicts of Interest: The authors declare no conflict of interest.

References

1. Tan, B.; Toman, E.; Li, Y.G.; Wu, Y.Y. Zinc stannate (Zn_2SnO_4) dye-sensitized solar cells. *J. Am. Chem. Soc.* **2007**, *129*, 4162–4163. [CrossRef] [PubMed]
2. Alpuche-Aviles, M.A.; Wu, Y. Photoelectrochemical study of the band structure of Zn_2SnO_4 prepared by the hydrothermal method. *J. Am. Chem. Soc.* **2009**, *131*, 3216–3224. [CrossRef] [PubMed]
3. Wang, K.; Huang, Y.; Huang, H.J.; Zhao, Y.; Qin, X.L.; Sun, X.; Wang, Y.L. Hydrothermal synthesis of flower-like Zn_2SnO_4 composites and their performance as anode materials for lithium-ion batteries. *Ceram. Int.* **2013**, *39*, 4569–4574. [CrossRef]
4. Wang, J.H.; Li, H.; Meng, S.G.; Zhang, L.; Fu, X.L.; Chen, S.F. One-pot hydrothermal synthesis of highly efficient SnO_x/Zn_2SnO_4 composite photocatalyst for the degradation of methyl orange and gaseous benzene. *Appl. Catal. B Environ.* **2017**, *200*, 19–30. [CrossRef]
5. Lou, X.; Jia, X.; Xu, J.; Liu, S.; Cao, Q. Hydrothermal synthesis, characterization and photocatalytic properties of Zn_2SnO_4 nanocrystal. *Mater. Sci. Eng. A* **2006**, *432*, 221–225. [CrossRef]
6. Zhao, Q.Q.; Deng, X.L.; Ding, M.; Huang, J.Z.; Ju, D.X; Xu, X.J. Synthesis of hollow cubic Zn_2SnO_4 sub-microstructures with enhanced photocatalytic performance. *J. Alloys Compd.* **2016**, *671*, 328–333. [CrossRef]
7. Junploy, P.; Phuruangrat, A.; Plubphon, N.; Thongtem, S.; Thongtem, T. Photocatalytic degradation of methylene blue by Zn_2SnO_4-SnO_2 system under UV visible radiation. *Mater. Sci. Semicond. Process.* **2017**, *66*, 56–61. [CrossRef]
8. Jia, T.K.; Zhao, J.W.; Fu, F.; Deng, Z.; Wang, W.M.; Fu, Z.Y.; Meng, F.C. Synthesis, characterization and photcatalytic activity Zn-doped SnO_2/Zn_2SnO_4 coupled nanocomposites. *Int. J. Photoenergy* **2014**, *2014*, 197824. [CrossRef]
9. Fu, X.L.; Wang, X.; Long, J.; Ding, Z.; Yan, T.; Zhang, G.; Zhang, Z.; Lin, H.; Fu, X.Z. Hydrothermal synthesis, characterization, and photocatalytic properties of Zn_2SnO_4. *J. Solid State Chem.* **2009**, *182*, 517–524. [CrossRef]
10. Foletto, E.L.; Jahn, S.L.; Moreira, R.d.F.P.M. Hydrothermal of preparation of Zn_2SnO_4 nanocrystals and photocatalytic degradation of a leather dye. *J. Appl. Electrochem.* **2010**, *40*, 59–63. [CrossRef]
11. Núñez, J.; Fresno, F.; Collado, L.; Jana, P.; Coronado, J.M.; Serrano, D.P.; Víctor, A. Photocatalytic H_2 production from aqueous methanol solutions using metal-co-catalysed Zn_2SnO_4 nanostructures. *Appl. Catal. B Environ.* **2016**, *191*, 106–115. [CrossRef]
12. Li, Z.; Zhou, Y.; Zhang, J.; Tu, W.; Liu, Q.; Yu, T.; Zou, Z. Hexagonal nanoplate-textured micro-octahedron Zn_2SnO_4: combined effects toward enhanced efficiencies of dye-sensitized solar cell and photoreduction of CO_2 into hydrocarbon fuels. *Cryst. Growth Des.* **2012**, *12*, 1476–1481. [CrossRef]
13. Jiang, Y.; Peng, Z.; Zhang, S.; Li, F.; Liu, Z.; Mhang, J.; Liu, Y.; Wang, K. Facile in-situ Solvothermal Method to synthesize double shell $ZnIn_2S_4$ nanosheets/TiO_2 hollow nanosphere with enhanced photocatalytic activities. *Ceram. Int.* **2018**, *44*, 6115–6126. [CrossRef]
14. Yu, H.G.; Liu, R.; Wang, X.; Wang, P.; Yu, J.G. Enhanced visible-light photocatalytic activity of Bi_2WO_6 nanoparticles by Ag_2O cocatalyst. *Appl. Catal. B Environ.* **2012**, *111–112*, 326–333. [CrossRef]
15. Yu, J.G.; Ran, J.R. Facile preparation and enhanced photocatalytic H_2-production activity of $Cu(OH)_2$ cluster modified TiO_2. *Energy Environ. Sci.* **2011**, *4*, 1364–1371. [CrossRef]
16. Jia, T.K.; Li, J.L.; Long, F.; Fu, F.; Deng, Z.; Wang, Z.X.; Zhang, Y. Ultrathin g-C_3N_4 Nanosheet-nodified BiOCl hierarchical flower-like plate heterostructure with enhanced photostability and photocatalytic performance. *Crystal* **2017**, *7*, 266. [CrossRef]
17. Hassan, S.M.; Ahmed, A.I.; Mannaa, M.A. Structural, photocatalytic, biological and catalytic properties of SnO_2/TiO_2 nanoparticles. *Ceram. Int.* **2018**, *44*, 6201–6211. [CrossRef]

18. Duo, F.F.; Wang, Y.W.; Mao, X.M.; Fan, C.M.; Zhang, H. Double Br sources fabrication and photocatalytic property of p-n junction BiOBr/ZnO composites for phenol removal. *Cryst. Res. Technol.* **2014**, *49*, 721–730. [CrossRef]
19. Li, X.W.; Wang, L.; Zhang, L.; Zhuo, S.P. A facile route to the synthesis of magnetically separable BiOBr/$NiFe_2O_4$ composites with enhanced photocatalytic performance. *Appl. Surf. Sci.* **2017**, *419*, 589–594. [CrossRef]
20. Gao, M.C.; Zhang, D.F.; Pu, X.P.; Ma, H.Y.; Su, C.H.; Gao, X.; Dou, J.M. Surface decoration of BiOBr with $BiPO_4$ nanoparticles to build heterostructure photocatalysts with enhanced visible-light photocatalytic activity. *Sep. Purif. Technol.* **2016**, *170*, 183–189. [CrossRef]
21. Wang, X.J.; Yang, W.Y.; Li, F.T.; Zhao, J.; Liu, R.H.; Liu, S.J.; Li, B. Construction of amorphous TiO_2/BiOBr heterojunctions via facets coupling for enhanced photocatalytic activity. *J. Hazard. Mater.* **2015**, *292*, 126–136. [CrossRef] [PubMed]
22. Song, X.C.; Li, W.T.; Huang, W.Z.; Zhou, H.; Zheng, Y.F.; Yin, H.Y. A novel p-n heterojunction BiOBr/$ZnWO_4$: preparation and its improved visible lightphotocatalytic activity. *Mater. Chem. Phys.* **2015**, *160*, 251–256. [CrossRef]
23. Zhang, Z.; Huang, J.; Zhang, M.; Yuan, Q.; Dong, B. Ultrathin hexagonal SnS_2 nanosheets coupled with g-C_3N_4 nanosheets as 2D/2D heterojunction photocatalysts toward high photocatalytic activity. *Appl. Catal. B Environ.* **2015**, *163*, 298–305. [CrossRef]
24. Zhang, Z.Y.; Shao, C.L.; Li, X.; Wang, C.; Zhang, M.; Liu, Y.C. Electrospun nanofibers of *p*-type NiO/*n*-type ZnO heterojunctions with Enhanced photocatalytic activity. *ACS Appl. Mater. Interfaces* **2010**, *2*, 2915–2923. [CrossRef] [PubMed]
25. Zhang, Z.Y.; Shao, C.L.; Li, X.; Sun, Y.; Zhang, M.; Mu, J.; Zhang, P.; Guo, Z.; Liu, Y.C. Hierarchical assembly of ultrathin hexagonal SnS_2 nanosheets onto electrospun TiO_2 nanofibers: enhanced photocatalytic activity based on photoinduced interfacial charge transfer. *Nanoscale* **2013**, *5*, 606–618. [CrossRef] [PubMed]
26. Li, X.H.; Xu, H.Y.; Zhang, X.T.; Liu, Y.C.; Sun, J.W.; Lu, Y.M. Local chemical states and thermal stabilities of nitrogen dopants in ZnO film studied by temperature-dependent X-ray photoelectron spectroscopy. *Appl. Phys. Lett.* **2009**, *95*, 191903. [CrossRef]
27. Zhu, A.; Zhao, Q.; Li, X.; Shi, Y. $BiFeO_3$/TiO_2 nanotube arrays composite electrode: Construction, characterization, and enhanced photoelectrochemical properties. *ACS Appl. Mater. Interfaces* **2014**, *6*, 671–679. [CrossRef] [PubMed]
28. Tang, C.; Liu, E.Z.; Wan, J.; Hu, X.; Fan, J. Co_3O_4 nanoparticles decorated Ag3PO4 tetrapods as an efficient visible-light-driven heterojunction photocatalysts. *Appl. Catal. B Environ.* **2016**, *181*, 707–315. [CrossRef]
29. Jia, T.K.; Fu, F.; Long, F.; Min, Z.Y.; Zhao, J.W.; Chen, J.; Li, J.L. Synthesis, characterization and enhanced visible-light photocatalytic activity of Zn_2SnO_4/C nanocomposites with truncated octahedron morphology. *Ceram. Int.* **2016**, *42*, 3808–3815. [CrossRef]
30. Li, Q.; Zhao, X.; Yang, J.; Jia, C.; Jin, Z.; Fan, W.L. Exploring the effects of nanocrystal facet orientations in g-C_3N_4/BiOCl heterostructures on photocatalytic performance. *Nanoscale* **2015**, *7*, 18971–18983. [CrossRef] [PubMed]
31. Li, F.T.; Wang, Q.; Ran, J.R.; Hao, Y.J.; Wang, X.J.; Zhao, D.S.; Qiao, S.Z. Ionic liquid self-combustion synthesis of BiOBr/$Bi_{24}O_{31}Br_{10}$ heterojunctions with exceptional visible-light photocatalytic performances. *Nanoscale* **2015**, *7*, 1116–1126. [CrossRef] [PubMed]
32. Lin, H.F.; Liao, S.C.; Hung, W.W.; Hu, C.T. Thermal plasma synthesis and optical properties of Zn_2SnO_4 nanopowders. *Mater. Chem. Phys.* **2009**, *117*, 9–13. [CrossRef]
33. Pang, C.; Yan, B.; Liao, L.; Liu, B.; Zheng, Z.; Wu, T.; Yu, T. Synthesis, characterization and opto-electrical properties of ternary Zn_2SnO_4 nanowires. *Nanotechnology* **2010**, *21*, 465706. [CrossRef] [PubMed]
34. Li, Y.; Wang, J.; Yang, Y.; Zhang, Y.; He, D.; An, Q.; Cao, G.Z. Seed-induced growing various TiO_2 on nanostructures on g-C_3N_4 nanosheets with much enhanced photocatalytic acitivity under visible light. *J. Hazard. Mater.* **2015**, *292*, 79–89. [CrossRef] [PubMed]
35. Fu, J.; Tian, Y.L.; Chang, B.B.; Xi, F.N.; Dong, X.P. BiOBr-carbon nitride heterojunctions: Synthesis, enhanced activity and photocatalytic mechanism. *J. Mater. Chem.* **2012**, *22*, 21159–21166. [CrossRef]
36. Jia, T.K.; Wang, W.M.; Long, F.; Fu, Z.Y.; Wang, H.; Zhang, Q.J. Synthesis, characterization, and photocatalytic activity of Zn-doped SnO_2 hierarchical architectures assembled by nanocones. *J. Phys. Chem. C* **2009**, *113*, 9071–9077. [CrossRef]

37. Jia, T.K.; Wang, W.M.; Long, F.; Fu, Z.Y.; Wang, H.; Zhang, Q.J. Fabrication, characterization and photocatalytic activity of La-doped ZnO nanowires. *J. Alloys Compd.* **2009**, *484*, 410–415. [CrossRef]
38. Jia, T.K.; Fu, F.; Yu, D.; Cao, J.; Sun, G. Facile synthesis and characterization of N-doped TiO_2/C nanocomposites with enhanced visible-light photocatalytic performance. *Appl. Surf. Sci.* **2018**, *430*, 438–447. [CrossRef]

© 2018 by the authors. Licensee MDPI, Basel, Switzerland. This article is an open access article distributed under the terms and conditions of the Creative Commons Attribution (CC BY) license (http://creativecommons.org/licenses/by/4.0/).

nanomaterials

MDPI

Article

Construction of g-C_3N_4-mNb_2O_5 Composites with Enhanced Visible Light Photocatalytic Activity

Meiyin Wang, Hui Wang, Yuanhang Ren, Cheng Wang, Zhewei Weng, Bin Yue * and Heyong He *

Department of Chemistry and Shanghai Key Laboratory of Molecular Catalysis and Innovative Materials, Collaborative Innovation Center of Chemistry for Energy Materials, Fudan University, Shanghai 200433, China; 14110220018@fudan.edu.cn (M.W.); 17210220053@fudan.edu.cn (H.W.); YuanhangRen@fudan.edu.cn (Y.R.); 15110220022@fudan.edu.cn (C.W.); 16210220004@fudan.edu.cn (Z.W)

* Correspondence: yuebin@fudan.edu.cn (B.Y.); heyonghe@fudan.edu.cn (H.H.); Tel.: +86-21-31242779 (B.Y.); +86-21-31243916 (H.H.)

Received: 4 May 2018; Accepted: 8 June 2018; Published: 12 June 2018

Abstract: A series of composites consisting of g-C_3N_4 sheet and mesoporous Nb_2O_5 (mNb_2O_5) microsphere were fabricated by in situ hydrolysis deposition of $NbCl_5$ onto g-C_3N_4 sheet followed by solvothermal treatment. The samples were characterized using powder X-ray diffraction (XRD), Fourier transform infrared spectroscopy (FT-IR), transmission electron microscopy (TEM), N_2 adsorption-desorption, X-ray photoelectron spectroscopy (XPS), UV-vis diffuse reflectance spectroscopy (DRS) and photoluminescence spectroscopy (PL). The photocatalytic activity of the composites was studied by degradation of rhodamine B (RhB) and tetracycline hydrochloride (TC-HCl) in aqueous solution under visible light irradiation (λ > 420 nm). Compared with g-C_3N_4 and mNb_2O_5, g-C_3N_4-mNb_2O_5 composites have higher photocatalytic activity due to synergistic effect between g-C_3N_4 and mNb_2O_5. Among these composites, 4% g-C_3N_4-mNb_2O_5 has the highest efficiency and good recyclability for degradation of both RhB and TC-HCl.

Keywords: photocatalytic degradation; g-C_3N_4; mesoporous Nb_2O_5; organic pollutant

1. Introduction

Organic dyes and antibiotics are two types of important products which are widely used in textile and pharmaceutical industries, respectively. The direct discharge of these chemical compounds along with sewage to environment would be seriously harmful to ecosystem and human health. It is, therefore, desirable to explore efficient ways to remove them from water [1–3]. In the past decades, the methods of biological treatment, physical adsorption and chemical transformation have been employed in removal of the organic pollutants [1,4–7]. Among them, much attention has been focused on photocatalytic degradation of organic pollutants over semiconductors such as TiO_2 [7,8], WO_3 [9], ZnO [10], MoS_2 [11], etc. due to their ability to oxidize organics through a redox process with low cost. Nevertheless, the majority of the above semiconductors still suffer from some problems such as limited absorption of visible light, difficult recycling of the catalyst, rapid recombination of photogenerated electrons, holes pairs, etc. n-Type transition metal oxide Nb_2O_5 attracts much interest in photocatalytic reactions due to its high chemical stability, water tolerance, and nontoxicity, although the light absorption of Nb_2O_5 with a wide band gap of ~3.4 eV limits its application only in the UV region [12–14]. Therefore, many efforts have been made to improve the absorption ability of Nb_2O_5 in the visible region, such as combination with other semiconductor material [15], ion doping [16] and deposition of noble metal [17].

Recently, graphitic carbon nitride (g-C_3N_4) has been widely used in photocatalytic reactions such as degradation of pollutants, hydrogen generation and selective oxidation of alcohols under visible light due to its excellent chemical/thermal stability, small band gap of 2.7 eV and low cost [18–20]. However, pure g-C_3N_4 shows low photocatalytic efficiency owing to fast recombination of photogenerated electrons and holes and small specific surface area [21,22]. It was reported that the composites of g-C_3N_4 and other semiconductors, such as TiO_2 [23–25], WO_3 [26], MoS_2 [27], and ZnO [28], could produce a certain number of heterojunction sites which is beneficial to promote electron-hole separation and restrain the recombination efficiently. In addition, some efforts have been paid to the Nb_2O_5/g-C_3N_4 composite which exhibits photocatalytic activity in degradation of tetracycline hydrochloride (TC-HCl) [29], methylene blue (MB) and rhodamine B (RhB) under UV and visible light [30], and production of H_2 under visible light [31]. However, it is still a challenge to prepare novel Nb_2O_5/g-C_3N_4 composites with strong interaction and high dispersion between mesoporous Nb_2O_5 microsphere and g-C_3N_4 layer which could be used as efficient photocatalysts.

Herein, a series of g-C_3N_4-mNb_2O_5 composites, prepared by in situ hydrolysis deposition of $NbCl_5$ onto g-C_3N_4 sheet followed by solvothermal treatment, have been used as photocatalysts in degradation of RhB and TC-HCl and characterized in details. The results indicate that 4% g-C_3N_4-mNb_2O_5 exhibits low photoluminescence (PL) intensity and narrow band gap which account for its high catalytic activity.

2. Materials and Methods

2.1. Materials

Niobium chloride ($NbCl_5$) was purchased from Strem Chemicals, Inc (Newburyport, MA, USA). Melamine (99%), absolute ethanol (analytical grade) and tetracycline hydrochloride (TC-HCl, 96%) were purchased from Shanghai Aladdin Bio-Chem Technology Co., Ltd. (Shanghai, China). Rhodamine B (RhB) was purchased from Sinopharm Chemical Reagent Co., Ltd. (Shanghai, China). Pluronic 123 (PEG-PPG-PEG, P123) was purchased from Sigma-Aldrich, Co. (ST. Louis, MO, USA). All the reagents were used without further purification.

2.2. Preparation

The bulk g-C_3N_4 was prepared by heating 2.50 g of melamine in an alumina crucible with a cover in air at 550 °C with ramp rate of 2 °C/min and maintained at 550 °C for 4 h [32]. After being cooled down to room temperature, the obtained product was ground into powder.

The g-C_3N_4-mNb_2O_5 composites were prepared through in situ hydrolysis of $NbCl_5$ onto g-C_3N_4 and then solvothermal treatment. Typically, 0.99 g of P123 was added into 20 g of absolute ethanol with vigorous stirring until complete dissolution of P123. Bulk g-C_3N_4 powder was dispersed into absolute ethanol and the suspension was sonicated for 1 h. Then, 1.49 g of $NbCl_5$ was added into a pear-shaped flask under N_2 atmosphere in a glovebox. The solution of P123 and the suspension of g-C_3N_4 were added and the mixture was stirred for 30 min. After that, 0.50 mL of distilled water was pumped into the flask with a rate of 0.0167 mL/min under stirring. After stirring for another 30 min, the mixture was transferred into a 40 mL Teflon-lined stainless steel autoclave and heated at 180 °C for 24 h. After being cooled down to room temperature, the precipitate was obtained by filtration and washing by ethanol for four times. The solid was dried at 30 °C overnight in a drying oven under vacuum and then transferred to a tube furnace to be calcined at 400 °C for 3 h with a rate of 2 °C/min. The g-C_3N_4-mNb_2O_5 composites prepared with the g-C_3N_4 weight ratios of 1%, 4%, 10%, 20% and 50% were denoted as 1% g-C_3N_4-mNb_2O_5, 4% g-C_3N_4-mNb_2O_5, 10% g-C_3N_4-mNb_2O_5, 20% g-C_3N_4-mNb_2O_5, 50% g-C_3N_4-mNb_2O_5, respectively. The pure mNb_2O_5 was prepared by the same procedure without adding g-C_3N_4. In addition, the corresponding mechanic mixture of g-C_3N_4 and mNb_2O_5 with the same weight ratio of 4% g-C_3N_4-mNb_2O_5 was prepared by simple grinding and was denoted as 4% g-C_3N_4/mNb_2O_5.

2.3. Characterization

Powder X-ray diffraction (XRD) patterns were carried out on a Bruker D8 Advance diffractometer (Karlsruhe, Germany) with Cu Kα radiation (λ = 0.15418 nm) operated at 40 kV and 40 mA in the 2θ range of 5–70°. The Fourier transform infrared (FT-IR) spectra were recorded on a Thermo Fisher Nicolet iS10 instrument (Waltham, MA, USA) with KBr pellet from 4000 to 400 cm^{-1}. X-ray photoelectron spectroscopy (XPS) was measured by a Perkin Elmer PHI 5000C spectroscope (Waltham, MA, USA). The spectra were recorded with Mg Kα line as the excitation source ($h\nu$ = 1253.6 eV) at 14 kV and 20 mA. UV-vis diffuse reflectance spectroscopy (DRS) was performed on a Perkin Elmer Lambda 650 spectrophotometer (Waltham, MA, USA) using $BaSO_4$ as reference. The transmission electron microscopic (TEM) images were obtained using a FEI Tecnai G^2 F20 S-Twin field emission transmission electron microscope (Hillsboro, Oregon, USA) with an accelerating voltage of 200 kV and a JEOL JEM-2011 transmission electron microscope with an accelerating voltage of 200 kV. The samples were prepared by dropping the ethanol suspension of samples onto the copper grid. The N_2 adsorption-desorption measurements were carried out at 77.3 K on a Micromeritics Tristar II 3020 analytical system. The specific surface area was calculated by the Brunauer-Emmett-Teller (BET) method. Pore size distributions were obtained from analysis of the desorption branch of the isotherms using the Barrett-Joyner-Halenda (BJH) model. Photoluminescence (PL) spectra were obtained on an Edinburgh Instruments FLS-980 spectrometer (Edinburgh, UK) with the excitation wavelength of 300 nm.

2.4. Photocatalytic Experiments

The photocatalytic activity of the as-prepared catalysts was performed by photocatalytic degradation of 15 mg/L RhB aqueous solution and 40 mg/L TC-HCl aqueous solution under visible light. The photocatalytic degradation tests were carried out in an instrument (CEL-HXF300, Beijing China Education Au-light Co., Ltd., Beijing, China) with a cylindrical glass reactor and a condensate water circulation equipment (Figure S1). The visible light was provided by a Xe lamp (14 V, 15 A) with a 420 nm cut-off filter (Figure S2). In a typical photocatalytic degradation of RhB or TC-HCl experiment, 30 mg of catalyst was added into 70 mL of RhB aqueous solution or 100 mL of TC-HCl aqueous solution. The suspension was firstly stirred in dark at room temperature to reach adsorption/desorption equilibrium. Then, the reaction was irradiated under visible light (λ > 420 nm) and the solution was vigorously stirred with air bubbling. During the process, 3 mL of suspension was taken at intervals of 20 min (for RhB) or intervals of 5 min (for TC-HCl). All suspensions were centrifuged to remove the catalyst and the concentration of the RhB and TC-HCl were analyzed by UV-vis spectrophotometer with the scanning ranges of 300–700 nm for RhB and 200–600 nm for TC-HCl. To be reused for next reaction cycle, the separated catalyst by centrifugation was washed by water and ethanol three times and then dried at 30 °C in a drying oven under vacuum.

3. Results and Discussion

3.1. Characterization of the Catalysts

3.1.1. XRD

XRD was employed to analyze mNb_2O_5, $g\text{-}C_3N_4$ and $g\text{-}C_3N_4\text{-}mNb_2O_5$ composites with different contents of $g\text{-}C_3N_4$ and the results are shown in Figure 1. For mNb_2O_5, the broad diffraction peak around 23.7° corresponds to partial crystallized Nb_2O_5 (JCPDS number: 19-0862), which results from the relatively low calcination temperature of 400 °C. The main diffractions of $g\text{-}C_3N_4$ are at 12.8° and 27.1°, which are assigned to (100) plane due to in-plane tris-s-triazine frameworks and (002) plane due to interlayer stacking of aromatic systems, respectively. The latter reflects interlayer stacking of aromatic systems and the calculated interlayer distance is 0.338 nm [33–35]. In the composites, the diffractions of mNb_2O_5 are retained and the characteristic (002) diffraction of $g\text{-}C_3N_4$ is observed when the content of $g\text{-}C_3N_4$ increases to 20% and 50%.

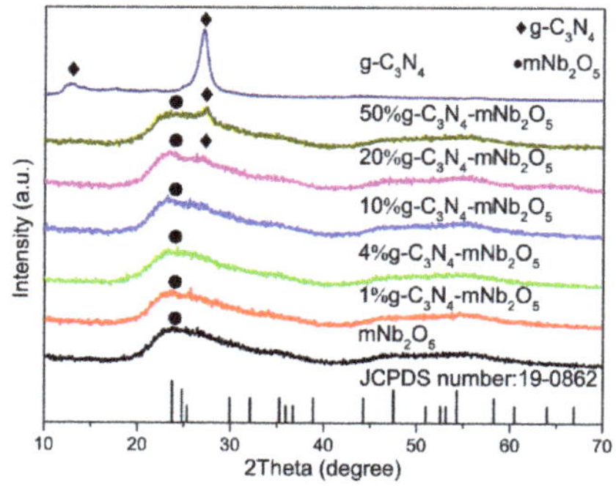

Figure 1. XRD patterns for mNb_2O_5, $g\text{-}C_3N_4$ and $g\text{-}C_3N_4\text{-}mNb_2O_5$ composites with different contents of $g\text{-}C_3N_4$.

3.1.2. FT-IR

Figure 2 shows the FT-IR spectra of mNb_2O_5, $g\text{-}C_3N_4$ and $g\text{-}C_3N_4\text{-}mNb_2O_5$ composites. For the mNb_2O_5 sample, a broad band around 3382 cm^{-1} is attributed to vibration of O-H of the water molecules adsorbed on the sample [27,30]. The broad band at 615 cm^{-1} is ascribed to Nb–O–Nb angular vibration [36]. As for $g\text{-}C_3N_4$, the broad band ranging from 3000 to 3500 cm^{-1} is due to the stretching vibration of N–H groups in $g\text{-}C_3N_4$ and O–H of adsorbed water [25]. Other characteristic bands arisen from the typical stretching modes of C=N and C–N in heterocycles of $g\text{-}C_3N_4$ are at 1237, 1317, 1405, 1461, 1574 and 1640 cm^{-1}. Additionally, the band at 803 cm^{-1} of $g\text{-}C_3N_4$ is assigned to the breathing mode of the triazine units [33,34]. For $g\text{-}C_3N_4\text{-}mNb_2O_5$ composites, all the characteristic bands of mNb_2O_5 and $g\text{-}C_3N_4$ exist except 1% $g\text{-}C_3N_4\text{-}mNb_2O_5$ due to small amount of $g\text{-}C_3N_4$, which confirm the existence of two components in the composites. With the increasing of $g\text{-}C_3N_4$, the band at 3382 cm^{-1} of mNb_2O_5 slightly shifts to smaller wavenumber for $g\text{-}C_3N_4\text{-}mNb_2O_5$ composites, which reveals the vibration of N-H. Moreover, the band of $g\text{-}C_3N_4$ at 1237 cm^{-1} slightly shifts to 1244 cm^{-1} for all $g\text{-}C_3N_4\text{-}mNb_2O_5$ composites except 1% $g\text{-}C_3N_4\text{-}mNb_2O_5$ (Figure 2b), revealing the interaction between $g\text{-}C_3N_4$ and mNb_2O_5.

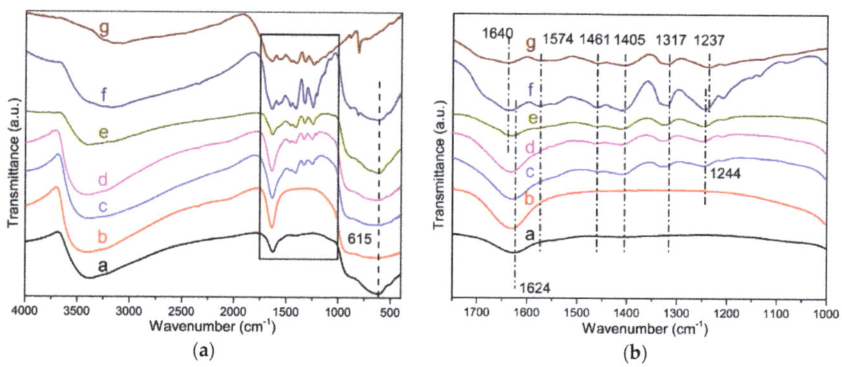

Figure 2. (**a**) FT-IR spectra of mNb_2O_5, $g\text{-}C_3N_4$ and $g\text{-}C_3N_4\text{-}mNb_2O_5$ composites with different contents of $g\text{-}C_3N_4$. (a, mNb_2O_5; b, 1% $g\text{-}C_3N_4\text{-}mNb_2O_5$; c, 4% $g\text{-}C_3N_4\text{-}mNb_2O_5$; d, 10% $g\text{-}C_3N_4\text{-}mNb_2O_5$; e, 20% $g\text{-}C_3N_4\text{-}mNb_2O_5$; f, 50% $g\text{-}C_3N_4\text{-}mNb_2O_5$; g, $g\text{-}C_3N_4$); and (**b**) enlarged FT-IR spectra corresponding to rectangle region from (**a**).

3.1.3. TEM

The prepared samples were investigated by field emission transmission electron microscope. Figure 3 shows the TEM images of g-C_3N_4, mNb_2O_5 and 4% g-C_3N_4-mNb_2O_5. The mNb_2O_5 sample is mesoporous materials with pore size of 3–5 nm and its lattice fringe spacing is 0.395 nm (Figure 3a). Figure 3b shows that g-C_3N_4 is veil-like with lamellar structure. In Figure 3c, we could measure that the interlayer distance g-C_3N_4 is 0.339 nm, which is in accordance with XRD result. For the 4% g-C_3N_4-mNb_2O_5 sample, it can be observed that mNb_2O_5 microspheres are deposited on the surface of g-C_3N_4 sheet in Figure 3d,e. Moreover, the corresponding EDS of Figure 3e indicates that the sample consists of C, N, O and Nb elements (Figure 3f). The relevant element contents are shown in Table S1.

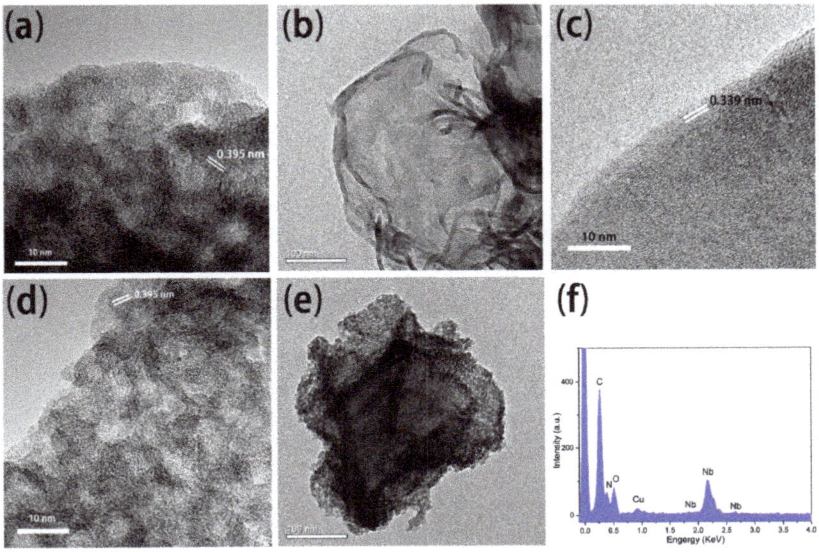

Figure 3. TEM images of: (**a**) mNb_2O_5; (**b**,**c**) g-C_3N_4; and (**d**,**e**) 4% g-C_3N_4-mNb_2O_5; (**f**) EDS analysis of 4% g-C_3N_4-mNb_2O_5 corresponding to (**e**).

3.1.4. N_2 Adsorption-Desorption

Figure 4 shows the nitrogen adsorption-desorption isotherms of g-C_3N_4, mNb_2O_5, 1% g-C_3N_4-mNb_2O_5, 4% g-C_3N_4-mNb_2O_5, 10% g-C_3N_4-mNb_2O_5 and 50% g-C_3N_4-mNb_2O_5. The isotherm of g-C_3N_4 belongs to the type IV isotherm with the H4 type hysteresis loop. All isotherms of mNb_2O_5, 1% g-C_3N_4-mNb_2O_5, 4% g-C_3N_4-mNb_2O_5, 10% g-C_3N_4-mNb_2O_5 and 50% g-C_3N_4-mNb_2O_5 are ascribed to the type V isotherm with the H3 type hysteresis loop, indicating the existence of mesoporous structure. The BET specific surface areas of g-C_3N_4 and mNb_2O_5 are 11.2 and 138 m^2/g, respectively. The surface area of the composites follows the sequence: 4% g-C_3N_4-mNb_2O_5 (190 m^2/g) > 10% g-C_3N_4-mNb_2O_5 (182 m^2/g) > 1% g-C_3N_4-mNb_2O_5 (152 m^2/g) > 50% g-C_3N_4-mNb_2O_5 (136 m^2/g). Among them, 4% g-C_3N_4-mNb_2O_5 exhibits the largest surface area, which is 1.4 times larger than that of mNb_2O_5 and 17 times larger than that of g-C_3N_4. It may originate from that, after ultrasonication of g-C_3N_4, the interlayer distance of g-C_3N_4 increases and the addition of g-C_3N_4 leads to partial disaggregation of mNb_2O_5 [37,38]. When the content of g-C_3N_4 is 50%, high content of g-C_3N_4 in the composites results in g-C_3N_4 stacking together, thus 50% g-C_3N_4-mNb_2O_5 sample shows almost the same specific surface as mNb_2O_5.

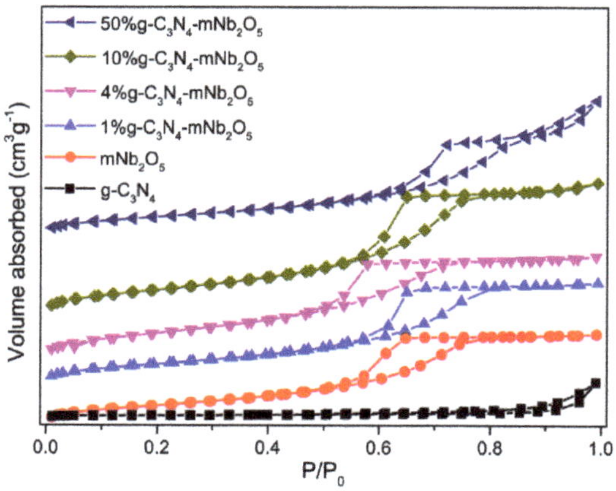

Figure 4. Nitrogen adsorption–desorption isotherms of g-C_3N_4, mNb_2O_5, 1% g-C_3N_4-mNb_2O_5, 4% g-C_3N_4-mNb_2O_5, 10% g-C_3N_4-mNb_2O_5 and 50% g-C_3N_4-mNb_2O_5.

3.1.5. UV-vis DRS

UV-vis diffuse reflectance spectroscopy (DRS) was performed to estimate the band gap of the catalysts, which is important to determine if the catalysts can be excited in the visible-light region [39]. Figure 5a shows the UV-vis diffuse reflectance spectra of mNb_2O_5, g-C_3N_4 and g-C_3N_4-mNb_2O_5 composites with different content of g-C_3N_4. Due to the high band energy value of mNb_2O_5 (3.34 eV), mNb_2O_5 only absorbs ultraviolet light with its fundamental absorption edge near 371 nm. However, g-C_3N_4 possesses a broad peak in the visible region with an absorption edge at ca. 458 nm. For g-C_3N_4-mNb_2O_5 composites, the absorption edge exhibits an obvious red shift relative to pristine mNb_2O_5, indicating the composites could absorb more visible light than mNb_2O_5. The band gap energy (Eg) of the samples was determined by UV-vis DRS with the Tauc model according to the following equation:

$$\alpha h\upsilon = A(h\upsilon - E_g)^{n/2} \quad (1)$$

where α, h, υ and A corresponds to absorption coefficient, Planck constant, light frequency and a constant, respectively, and the constant n depends on whether the transition is direct (n = 1) or indirect (n = 4) [40]. For both g-C_3N_4 and mNb_2O_5, the values of n are 1 [41,42]. Thus, the band gap values of 1% g-C_3N_4-mNb_2O_5, 4% g-C_3N_4-mNb_2O_5, 10% g-C_3N_4-mNb_2O_5, 20% g-C_3N_4-mNb_2O_5 and 50% g-C_3N_4-mNb_2O_5 are estimated as 3.34, 3.10, 3.19, 3.15 and 2.88 eV, respectively (Figure 5b). The narrowed band gaps of 4% g-C_3N_4-mNb_2O_5, 10% g-C_3N_4-mNb_2O_5 and 20% g-C_3N_4-mNb_2O_5 results from the formation of heterostructure between g-C_3N_4 and mNb_2O_5 [29]. It implies that 4% g-C_3N_4-mNb_2O_5 can harvest more visible light, which is beneficial to improve the visible-light photocatalytic activity of catalysts. Interestingly, the band gap of 50% g-C_3N_4-mNb_2O_5 is lower than that of other composites, due to high content of g-C_3N_4 in 50% g-C_3N_4-mNb_2O_5.

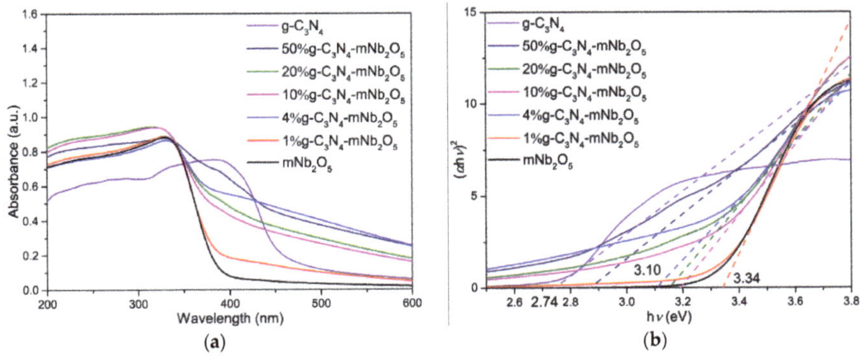

Figure 5. (**a**) UV-vis diffuse reflectance spectra; and (**b**) plots of the $(\alpha h\upsilon)^2$ vs. $(h\upsilon)$ of mNb_2O_5, $g\text{-}C_3N_4$, and $g\text{-}C_3N_4\text{-}mNb_2O_5$ composites with different content of $g\text{-}C_3N_4$.

3.1.6. XPS

XPS was used to characterize the surface chemical composition and elemental valence states of the samples. As shown in Figure 6a, C and N elements are detected from $g\text{-}C_3N_4$ and Nb and O elements are detected from mNb_2O_5. As for the spectrum of 4% $g\text{-}C_3N_4\text{-}mNb_2O_5$, a very weak N 1s peak is ascribed to small content of $g\text{-}C_3N_4$. Figure 6b presents the high resolution XPS spectra of Nb 3d for 4% $g\text{-}C_3N_4\text{-}mNb_2O_5$ and mNb_2O_5. For mNb_2O_5, the signals of Nb $3d_{5/2}$ and $3d_{3/2}$ locate at 207.0 and 209.6 eV, respectively, whereas the signals of Nb $3d_{5/2}$ and $3d_{3/2}$ shift slightly (ΔBE ~0.4 eV) to a higher binding energy for 4% $g\text{-}C_3N_4\text{-}mNb_2O_5$. The upshift may be attributed to band bending.

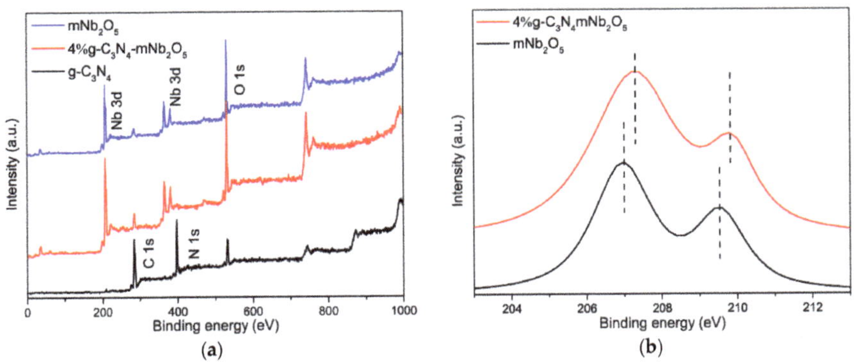

Figure 6. (**a**) XPS survey spectra of $g\text{-}C_3N_4$, mNb_2O_5 and 4% $g\text{-}C_3N_4\text{-}mNb_2O_5$; and (**b**) Nb 3d spectra for mNb_2O_5 and 4% $g\text{-}C_3N_4\text{-}mNb_2O_5$.

3.1.7. PL Spectroscopy

To test the generation and recombination efficiency of photogenerated electrons and holes in semiconductors, PL spectroscopy is often conducted. The PL spectra of mNb_2O_5, 4% $g\text{-}C_3N_4\text{-}mNb_2O_5$, 10% $g\text{-}C_3N_4\text{-}mNb_2O_5$, 20% $g\text{-}C_3N_4\text{-}mNb_2O_5$, 50% $g\text{-}C_3N_4\text{-}mNb_2O_5$ and $g\text{-}C_3N_4$ with an excitation wavelength of 300 nm are shown in Figure 7. It can be observed that mNb_2O_5 has a strong emission peak at 474 nm and $g\text{-}C_3N_4$ has a strong emission peak at 472 nm. Compared to mNb_2O_5, the PL emission intensity of $g\text{-}C_3N_4\text{-}mNb_2O_5$ samples is significantly reduced, suggesting that the $g\text{-}C_3N_4\text{-}mNb_2O_5$ composite has a lower recombination rate of photogenerated electrons and holes than mNb_2O_5. Among the composites, the 4% $g\text{-}C_3N_4\text{-}mNb_2O_5$ sample shows the lowest emission intensity at the similar emission peak position which means the lowest recombination efficiency of photogenerated electrons and holes [27,43,44]. Combined with the UV-vis DRS analysis result that 4% g-C_3N_4-mNb_2O_5 has the lowest band gap, 4% g-C_3N_4-mNb_2O_5 demonstrates best potential in photocatalytic degradation reactions among the catalysts studied in this work.

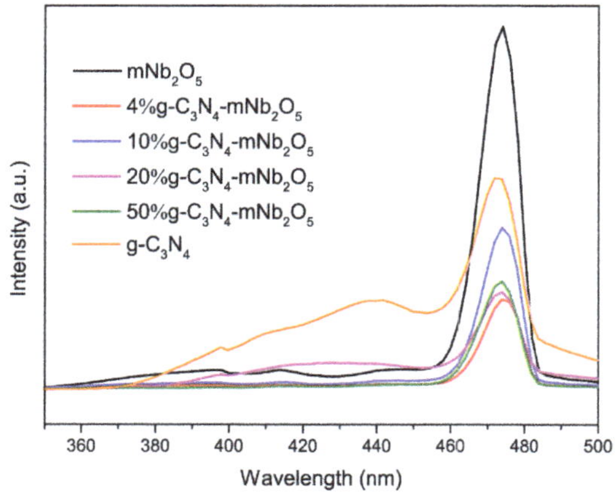

Figure 7. PL spectra of mNb_2O_5, 4% g-C_3N_4-mNb_2O_5, 10% g-C_3N_4-mNb_2O_5, 20% g-C_3N_4-mNb_2O_5, 50% g-C_3N_4-mNb_2O_5 and g-C_3N_4.

3.2. Visible-Light Photocatalytic Performance and Stability of the Catalysts

The photocatalytic test of mNb_2O_5, g-C_3N_4 and g-C_3N_4-mNb_2O_5 composites with different content of g-C_3N_4 in degradation of RhB and TC-HCl were carried out under visible light irradiation ($\lambda > 420$ nm).

3.2.1. Photodegradation of RhB under Visible Light Irradiation

The photocatalytic degradation results are shown in Figure 8a. The standard curve of the absorbance intensity (A) vs. concentration (C) of RhB solution is shown in Figure S4a. The RhB suspension was firstly stirred in dark for 2 h at room temperature to reach adsorption/desorption equilibrium (Figure S3a) [45,46]. No evident decrease of RhB concentration is observed in the absence of catalyst. Therefore, the direct photolysis of RhB could be ignored. Pure mNb_2O_5 and g-C_3N_4 show low photocatalytic activities with only about 27.3% and 25.4% of RhB being degraded after 180 min of irradiation, respectively. The g-C_3N_4-mNb_2O_5 composites show higher degradation efficiency than mNb_2O_5 and g-C_3N_4. For the catalysts containing different content of g-C_3N_4, the sample 4% g-C_3N_4-mNb_2O_5 exhibits the highest degradation efficiency for RhB (97.5%) within 180 min, which is 3.6 and 3.9 times higher than that of mNb_2O_5 and g-C_3N_4, respectively. Furthermore, the corresponding mechanic mixture 4% g-C_3N_4/mNb_2O_5 shows 51% of degradation efficiency for RhB, which is much smaller than that of 4% g-C_3N_4-mNb_2O_5 composite. Therefore, the high degradation efficiency of 4% g-C_3N_4-mNb_2O_5 is ascribed to the formation of heterostructure between g-C_3N_4 and mNb_2O_5, as found in the UV-vis DRS and PL studies.

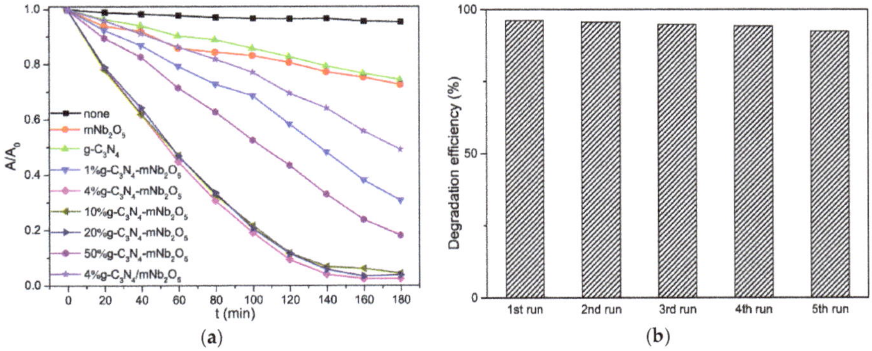

Figure 8. (**a**) Photolysis of RhB and photocatalytic activity over as-prepared photocatalysts for RhB; and (**b**) recyclability for the photodegradation of RhB in the presence of 4% g-C_3N_4-mNb_2O_5 under visible light irradiation.

In addition to photocatalytic efficiency, stability and recyclability of the catalysts are also important for application of the catalysts. The 4% g-C_3N_4-mNb_2O_5 sample for photodegradation of RhB was further tested for five cycles. As shown in Figure 8b, after five cycles, the high efficiency for photocatalytic degradation of RhB by 4% g-C_3N_4-mNb_2O_5 is maintained, demonstrating good reusability and stability of 4% g-C_3N_4-mNb_2O_5.

3.2.2. Photodegradation of TC-HCl under Visible Light Irradiation

The photocatalytic degradation of TC-HCl results are shown in Figure 9a. The standard curve of the absorbance intensity (A) vs. concentration (C) of TC-HCl solution can be seen in Figure S4b. The TC-HCl suspension was firstly stirred in dark for 1 h at room temperature to reach adsorption/desorption equilibrium (Figure S3b) [45,46]. No evident decrease of TC-HCl concentration is observed in the absence of catalyst. The pure mNb_2O_5 and g-C_3N_4 show low photocatalytic efficiency in degradation of TC-HCl with about 49.7% and 5.3% after 60 min of irradiation, respectively. The degradation efficiency of g-C_3N_4-mNb_2O_5 composites within 60 min could be listed as the following order: 4% g-C_3N_4-mNb_2O_5 > 10% g-C_3N_4-mNb_2O_5 ≈ 20% g-C_3N_4-mNb_2O_5 > 50% g-C_3N_4-mNb_2O_5 > 1% g-C_3N_4-mNb_2O_5. All g-C_3N_4-mNb_2O_5 composites display higher degradation efficiency than that of mNb_2O_5 and g-C_3N_4. For comparison, the corresponding mechanic mixture 4% g-C_3N_4/mNb_2O_5 shows its degradation efficiency of 52%. Overall, 4% g-C_3N_4-mNb_2O_5 is the optimal catalyst, for which the degradation efficiency of TC-HCl in 60 min is 75.7%. After 30-min experimental run, for 4% g-C_3N_4-mNb_2O_5, 10% g-C_3N_4-mNb_2O_5 and 20% g-C_3N_4-mNb_2O_5, degradation reaction of TC-HCl solution has approached the endpoint. Further degradation is hard to proceed so that the degradation rate is slow and the change of absorbance change is not evident. An appropriate content of g-C_3N_4 leads to this composite with large surface area, narrow band gap and low PL intensity, which favors the titled photocatalytic reactions [41,44,47].

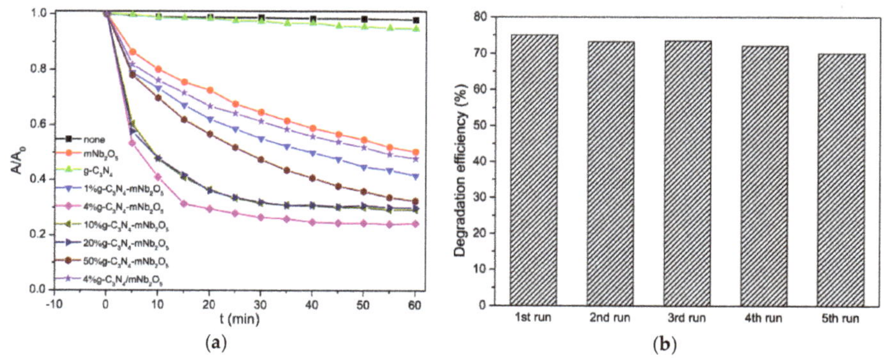

Figure 9. (**a**) Photolysis of TC-HCl and photocatalytic activity over as-prepared photocatalysts for TC-HCl; and (**b**) recyclability for the photodegradation of TC-HCl in the presence of 4% g-C_3N_4-mNb_2O_5 under visible light irradiation.

The stability and recyclability of 4% g-C_3N_4-mNb_2O_5 catalysts for photocatalytic degradation of TC-HCl were also tested for five cycles. As shown in Figure 9b, after five cycles, photodegradation efficiency of 4% g-C_3N_4-mNb_2O_5 for TC-HCl is almost unchanged. Thus, 4% g-C_3N_4-mNb_2O_5 sample could maintain good stability and it is able to be reused in the photodegradation reaction of TC-HCl as well as in photodegradation reaction of RhB.

In other words, in two photocatalytic reactions of RhB solution and TC-HCl solution under visible-light irradiation, g-C_3N_4-mNb_2O_5 composites showed enhanced photocatalytic activity than mNb_2O_5. The 4% g-C_3N_4-mNb_2O_5 sample shows the highest photocatalytic activity. The other photocatalytic performance of Nb_2O_5/g-C_3N_4 composites were also tested by Ribeiro et al., Shi et al., and Li et al. [29–31]. Compared to the reported work, our work focus on the stability of catalysts, which is greatly distinct. The strategy based on the in situ hydrolysis can efficiently avoid the aggregation of Nb_2O_5, thus further facilitating the full exertion of catalyst activity. Besides, the interlayer of g-C_3N_4 sheets can be enlarged by the existence of Nb_2O_5, increasing the surface area of catalyst. The stability of our catalyst is superior to many reported literatures when evaluated in the photocatalytic of RhB and TC-HCl, which can be attributed to highly uniformity and tightly anchoring of Nb_2O_5 on g-C_3N_4 sheets.

4. Conclusions

In summary, we have prepared a series of g-C_3N_4-mNb_2O_5 composites by in situ $NbCl_5$ hydrolysis, deposition and solvothermal treatment. It is found that the g-C_3N_4-mNb_2O_5 composites exhibit lower PL intensity than pure mNb_2O_5 and g-C_3N_4 and narrower band gap than pure mNb_2O_5, leading to a significant enhancement of photocatalytic activity with excellent stability and recyclability for degradation of RhB and TC-HCl. Among the composites, 4% g-C_3N_4-mNb_2O_5 sample shows the highest photocatalytic activity under visible-light irradiation (λ > 420 nm), which is ascribed to the strong interaction between g-C_3N_4 and Nb_2O_5, its smallest band gap and lowest PL intensity. Based on the above characterization and photocatalytic results, the excellent photocatalytic activity of 4% g-C_3N_4-mNb_2O_5 can be attributed to combined effects between g-C_3N_4 and mNb_2O_5 as follows: (1) the induction of g-C_3N_4 into mNb_2O_5 leads to extension the absorption into visible light region; (2) the recombination of photogenerated electrons and holes is inhibited to some extent; (3) its highest surface area among all composites leads to more active sites; and (4) interaction exists between mNb_2O_5 and g-C_3N_4, as observed in XPS and FT-IR. These also account for the remarkable stability and recyclability of the catalyst 4% g-C_3N_4-mNb_2O_5 in the degradation reaction.

Highlights:

1. g-C_3N_4-mNb_2O_5 composites were obtained by in situ hydrolysis deposition and solvothermal treatment.
2. Under the visible light irradiation, 4% g-C_3N_4-mNb_2O_5 exhibited the highest photocatalytic activities to degrade rhodamine B (97.5% degradation within 180 min) and tetracycline hydrochloride (75.7% degradation within 60 min) solution with excellent stability and recyclability.
3. The g-C_3N_4-mNb_2O_5 composites showed narrower band gap, lower PL intensity and larger surface area than pure mNb_2O_5 and g-C_3N_4, leading to a significant enhancement of their photocatalytic activity.

Supplementary Materials: The following are available online at http://www.mdpi.com/2079-4991/8/6/427/s1, Figure S1: A schematic diagram of photocatalytic equipment, Figure S2: Emission spectrum of Xe lamp with 420 nm filter, Figure S3: Influence of adsorption process on: (a) RhB solution (15 mg/L, 70 mL); and (b) TC-HCl solution (40 mg/L, 100 mL) containing 30 mg 4% g-C_3N_4-mNb_2O_5 powder under dark, Figure S4: Absorption changes of: (a) RhB solution (15 mg/L, 70 mL); and (b) TC-HCl solution (40 mg/L, 100 mL) containing 30 mg 4% g-C_3N_4-mNb_2O_5 powder under visible light irradiation, Figure S5: Standard curves of absorbance intensity vs. concentration of: (a) RhB solution; and (b) TC-HCl solution, Table S1: The elements content of 4% g-C_3N_4-mNb_2O_5 by EDS analysis.

Author Contributions: M.W. managed all the experiments and initial writing; H.W., Y.R., C.W. and Z.W. contributed greatly to data analysis, discussion and manuscript preparation. B.Y. and H.H. supervised and managed the project.

Acknowledgments: This work was supported by the National Natural Science Foundation of China (21673046, 21473036, 91645201 and 21371035).

Conflicts of Interest: The authors declare no conflict of interest.

References

1. Barbosa, M.O.; Moreira, N.F.F.; Ribeiro, A.R.; Pereira, M.F.R.; Silva, A.M.T. Occurrence and removal of organic micropollutants: An overview of the watch list of EU Decision 2015/495. *Water Res.* **2016**, *94*, 257–279. [CrossRef] [PubMed]
2. Zhang, Q.Q.; Ying, G.G.; Pan, C.G.; Liu, Y.S.; Zhao, J.L. Comprehensive evaluation of antibiotics emission and fate in the river basins of China: Source analysis, multimedia modeling, and linkage to bacterial resistance. *Environ. Sci. Technol.* **2015**, *49*, 6772–6782. [CrossRef] [PubMed]
3. Molla, A.; Sahu, M.; Hussain, S. Synthesis of tunable band gap semiconductor nickel sulphide nanoparticles: Rapid and round the clock degradation of organic dyes. *Sci. Rep.* **2016**, *6*, 1–11. [CrossRef] [PubMed]
4. Gao, P.; Munir, M.; Xagoraraki, I. Correlation of tetracycline and sulfonamide antibiotics with corresponding resistance genes and resistant bacteria in a conventional municipal wastewater treatment plant. *Sci. Total Environ.* **2012**, *421*, 173–183. [CrossRef] [PubMed]
5. Lv, Z.Y.; Liu, X.J.; Jia, B.; Wang, H.; Wu, Y.; Lu, Z.P. Development of a novel high-entropy alloy with eminent efficiency of degrading azo dye solutions. *Sci. Rep.* **2016**, *6*, 1–11. [CrossRef] [PubMed]
6. Ahmed, M.B.; Zhou, J.L.; Ngo, H.H.; Guo, W.S. Adsorptive removal of antibiotics from water and wastewater: Progress and challenges. *Sci. Total Environ.* **2015**, *532*, 112–126. [CrossRef] [PubMed]
7. Gaya, U.I.; Abdullah, A.H. Heterogeneous photocatalytic degradation of organic contaminants over titanium dioxide: A review of fundamentals, progress and problems. *J. Photochem. Photobiol. C* **2008**, *9*, 1–12. [CrossRef]
8. Sorathiya, K.; Mishra, B.; Kalarikkal, A.; Reddy, K.P.; Gopinath, C.S.; Khushalani, D. Enhancement in rate of photocatalysis upon catalyst recycling. *Sci. Rep.* **2016**, *6*, 1–9. [CrossRef] [PubMed]
9. Szilagyi, I.M.; Forizs, B.; Rosseler, O.; Szegedi, A.; Nemeth, P.; Kiraly, P.; Tarkanyi, G.; Vajna, B.; Varga-Josepovits, K.; Laszlo, K.; et al. WO_3 photocatalysts: Influence of structure and composition. *J. Catal.* **2012**, *294*, 119–127. [CrossRef]
10. Daneshvar, N.; Salari, D.; Khataee, A.R. Photocatalytic degradation of azo dye acid red 14 in water on ZnO as an alternative catalyst to TiO_2. *J. Photochem. Photobiol. A* **2004**, *162*, 317–322. [CrossRef]

11. Zhou, Z.H.; Lin, Y.L.; Zhang, P.A.; Ashalley, E.; Shafa, M.; Li, H.D.; Wu, J.; Wang, Z.M. Hydrothermal fabrication of porous MoS_2 and its visible light photocatalytic properties. *Mater. Lett.* **2014**, *131*, 122–124. [CrossRef]
12. Zhao, Y.; Eley, C.; Hu, J.P.; Foord, J.S.; Ye, L.; He, H.Y.; Tsang, S.C.E. Shape-dependent acidity and photocatalytic activity of Nb_2O_5 nanocrystals with an active TT (001) surface. *Angew. Chem. Int. Ed.* **2012**, *51*, 3846–3849. [CrossRef] [PubMed]
13. Liu, H.; Gao, N.; Liao, M.Y.; Fang, X.S. Hexagonal-like Nb_2O_5 nanoplates-based photodetectors and photocatalyst with high performances. *Sci. Rep.* **2015**, *5*, 1–9. [CrossRef] [PubMed]
14. Zhao, W.L.; Zhao, W.; Zhu, G.L.; Lin, T.Q.; Xu, F.F.; Huang, F.Q. Black Nb_2O_5 nanorods with improved solar absorption and enhanced photocatalytic activity. *Dalton Trans.* **2016**, *45*, 3888–3894. [CrossRef] [PubMed]
15. Oliveira, L.C.A.; Oliveira, H.S.; Mayrink, G.; Mansur, H.S.; Mansur, A.A.P.; Moreira, R.L. One-pot synthesis of CdS@Nb_2O_5 core shell nanostructures with enhanced photocatalytic activity. *Appl. Catal. B* **2014**, *152*, 403–412. [CrossRef]
16. Hu, C.Y.; Teoh, W.Y.; Ji, S.L.; Ye, C.H.; Iwase, A. In situ metal doping during modified anodization synthesis of Nb_2O_5 with enhanced photoelectrochemical water splitting. *AIChE J.* **2016**, *62*, 352–358. [CrossRef]
17. Zhang, H.L.; Lin, Q.Y.; Ning, S.B.; Zhou, Y.G.; Lin, H.X.; Long, J.L.; Zhang, Z.Z.; Wang, X.X. One-step synthesis of mesoporous Pt-Nb_2O_5 nanocomposites with enhanced photocatalytic hydrogen production activity. *RSC Adv.* **2016**, *6*, 96809–96815. [CrossRef]
18. Zhao, Z.W.; Sun, Y.J.; Dong, F. Graphitic carbon nitride based nanocomposites: A review. *Nanoscale* **2015**, *7*, 15–37. [CrossRef] [PubMed]
19. Su, F.Z.; Mathew, S.C.; Lipner, G.; Fu, X.Z.; Antonietti, M.; Blechert, S.; Wang, X.C. mpg-C_3N_4-catalyzed selective oxidation of alcohols using O_2 and visible Light. *J. Am. Chem. Soc.* **2010**, *132*, 16299–16301. [CrossRef] [PubMed]
20. Yang, X.L.; Qian, F.F.; Zou, G.J.; Li, M.L.; Lu, J.R.; Li, Y.M.; Bao, M.T. Facile fabrication of acidified g-C_3N_4/g-C_3N_4 hybrids with enhanced photocatalysis performance under visible light irradiation. *Appl. Catal. B* **2016**, *193*, 22–35. [CrossRef]
21. Wang, X.C.; Maeda, K.; Thomas, A.; Takanabe, K.; Xin, G.; Carlsson, J.M.; Domen, K.; Antonietti, M. A metal-free polymeric photocatalyst for hydrogen production from water under visible light. *Nat. Mater.* **2009**, *8*, 76–80. [CrossRef] [PubMed]
22. Wang, X.C.; Maeda, K.; Chen, X.F.; Takanabe, K.; Domen, K.; Hou, Y.D.; Fu, X.Z.; Antonietti, M. Polymer semiconductors for artificial photosynthesis: Hydrogen evolution by mesoporous graphitic carbon nitride with visible Light. *J. Am. Chem. Soc.* **2009**, *131*, 1680–1681. [CrossRef] [PubMed]
23. Yan, J.Q.; Wu, H.; Chen, H.; Zhang, Y.X.; Zhang, F.X.; Liu, S.F. Fabrication of TiO_2/C_3N_4 heterostructure for enhanced photocatalytic Z-scheme overall water splitting. *Appl. Catal. B* **2016**, *191*, 130–137. [CrossRef]
24. Tong, Z.W.; Yang, D.; Xiao, T.X.; Tian, Y.; Jiang, Z.Y. Biomimetic fabrication of g-C_3N_4/TiO_2 nanosheets with enhanced photocatalytic activity toward organic pollutant degradation. *Chem. Eng. J.* **2015**, *260*, 117–125. [CrossRef]
25. Jiang, Y.H.; Li, F.; Liu, Y.; Hong, Y.Z.; Liu, P.P.; Ni, L. Construction of TiO_2 hollow nanosphere/g-C_3N_4 composites with superior visible-light photocatalytic activity and mechanism insight. *J. Ind. Eng. Chem.* **2016**, *41*, 130–140. [CrossRef]
26. Li, Y.G.; Wei, X.L.; Yan, X.Y.; Cai, J.T.; Zhou, A.N.; Yang, M.R.; Liu, K.Q. Construction of inorganic-organic 2D/2D WO_3/g-C_3N_4 nanosheet arrays toward efficient photoelectrochemical splitting of natural seawater. *Phys. Chem. Chem. Phys.* **2016**, *18*, 10255–10261. [CrossRef] [PubMed]
27. Wang, X.; Hong, M.Z.; Zhang, F.W.; Zhuang, Z.Y.; Yu, Y. Recyclable nanoscale zero valent iron doped g-C_3N_4/MoS_2 for efficient photocatalysis of RhB and Cr(VI) driven by visible light. *ACS Sustain. Chem. Eng.* **2016**, *4*, 4055–4062. [CrossRef]
28. Wang, J.; Xia, Y.; Zhao, H.Y.; Wang, G.F.; Xiang, L.; Xu, J.L.; Komarneni, S. Oxygen defects-mediated Z-scheme charge separation in g-C_3N_4/ZnO photocatalysts for enhanced visible-light degradation of 4-chlorophenol and hydrogen evolution. *Appl. Catal. B* **2017**, *206*, 406–416. [CrossRef]
29. Hong, Y.Z.; Li, C.S.; Zhang, G.Y.; Meng, Y.D.; Yin, B.X.; Zhao, Y.; Shi, W.D. Efficient and stable Nb_2O_5 modified g-C_3N_4 photocatalyst for removal of antibiotic pollutant. *Chem. Eng. J.* **2016**, *299*, 74–84. [CrossRef]

30. Carvalho, K.T.G.; Nogueira, A.E.; Lopes, O.F.; Byzynski, G.; Ribeiro, C. Synthesis of g-C_3N_4/Nb_2O_5 heterostructures and their application in the removal of organic pollutants under visible and ultraviolet irradiation. *Ceram. Int.* **2017**, *43*, 3521–3530. [CrossRef]
31. Huang, Q.Z.; Wang, J.C.; Wang, P.P.; Yao, H.C.; Li, Z.J. In-situ growth of mesoporous Nb_2O_5 microspheres on g-C_3N_4 nanosheets for enhanced photocatalytic H_2 evolution under visible light irradiation. *Int. J. Hydrogen Energy* **2017**, *42*, 6683–6694. [CrossRef]
32. Niu, P.; Zhang, L.L.; Liu, G.; Cheng, H.M. Graphene-like carbon nitride nanosheets for improved photocatalytic activities. *Adv. Funct. Mater.* **2012**, *22*, 4763–4770. [CrossRef]
33. Zheng, Y.; Lin, L.H.; Ye, X.J.; Guo, F.S.; Wang, X.C. Helical graphitic carbon nitrides with photocatalytic and optical activities. *Angew. Chem. Int. Ed.* **2014**, *53*, 11926–11930. [CrossRef] [PubMed]
34. Yan, S.C.; Li, Z.S.; Zou, Z.G. Photodegradation performance of g-C_3N_4 fabricated by directly heating melamine. *Langmuir* **2009**, *25*, 10397–10401. [CrossRef] [PubMed]
35. Dong, F.; Wu, L.W.; Sun, Y.J.; Fu, M.; Wu, Z.B.; Lee, S.C. Efficient synthesis of polymeric g-C_3N_4 layered materials as novel efficient visible light driven photocatalysts. *J. Mater. Chem.* **2011**, *21*, 15171–15174. [CrossRef]
36. Yue, Z.K.; Chu, D.M.; Huang, H.; Huang, J.; Yang, P.; Du, Y.K.; Zhu, M.S.; Lu, C. A novel heterogeneous hybrid by incorporation of Nb_2O_5 microspheres and reduced graphene oxide for photocatalytic H_2 evolution under visible light irradiation. *RSC Adv.* **2015**, *5*, 47117–47124. [CrossRef]
37. Ding, J.; Liu, Q.Q.; Zhang, Z.Y.; Liu, X.; Zhao, J.Q.; Cheng, S.B.; Zong, B.N.; Dai, W.L. Carbon nitride nanosheets decorated with WO_3 nanorods: Ultrasonic-assisted facile synthesis and catalytic application in the green manufacture of dialdehydes. *Appl. Catal. B* **2015**, *165*, 511–518. [CrossRef]
38. Yang, S.B.; Gong, Y.J.; Zhang, J.S.; Zhan, L.; Ma, L.L.; Fang, Z.Y.; Vajtai, R.; Wang, X.C.; Ajayan, P.M. Exfoliated Graphitic Carbon Nitride Nanosheets as Efficient Catalysts for Hydrogen Evolution Under Visible Light. *Adv. Mater.* **2013**, *25*, 2452–2456. [CrossRef] [PubMed]
39. Aguirre, M.E.; Zhou, R.X.; Eugene, A.J.; Guzman, M.I.; Grela, M.A. Cu_2O/TiO_2 heterostructures for CO_2 reduction through a direct Z-scheme: Protecting Cu_2O from photocorrosion. *Appl. Catal. B* **2017**, *217*, 485–493. [CrossRef]
40. Butler, M.A. Photoelectrolysis and physical-properties of semiconducting electrode WO_3. *J. Appl. Phys.* **1977**, *48*, 1914–1920. [CrossRef]
41. Shang, Y.Y.; Chen, X.; Liu, W.W.; Tan, P.F.; Chen, H.Y.; Wu, L.D.; Ma, C.; Xiong, X.; Pan, J. Photocorrosion inhibition and high-efficiency photoactivity of porous g-C_3N_4/Ag_2CrO_4 composites by simple microemulsion-assisted co-precipitation method. *Appl. Catal. B* **2017**, *204*, 78–88. [CrossRef]
42. Lopes, O.F.; Paris, E.C.; Ribeiro, C. Synthesis of Nb_2O_5 nanoparticles through the oxidant peroxide method applied to organic pollutant photodegradation: A mechanistic study. *Appl. Catal. B* **2014**, *144*, 800–808. [CrossRef]
43. Liu, B.S.; Zhao, X.J.; Zhao, Q.N.; He, X.; Feng, J.Y. Effect of heat treatment on the UV-vis-NIR and PL spectra of TiO_2 films. *J. Electron Spectrosc. Relat. Phenom.* **2005**, *148*, 158–163. [CrossRef]
44. Xiang, Q.J.; Lv, K.L.; Yu, J.G. Pivotal role of fluorine in enhanced photocatalytic activity of anatase TiO_2 nanosheets with dominant (001) facets for the photocatalytic degradation of acetone in air. *Appl. Catal. B* **2010**, *96*, 557–564. [CrossRef]
45. Zhou, R.X.; Guzman, M.I. CO_2 Reduction under Periodic Illumination of ZnS. *J. Phys. Chem. C* **2014**, *118*, 11649–11656. [CrossRef]
46. Zhou, R.X.; Guzman, M.I. Photocatalytic Reduction of Fumarate to Succinate on ZnS Mineral Surfaces. *J. Phys. Chem. C* **2016**, *120*, 7349–7357. [CrossRef]
47. Liu, D.; Fernandez, Y.; Ola, O.; Mackintosh, S.; Maroto-Valer, M.; Parlett, C.M.A.; Lee, A.F.; Wu, J.C.S. On the impact of Cu dispersion on CO_2 photoreduction over Cu/TiO_2. *Catal. Commun.* **2012**, *25*, 78–82. [CrossRef]

© 2018 by the authors. Licensee MDPI, Basel, Switzerland. This article is an open access article distributed under the terms and conditions of the Creative Commons Attribution (CC BY) license (http://creativecommons.org/licenses/by/4.0/).

nanomaterials

MDPI

Article

Facile Synthesis of Novel $CaIn_2S_4/ZnIn_2S_4$ Composites with Efficient Performance for Photocatalytic Reduction of Cr(VI) under Simulated Sunlight Irradiation

Siyu Xu [1], Jun Dai [1,2], Juan Yang [1,2,3,*], Jun You [3] and Jingyi Hao [3]

1 Institute of Chemical Safety, School of Safety Science and Engineering, Henan Polytechnic University, Jiaozuo 454003, China; xusiyu2017@163.com (S.X.); daijun@hpu.edu.cn (J.D.)
2 The Collaborative Innovation Center of Coal Safety Production of Henan, Henan Polytechnic University, Jiaozuo 454003, China
3 Institute of Applied Chemistry, College of Chemistry and Chemical Engineering, Henan Polytechnic University, Jiaozuo 454003, China; youjunhpu@163.com (J.Y.); haojingyi2018@163.com (J.H.)
* Correspondence: yangjuanhpu@163.com; Tel.: +86-391-398-7881

Received: 29 May 2018; Accepted: 25 June 2018; Published: 27 June 2018

Abstract: A series of novel and efficient heterostructured composites $CaIn_2S_4/ZnIn_2S_4$ have been synthesized using a facile hydrothermal method. XRD patterns indicate the as-prepared catalysts are two-phase composites of cubic phase $CaIn_2S_4$ and hexagonal phase $ZnIn_2S_4$. FESEM (field emission scanning electron microscope) images display that the synthesized composites are composed of flower-like microspheres with wide diameter distribution. UV–Vis diffuse reflectance spectra (DRS) show that the optical absorption edges of the $CaIn_2S_4/ZnIn_2S_4$ composites shift toward longer wavelengths with the increase of the $CaIn_2S_4$ component. The photocatalytic activities of the as-synthesized composites are investigated by using the aqueous-phase Cr(VI) reduction under simulated sunlight irradiation. This is the first report on the application of the $CaIn_2S_4/ZnIn_2S_4$ composites as stable and efficient photocatalysts for the Cr(VI) reduction. The fabricated $CaIn_2S_4/ZnIn_2S_4$ composites possess higher photocatalytic performance in comparison with pristine $CaIn_2S_4$ or $ZnIn_2S_4$. The $CaIn_2S_4/ZnIn_2S_4$ composite with a $CaIn_2S_4$ molar content of 30% exhibits the optimum photocatalytic activity. The primary reason for the significantly enhanced photoreduction activity is proved to be the substantially improved separation efficiency of photogenerated electrons/holes caused by forming the $CaIn_2S_4/ZnIn_2S_4$ heterostructured composites. The efficient charge separation can be evidenced by steady-state photoluminescence spectra (PLs) and transient photocurrent response. Based on the charge transfer between $CaIn_2S_4$ and $ZnIn_2S_4$, an enhancement mechanism of photocatalytic activity and stability for the Cr(VI) reduction is proposed.

Keywords: $CaIn_2S_4/ZnIn_2S_4$ composites; Cr(VI); photocatalysis; simulated sunlight; hydrothermal method

1. Introduction

Semiconductor-based photocatalytic technology has exhibited great potential in green controlling of environmental contaminants and converting CO_2 to valuable chemicals [1–6]. However, the most studied photocatalysts (such as TiO_2, ZnO, and ZnS) are active only under ultraviolet light irradiation. In view of the practical applications, visible light active photocatalysts with narrow bandgaps are greatly desirable. A great deal of research and efforts have been devoted to fabricating and synthesizing visible-light-driven catalysts, [7–10]. For instance, visible-light photocatalysts based on metal sulfide, including the doped or functionalized ZnS [7,8], $CdIn_2S_4$ [9], $ZnIn_2S_4$ [10,11], $CdIn_2S_4/ZnIn_2S_4$ [12], and Cu_2ZnSnS_4 [13] have received broad attention in recent decades.

Among these metal sulfides, $ZnIn_2S_4$, as a ternary chalcogenide semiconductor possessing a narrow bandgap (2.34–2.48 eV), has aroused great interest in the field of visible-light-driven contaminants' degradation and hydrogen production by water splitting [10,11]. Although the studies have revealed that $ZnIn_2S_4$ has strong visible-light absorption, the catalytic activity of pristine $ZnIn_2S_4$ is lower than expected due to fast recombination of photogenerated charge carriers. Many efforts have been made to solve these drawbacks, including morphologies control, metal doping, and heterostructured composites construction [14–17]. The fabrication of heterojunction catalysts by coupling two kinds of semiconductor particles with a suitable energy band structure is identified as a valid method to promote photocatalytic activity because of the resulting effective separation of the photoinduced charge [18,19]. To improve the photocatalytic activity of pure $ZnIn_2S_4$, the heterojunction photocatalysts have been reported by hybridization with TiO_2, CdS, and C_3N_4 [20–22].

On the other hand, the metal sulfide semiconductors with narrow bandgaps generally undergo photocorrosion induced by the oxidative process of photogenerated holes, which result in poor photostability or deactivation during long-term recycling use under sunlight irradiation. For instance, CdS particles possess excellent catalytic activity, acceptable stability, and photocorrosion resistance under visible-light illumination, whereas the catalyst becomes unstable when exposed to sunlight irradiation [23]. As a visible-light photocatalyst, $ZnIn_2S_4$ also suffers from the critical drawback of high photocorrosion under sunlight illumination [24]. Very recently, Zhao et al. have synthesized chemically cross-linked $ZnIn_2S_4$/RGO heterostructured catalysts and investigated the sunlight-driven photocatalytic performance of 4-nitrophenol degradation. The experimental results demonstrate the as-fabricated $ZnIn_2S_4$/RGO possessed not only the enhanced visible-light catalytic performance but also prominently improved sunlight stability [24]. Therefore, to practically utilize sunlight-driven photocatalytic technology to control contaminants, the employed catalysts must be not only visible-light active but also stable when subjected to sunlight illumination.

$CaIn_2S_4$ is also an attractive ternary chalcogenide with a narrow bandgap (~1.9 eV) and meanwhile, is the cheapest alkaline earth metal-based semiconductor, which can serve as a potential visible-light catalyst. However, only a few of the investigations on the construction and synthesis of $CaIn_2S_4$-based composite photocatalysts have been reported to date [25–27]. Similar to other narrow-bandgap catalysts, the photocatalytic degradation efficiency of $CaIn_2S_4$ alone is very limited owing to the poor separation efficiency of photogenerated electrons and holes. By constructing the composite catalysts, such as $CaIn_2S_4$-RGO, $CaIn_2S_4$/g-C_3N_4, and $CaIn_2S_4/TiO_2$, the photocatalytic activity of $CaIn_2S_4$ has been significantly improved. For instance, direct Z-scheme $CaIn_2S_4/TiO_2$ catalysts with different $CaIn_2S_4$ weight percentages were prepared, and the photocatalytic performance of these composite catalysts was investigated by the degradation of isoniazid and metronidazole in the pharmaceutical wastewater. The improved catalytic performance can be ascribed to the significantly suppressed recombination of photoinduced charge carriers based on the Z-scheme charge transfer over $CaIn_2S_4/TiO_2$ catalysts. The aforementioned $CaIn_2S_4$-based composites show higher photocatalytic performance than the pure $CaIn_2S_4$ or TiO_2; nevertheless, the preparation processes are relatively complicated, and these composites cannot be obtained through one-step synthesis. Further research on the synthesis of the $CaIn_2S_4$-based composite photocatalysts by one-step hydrothermal methods is highly necessary to improve the photocatalytic activity of $CaIn_2S_4$.

Hexavalent Cr(VI) ions have been listed as one of the priority pollutants by the United States Environmental Protection Agency (US EPA) [28], due to the high toxicity, mutagenicity, and teratogenicity to the aquatic environment and human beings. The elimination of Cr(VI) ions have received increasing attention in the field of wastewater purification. The traditional techniques for removing Cr(VI) ions generally include precipitation, adsorption, ion-exchange, electro-coagulation [29], membrane separation, and photocatalytic reduction [30,31]. Considering solar energy conversion, photocatalytic reduction is postulated to be an efficient and green technology for the elimination of Cr(VI) ions from contaminated water. The highly toxic Cr(VI) can be photoreduced to less harmful Cr(III) by means of a certain photocatalyst and reaction system.

In this study, flower-like $CaIn_2S_4/ZnIn_2S_4$ heterojunction composites are successfully synthesized by using a one-step hydrothermal method, and the photocatalytic performance of the Cr(VI) reduction is investigated under simulated sunlight illumination. To our knowledge, this is the first report about the $CaIn_2S_4/ZnIn_2S_4$ heterojunction composites for the photocatalytic reduction of Cr(VI). The heterojunction composites exhibit much higher reduction efficiency of Cr(VI) than the pure $CaIn_2S_4$ or $ZnIn_2S_4$. Meanwhile, the $CaIn_2S_4/ZnIn_2S_4$ composites present excellent solar stability, and the photocorrosion of $ZnIn_2S_4$ is dramatically inhibited by constructing the $CaIn_2S_4/ZnIn_2S_4$ composites. The detailed mechanism of enhanced photocatalytic performance for the Cr(VI) reduction over the $CaIn_2S_4/ZnIn_2S_4$ composites is also proposed.

2. Materials and Methods

2.1. Materials

All the chemicals were of analytical grade and used as received without further purification. Calcium nitrate ($Ca(NO_3)_2$), zinc sulfate ($ZnSO_4$), indium chloride ($InCl_3$), thioacetamide (TAA), potassium dichromate, sulfuric acid (H_2SO_4), and ammonium oxalate were obtained from Aladdin Industrial Inc. (Shanghai, China). Diphenylcarbazide (DPC) was purchased from J & K Scientific Ltd. (Beijing, China). Double distilled water was used throughout this study.

2.2. Synthesis of Composite Photocatalysts

The $CaIn_2S_4/ZnIn_2S_4$ composite photocatalysts were prepared by hydrothermal route. Taking 20% $CaIn_2S_4/ZnIn_2S_4$ as an example, in a typical synthesis process, 0.2 mmol $Ca(NO_3)_3 \cdot 4H_2O$ and 0.8 mmol $ZnSO_4 \cdot 7H_2O$ were added into 40 mL of distilled water, followed by vigorous stirring for 30 min to form a clear solution. After that, 2 mmol $InCl_3 \cdot 4H_2O$ and 8 mmol TAA were put into the above-obtained mixed solution and stirred for additional 60 min at room temperature. The reaction mixture was then transferred into a Teflon-lined steel autoclave (Zhengxin Instrument, Yancheng, China), which was heated at 180 °C for 12 h. Finally, the obtained yellow solid was collected by filtration, washed with distilled water several times, and dried at 60 °C for 8 h. On that basis, different $CaIn_2S_4/ZnIn_2S_4$ composites with $CaIn_2S_4$ molar ratios of 5%, 10%, 20%, 30%, and 50% were synthesized and denoted as 5% $CaIn_2S_4/ZnIn_2S_4$, 10% $CaIn_2S_4/ZnIn_2S_4$, 20% $CaIn_2S_4/ZnIn_2S_4$, 30% $CaIn_2S_4/ZnIn_2S_4$, and 50% $CaIn_2S_4/ZnIn_2S_4$, respectively.

For comparison, the pure $ZnIn_2S_4$ was prepared by using a similar process without $Ca(NO_3)_3$. 1.0 mmol $ZnSO_4 \cdot 7H_2O$ was put into 40 mL of distilled water and stirred for 30 min. Then, 2 mmol $InCl_3 \cdot 4H_2O$ and 8 mmol TAA were added to the above-obtained solution and stirred for another 60 min. This mixture was subsequently transferred into a Teflon-lined steel autoclave and heated at 180 °C for 12 h. The precipitate was finally washed with deionized water and dried at 60 °C for 8 h to obtain pure $ZnIn_2S_4$. For the preparation of the pure $CaIn_2S_4$, 1.0 mmol $Ca(NO_3)_3 \cdot 4H_2O$ was added into 40 mL of distilled water and stirred for 30 min. After that, 2 mmol $InCl_3 \cdot 4H_2O$ and 8 mmol TAA were put into the above solution and then stirred for another 60 min. The obtained mixture was transferred into a Teflon-lined steel autoclave and heated at 180 °C for 12 h. The orange precipitate was finally washed with deionized water and dried at 60 °C for 8 h to obtain pure $CaIn_2S_4$.

2.3. Material Characterization

The crystal structures of the as-prepared composites were investigated by using an X-ray diffractometer (XRD, Bruker D8 Advance, Karlsruhe, Germany), with Cu Kα radiation (λ = 0.15405 nm). The morphologies of the synthesized samples were observed by a field emission scanning electron microscope (FEI, Quanta250, Hillsboro, OR, USA). The X-ray photoelectron spectroscopy (XPS) experiment was carried out by using a Thermo Scientific ESCALAB 250xi system (Waltham, MA, USA), equipped with an Al anode. The BET (Brunauer-Emmett-Teller) surface areas of the composite samples were determined on a surface area analyzer (Autosorb-IQ, Quantachrome, Boynton Beach, FL, USA). UV–visible diffuse reflectance spectra (DRS) were collected on a Shimadzu UV-2550 spectrometer (Kyoto, Japan) using $BaSO_4$ as the reflectance standard. Photoluminescence spectra (PLs) were measured by using an Edinburgh FLS 980 fluorescence spectrometer (Livingston, UK), with a 330 nm excitation wavelength. The transient photocurrent measurement was performed on a CHI electrochemical workstation (CHI 760D, Shanghai, China) in the standard three-electrode system. Ag/AgCl electrode and a platinum wire were employed as the reference electrode and the counter electrode, respectively. The working electrodes were prepared according to our previous report [32]. The photocurrent was measured in the electrolyte of a 0.5 mol/L Na_2SO_4 aqueous solution (pH ~6.8) at a bias of +0.5 V. The light source was the same as that used in the photocatalytic experiments.

2.4. Photocatalytic Reduction of Cr(VI)

For the photocatalytic reduction of Cr(VI) under simulated sunlight irradiation, a 300 W xenon lamp (PLS-SXE 300, Perfect Light Co. Ltd., Beijing, China) was used as the light source. The incident light intensity was 75 mW/cm^2. In the typical photocatalytic test, 50 mg of photocatalyst was suspended in 50 mL of 20 mg/L Cr(VI) aqueous solution. After adding 5 mg of ammonium oxalate (a scavenger for photo-hole), the suspension was magnetically stirred in the dark for 30 min to establish an adsorption–desorption equilibrium. As the photocatalytic reduction proceeded, 3 mL of the reaction solution was taken out at a given time interval and centrifuged to remove the catalyst particles. The concentrations of Cr(VI) were determined colorimetrically at 540 nm using the diphenylcarbazide (DPC) method on a Shimadzu UV-160A UV–Vis spectrophotometer [33]. In addition, the total Cr ions concentrations were measured by an inductively coupled plasma-optical emission spectrophotometer (Agilent 725 ICP-OES, Palo Alto, CA, USA).

The apparent quantum efficiency (AQE) of Cr(VI) the photocatalytic reduction was measured under the same photocatalytic reaction conditions, except by using a 420 nm bandpass filter. The apparent quantum efficiencies were calculated according to the following equation:

$$\mathrm{AQE}(\%) = \frac{\text{Number of reacted electrons}}{\text{Number of incident photons}} \times 100 = \frac{3 \times \text{number of reduced Cr(VI)}}{\text{Number of incident photons}} \times 100$$

$$\text{Number of incident photons} = \frac{I \times A \times t}{E(\lambda = 420\ \text{nm})}$$

$$E(\lambda = 420\ \text{nm}) = \frac{h \times c}{\lambda}$$

where I, A, t, c, λ, and h are light intensity, light exposure area, irradiation time, light velocity, light wavelength, and Planck constant, respectively.

To evaluate the catalytic stability of the $CaIn_2S_4/ZnIn_2S_4$ composites, the photocatalysts—after the first run for the Cr(VI) reduction—were separated by centrifugation from the suspension and washed with 1 M nitrite acid solution and deionized water. After being dried at 60 °C, the recovered photocatalysts were reused for the next run of the photocatalytic Cr(VI) reductions under the same experimental conditions.

3. Results and Discussion

3.1. XRD Analysis and BET Surface Area

The phase composition and crystalline properties of the $CaIn_2S_4/ZnIn_2S_4$ composite samples were analyzed by XRD. Figure 1 displays the XRD patterns of the pure $ZnIn_2S_4$, $CaIn_2S_4$, and $CaIn_2S_4/ZnIn_2S_4$ composites. As indicated in Figure 1, the diffraction peaks of the pure $ZnIn_2S_4$ at 2θ = 21.1°, 27.6°, 30.3°, 47.4°, 52.0°, and 55.6° corresponded to (006), (102), (104), (110), (116), and (200) crystal planes of hexagonal phase $ZnIn_2S_4$ (JCPDS NO. 65-2023), respectively [24]. No other diffraction peaks were detected in the XRD pattern of $ZnIn_2S_4$, indicating that the obtained $ZnIn_2S_4$ is highly pure. With the addition of $Ca(NO_3)_3$ in the preparation process, new diffraction peaks appeared in the XRD patterns, which belonged to the characteristic peaks of the cubic phase $CaIn_2S_4$ (JCPDS No. 16-0341) [25–27]. It can be also observed from Figure 1 that the intensity of the diffraction peaks belonging to the cubic $CaIn_2S_4$ phase increased gradually by increasing the mole proportion of $CaIn_2S_4$, which indicated the existence of both $CaIn_2S_4$ and $ZnIn_2S_4$ in the as-synthesized composites. In addition, the diffraction peaks corresponding to binary sulfides, oxides, and other new compounds were not observed, indicating that $CaIn_2S_4$ and $ZnIn_2S_4$ maintained the pure phase and no impurities were formed in the obtained $CaIn_2S_4/ZnIn_2S_4$ composites.

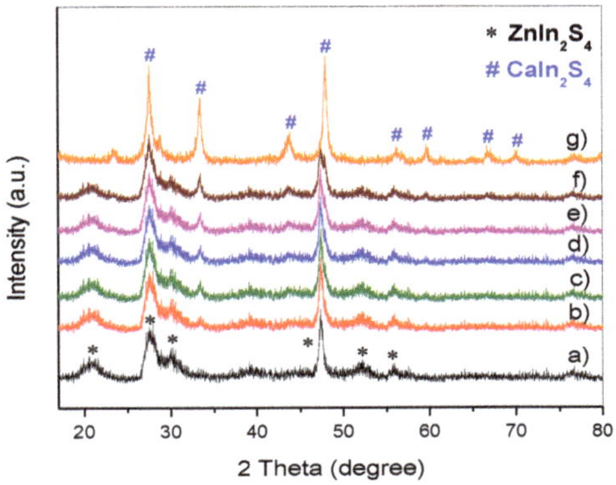

Figure 1. XRD patterns of pure $ZnIn_2S_4$, $CaIn_2S_4$, and $CaIn_2S_4/ZnIn_2S_4$ composites: (**a**) pure $ZnIn_2S_4$, (**b**) 5% $CaIn_2S_4/ZnIn_2S_4$, (**c**) 10% $CaIn_2S_4/ZnIn_2S_4$, (**d**) 20% $CaIn_2S_4/ZnIn_2S_4$, (**e**) 30% $CaIn_2S_4/ZnIn_2S_4$, (**f**) 50% $CaIn_2S_4/ZnIn_2S_4$, and (**g**) pure $CaIn_2S_4$.

The BET surface area of the pure $ZnIn_2S_4$, $CaIn_2S_4$, and $CaIn_2S_4/ZnIn_2S_4$ composites was measured, and the results are summarized in Table 1. As can be noted from Table 1, the specific surface area of the $CaIn_2S_4/ZnIn_2S_4$ composites decreased slightly with the continuous increment of $CaIn_2S_4$ component. The actual molar ratios of Ca–Zn in the $CaIn_2S_4/ZnIn_2S_4$ composites were determined by inductively coupled plasma elemental analysis, as presented in Table 1. It can be observed from Table 1 that the experimentally measured molar ratios of Ca–Zn in the synthesized composites were close to those of the added proportions in the preparation process.

Table 1. BET (Brunauer-Emmett-Teller) specific surface area, molar ratios of Ca–Zn in the synthesized $CaIn_2S_4/ZnIn_2S_4$ and apparent quantum efficiency (AQE) of the Cr(VI) reduction over these composites.

Samples	S_{BET} (m^2/g)	Molar Ratios of Ca:Zn (%)	AQE (%)
pure $ZnIn_2S_4$	59.2	0	3.7
5% $CaIn_2S_4/ZnIn_2S_4$	56.3	5.05	4.1
10% $CaIn_2S_4/ZnIn_2S_4$	54.7	9.67	4.5
20% $CaIn_2S_4/ZnIn_2S_4$	53.4	18.30	5.2
30% $CaIn_2S_4/ZnIn_2S_4$	52.1	26.52	6.6
50% $CaIn_2S_4/ZnIn_2S_4$	50.5	43.29	5.5
pure $CaIn_2S_4$	46.0	–	2.3

3.2. SEM and Elemental Mapping Analysis

The SEM images of the as-synthesized pure $ZnIn_2S_4$, $CaIn_2S_4$, and $CaIn_2S_4/ZnIn_2S_4$ composites are indicated in Figure 2. As can be observed from Figure 2, the pure $ZnIn_2S_4$ was composed of hierarchical microspheres with a wide distribution of diameter, which was consistent with the previous reports [17,24]. Introducing the Ca component had almost no influence on the morphologies of the $CaIn_2S_4/ZnIn_2S_4$ composites, which also exhibited flower-like microspheres constructed by numerous nanosheets in the form of random self-assembly. As a consequence, the porous structures with wide pore-size distribution could be expected. This would benefit the photocatalytic reaction by increasing the specific surface area. Moreover, the formation of the $CaIn_2S_4/ZnIn_2S_4$ heterojunction was evidenced by the elemental mapping of the as-synthesized composites (Figure 3). Maps of Zn–K, Ca–K, In–L, and S–K display the same shape and location, indicating the coexistence of $ZnIn_2S_4$ and $CaIn_2S_4$ components in the obtained composites. This provided solid evidence for the formation of $CaIn_2S_4/ZnIn_2S_4$ heterostructured composites.

Figure 2. FE-SEM (field emission scanning electron microscope) images of (**a**) pure $ZnIn_2S_4$, (**b**) 10% $CaIn_2S_4/ZnIn_2S_4$, (**c**) 30% $CaIn_2S_4/ZnIn_2S_4$, and (**d**) pure $CaIn_2S_4$.

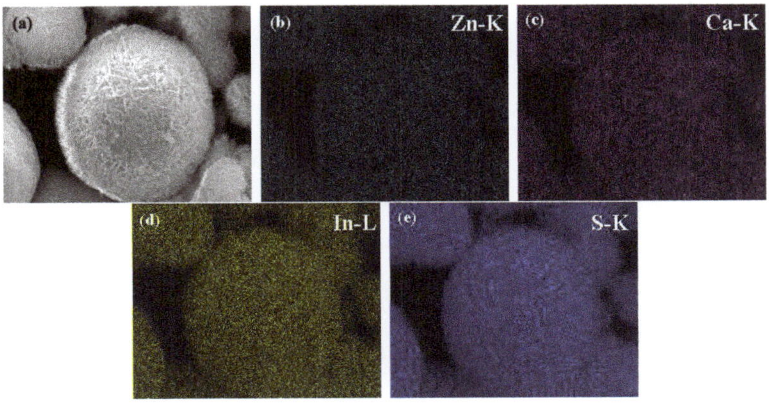

Figure 3. The EDS (Energy Dispersive Spectroscopy) elemental mapping images of the 30% $CaIn_2S_4/ZnIn_2S_4$ sample: (**a**) SEM image, (**b**) Zn-K, (**c**) Ca-K, (**d**) In-L, and (**e**) S-K, respectively.

3.3. *XPS Analysis and Optical Properties*

To determine the elemental composition and the corresponding chemical states of the synthesized composites, the XPS spectra of the 30% $CaIn_2S_4/ZnIn_2S_4$ sample are indicated in Figure 4. As can be seen from Figure 4a, the survey spectrum indicated the presence of Zn, Ca, In, and S elements in the 30% $CaIn_2S_4/ZnIn_2S_4$ sample. The high-resolution XPS spectrum for Zn is presented in Figure 4b. Two characteristic XPS signals were observed at binding energies of 1022.4 and 1045.3 eV, which were ascribed to Zn^{2+} $2p_{1/2}$ and Zn^{2+} $2p_{3/2}$, respectively [33,34]. The high-resolution XPS spectrum of In 3d is displayed in Figure 4c. The characteristic peaks centered at 444.7 and 452.3 eV can be attributed to the In $3d_{3/2}$ and In $3d_{5/2}$ signals of In^{3+} species, respectively [27,34]. In Figure 4d, the two peaks at the binding energies of 351.1 and 347.6 eV corresponded to the $2p_{3/2}$ and $2p_{1/2}$ levels of Ca^{2+}. Figure 4e shows an XPS signal centered at 162.5 eV, which can be assigned to the $2p_{1/2}$ level of S^{2-} in the as-prepared $CaIn_2S_4/ZnIn_2S_4$ composites [34]. These results further indicate that $CaIn_2S_4/ZnIn_2S_4$ composites can be successfully synthesized through a one-step hydrothermal reaction process.

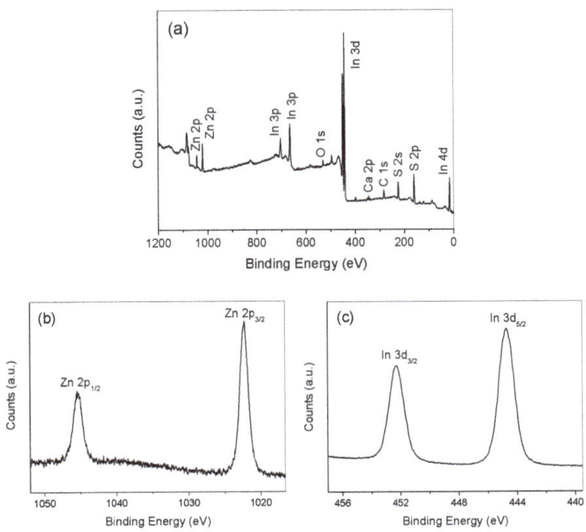

Figure 4. *Cont.*

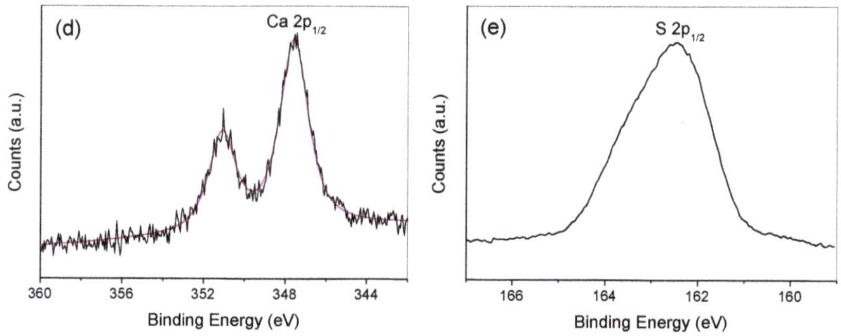

Figure 4. Typical XPS survey spectrum of 30% $CaIn_2S_4/ZnIn_2S_4$ (**a**), high-resolution XPS spectra of Zn 2p (**b**), In 3d (**c**), Ca 2p (**d**), and S 2p (**e**), respectively.

The optical absorption properties of the as-obtained $CaIn_2S_4/ZnIn_2S_4$ composites were analyzed by UV–Vis DRS, and the results are depicted in Figure 5a. The absorption edges of the pure $ZnIn_2S_4$ and $CaIn_2S_4$ samples were at around 537 nm and 638 nm, respectively. As can be also noted from Figure 5a, the absorption edges of the $CaIn_2S_4/ZnIn_2S_4$ samples were gradually red-shifted from 537 to 570 nm as the molar percentage of the $CaIn_2S_4$ component increased to 50%. The photoresponse of the $CaIn_2S_4/ZnIn_2S_4$ composites in visible-light region was significantly improved by comparison with that of the pure $ZnIn_2S_4$. Moreover, the UV–Vis DRS shown in Figure 5a were all very steep, demonstrating the visible-light absorption was ascribed to the intrinsic band transition instead of the transition from impurity levels [35]. Based on the optical absorption theory of the bandgap semiconductor, the bandgap energy of the pure $CaIn_2S_4$ and $ZnIn_2S_4$ can be calculated by the Equation (1) [36] as follows:

$$\alpha hv = A\ (hv - E_g)^{n/2} \tag{1}$$

where α, v, E_g, and A represent the absorption coefficient, light frequency, bandgap energy, and a constant, respectively. The value of n depends on the type of optical transition of a semiconductor ($n = 1$ for the direct transition and $n = 4$ for the indirect transition). According to the previous literature [27,37], $ZnIn_2S_4$ and $CaIn_2S_4$ are direct-transition semiconductors, and thus, the bandgap energies of $CaIn_2S_4$ and $ZnIn_2S_4$ can be estimated from the plots of $(\alpha hv)^2$ versus light energy (hv). As illustrated in Figure 5b, the estimated bandgaps were 2.03 and 2.43 eV for the pure $CaIn_2S_4$ and $ZnIn_2S_4$, respectively, which matched well with the literature values [15,27]. For the composite catalysts, the photocatalytic performance was primarily determined by the valence band (VB) and conduction band (CB) energy levels of the constituent semiconductors. Based on the following equations, the VB and CB positions of the pristine $ZnIn_2S_4$ and $CaIn_2S_4$ can be obtained:

$$E_{VB} = \chi - E_e + 0.5\ E_g \tag{2}$$

$$E_{CB} = E_{VB} - E_g \tag{3}$$

where E_{VB} and E_{CB} are the potential of the VB and CB edge, E_g is the bandgap energy, χ is the geometric mean of the absolute electronegativity of the constituent atoms in the semiconductor, and E_e is the energy of free electrons on the hydrogen scale with a value of 4.5 eV. The χ values of $ZnIn_2S_4$ and $CaIn_2S_4$ were 4.86 and 4.39 eV, respectively. Based on the above empirical equations, the E_{VB} values of $ZnIn_2S_4$ and $CaIn_2S_4$ were estimated to be +1.58 and +0.91 eV. The corresponding E_{CB} values were also calculated to be −0.85 and −1.12 eV, respectively.

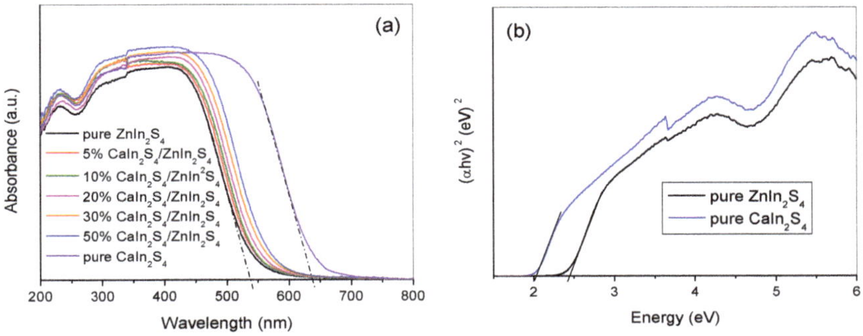

Figure 5. (**a**) UV–Vis DRS of the pure $ZnIn_2S_4$, $CaIn_2S_4$, and $CaIn_2S_4/ZnIn_2S_4$ composites, (**b**) Plots of $(\alpha hv)^2$ vs. light energy (hv) of the pure $ZnIn_2S_4$ and $CaIn_2S_4$.

3.4. Photocatalytic Activity

The photocatalytic performances of the $ZnIn_2S_4$, $CaIn_2S_4$, and $CaIn_2S_4/ZnIn_2S_4$ composites were investigated by the aqueous-phase Cr(VI) reduction under simulated sunlight irradiation (Figure 6A). Cr(VI) cannot be reduced in the absence of light illumination or photocatalysts. The pure $ZnIn_2S_4$ displayed a relatively higher photocatalytic activity of the Cr(VI) reduction than the $CaIn_2S_4$ sample. About 42% of Cr(VI) was reduced over the pure $CaIn_2S_4$, while $ZnIn_2S_4$ can reduce 63% of Cr(VI) after irradiation of 30 min. All the $CaIn_2S_4/ZnIn_2S_4$ composites exhibited higher photocatalytic efficiency than the pristine $CaIn_2S_4$ and $ZnIn_2S_4$, indicating that the combination of $CaIn_2S_4$ and $ZnIn_2S_4$ can improve the photocatalytic reduction performance of these composites.

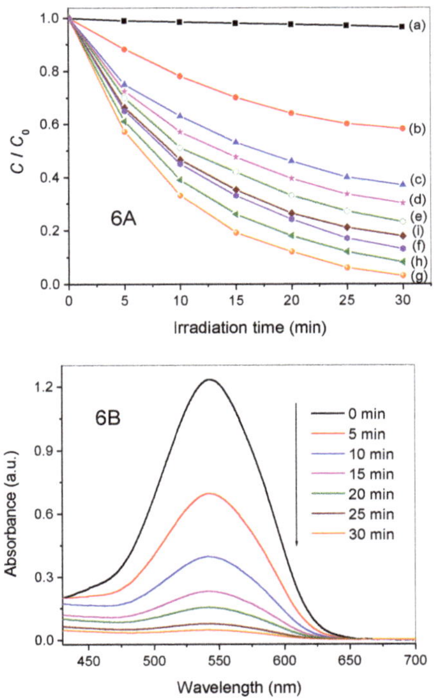

Figure 6. *Cont.*

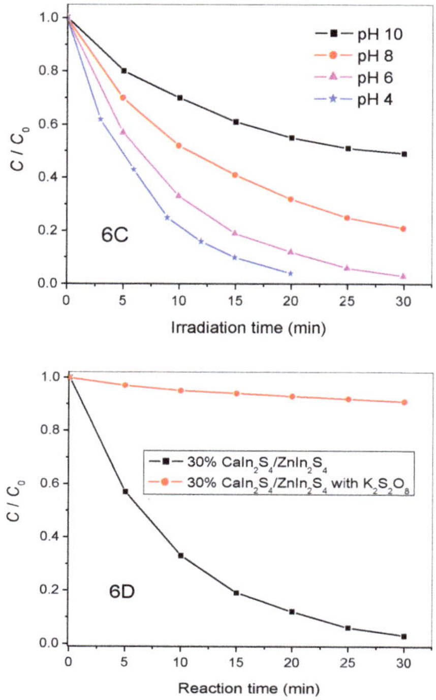

Figure 6. (**A**) Photocatalytic reduction of Cr(VI) as a function of irradiation time over different catalysts: (a) no catalyst, (b) pure $CaIn_2S_4$, (c) pure $ZnIn_2S_4$, (d) 5% $CaIn_2S_4/ZnIn_2S_4$, (e) 10% $CaIn_2S_4/ZnIn_2S_4$, (f) 20% $CaIn_2S_4/ZnIn_2S_4$, (g) 30% $CaIn_2S_4/ZnIn_2S_4$, (h) 50% $CaIn_2S_4/ZnIn_2S_4$, and (i) mechanically mixed 30%$CaIn_2S_4$ + 70%$ZnIn_2S_4$; (**B**) time-dependent absorption spectral pattern of diphenylcarbazide (DPC)-Cr(VI) complex solutions after the reduction over 30% $CaIn_2S_4/ZnIn_2S_4$ (pH = 6); (**C**) photocatalytic reduction of Cr(VI) under different pH values over 30% $CaIn_2S_4/ZnIn_2S_4$; (**D**) the controlled experiment for photocatalytic reduction of Cr(VI) over the 30% $CaIn_2S_4/ZnIn_2S_4$ composites with the addition of $K_2S_2O_8$ (0.1 mmol) as a scavenger for photoinduced electrons.

For the synthesized $CaIn_2S_4/ZnIn_2S_4$ composite catalysts, the photocatalytic were closely associated with the contents of the $CaIn_2S_4$ component. As can be noted from Figure 6A, the photocatalytic activities of the $CaIn_2S_4/ZnIn_2S_4$ composite catalysts increased with the increment of component $CaIn_2S_4$. The 30% $CaIn_2S_4/ZnIn_2S_4$ composite photocatalyst exhibited the highest activity for the Cr(VI) reduction, whereas the greater increase in the amount of $CaIn_2S_4$ resulted in a decrease in the Cr(VI) reduction rates. It can be ascribed to the low photocatalytic activity of the pure $CaIn_2S_4$ because of the slow separation of the photogenerated charge carriers. Moreover, the apparent quantum efficiencies (AQE) of the Cr(VI) photocatalytic reduction over the synthesized composites were also calculated, and the corresponding results are summarized in Table 1. As indicated in Table 1, the AQE of the 30% $CaIn_2S_4/ZnIn_2S_4$ nanocomposite catalyst reached 6.6%, which presented higher than that of the pure $ZnIn_2S_4$ (3.7%) or pure $CaIn_2S_4$ (2.3%). In addition, the AQE for the Cr(VI) reduction increased gradually with the increase of the $CaIn_2S_4$ constituent when the addition ratio of the $CaIn_2S_4$ precursor was no more than 30%. However, further increasing the molar ratio of $CaIn_2S_4$ led to a decrease in the AQE. This suggests that the heterostructured composites containing a suitable amount of $CaIn_2S_4$ and $ZnIn_2S_4$ contributed to improving optimally the photoactivity for the Cr(VI) reduction. Therefore, the optimal 30% $CaIn_2S_4/ZnIn_2S_4$ has been chosen as a representative catalyst for the following studies.

To further clarify the influence of the heterostructure on the photocatalytic performance of the Cr(VI) reduction, the 30% $CaIn_2S_4/ZnIn_2S_4$ sample was compared to its mechanical mixing counterpart sample 30% $CaIn_2S_4$ + 70% $ZnIn_2S_4$. The photocatalytic activity of the Cr(VI) reduction over the physical mixture was much lower than that of the 30% $CaIn_2S_4/ZnIn_2S_4$ composite obtained via the one-step hydrothermal method. This result shows that the heterojunction formed between the $CaIn_2S_4$ and $ZnIn_2S_4$ contributed to improving photocatalytic efficiency of the Cr(VI) reduction.

When using the optimal 30% $CaIn_2S_4/ZnIn_2S_4$ as a photocatalyst, the change in the temporal absorption spectra of the DPC-Cr(VI) complex solution is illustrated in Figure 6B. The absorption peak at 540 nm belonging to the DPC-Cr(VI) complex decreased rapidly with the increase of light irradiation time, and it almost vanished after light illumination for 30 min. To gain more insight into the photocatalytic process, the total Cr ions concentrations over the 30% $CaIn_2S_4/ZnIn_2S_4$ photocatalyst after the treated samples were measured by ICP emission spectrometer (Figure S1, Supplementary Materials). As indicated in Figure S1, the initial concentration of Cr(VI) is 19.6 ppm. The measured total Cr ions concentration after 30 min light irradiation were found to be 19.1 ppm. This result demonstrates that almost no Cr(0) was produced in the present photocatalytic system. That is to say, Cr(VI) was primarily reduced to Cr(III) by the photogenerated electrons of the $CaIn_2S_4/ZnIn_2S_4$ composites.

In addition, the photocatalytic experiments for the Cr(VI) reduction under different pH conditions with 30% $CaIn_2S_4/ZnIn_2S_4$ were also carried out, and the corresponding results are presented in Figure 6C. In a photocatalytic system, the reduction efficiency of the aqueous Cr(VI) was greatly influenced by the pH value according to the previous reports [30]. Cr(VI) existed in two forms in alkaline and acid medium, respectively. $CrO_4{}^{2-}$ is predominant in the alkaline medium, whereas $Cr_2O_7{}^{2-}$ plays a major role in the acid medium. The chemical redox reaction can be outlined as follows:

$$CrO_4{}^{2-} + 4H_2O + 3e^- \rightarrow Cr(OH)_3\downarrow + 5OH^- \text{ (alkaline)} \tag{4}$$

$$14H^+ + Cr_2O_7{}^{2-} + 6e^- \rightarrow 2Cr^{3+} + 7H_2O \text{ (acid)} \tag{5}$$

As depicted in Figure 6C, the reduction rate of Cr(VI) increased with the decrease of the pH value. About 38.7%, 58.8%, 80.6%, and 89.0% of Cr(VI) are reduced at pH 10, 8, 6, and 4 in the first 15 min, respectively. This could be due to the fact that the $Cr(OH)_3$ precipitate produced in the alkaline medium covers the activity sites of the photocatalysts [38]. The results demonstrate the acidic condition was more beneficial to the photocatalytic reduction of Cr(VI) over the synthesized $CaIn_2S_4/ZnIn_2S_4$ composites.

Additionally, the controlled experiment by adding $K_2S_2O_8$ (the trapping agent of photo-generated electrons, 0.1 mmol) [39] into the photocatalytic system of the Cr(VI) reduction was performed, and the corresponding result is depicted in Figure 6D. It can be seen clearly that the Cr(VI) reduction over the optimal photocatalyst 30% $CaIn_2S_4/ZnIn_2S_4$ hardly occurs in the presence of $K_2S_2O_8$, which indicates that the reduction of Cr(VI) is conducted by photogenerated electrons under simulated sunlight irradiation. These results also indicate that a suitable amount of $CaIn_2S_4$ can effectively improve the photocatalytic performance of $ZnIn_2S_4$ toward the Cr(VI) reduction.

3.5. Catalytic Stability

The stability of a given photocatalyst was also an important factor in the practical application [40]. Furthermore, narrow bandgap semiconductors are generally more unstable when exposed to the sunlight illumination [24]. To study the effect of the $CaIn_2S_4/ZnIn_2S_4$ heterostructure on the catalytic stability, the cycle experiments of the Cr(VI) reduction were carried out and compared by using $ZnIn_2S_4$ and 30% $CaIn_2S_4/ZnIn_2S_4$ as photocatalysts under simulated sunlight irradiation (Figure 7). The results indicate that for the pristine $ZnIn_2S_4$, the reduction efficiency of Cr(VI) decreased about 20.9% after five repeated uses, demonstrating that $ZnIn_2S_4$ was unstable to some extent under simulated sunlight illumination. In comparison, a relatively low decrease in the reduction efficiency of Cr(VI) over 30% $CaIn_2S_4/ZnIn_2S_4$ was observed, and only a ca. 2.6% decrease of the reduction efficiency is obtained after reusing five cycles. This suggests that the sunlight stability of $ZnIn_2S_4$ can be improved through forming the heterojunction composites with $CaIn_2S_4$. Additionally, the XPS analysis of the used 30% $CaIn_2S_4/ZnIn_2S_4$ catalyst was also performed, and the obtained results are displayed in Figure S2 of the Supplementary Materials. As can be found from Figure S2, the surface element composition and the chemical state of the 30% $CaIn_2S_4/ZnIn_2S_4$ sample before and after the photocatalytic reaction show no obvious difference. The results confirm the superior stability of the $CaIn_2S_4/ZnIn_2S_4$ composite photocatalysts under simulated sunlight, which is more promising for the practical photocatalytic applications in environmental restoration.

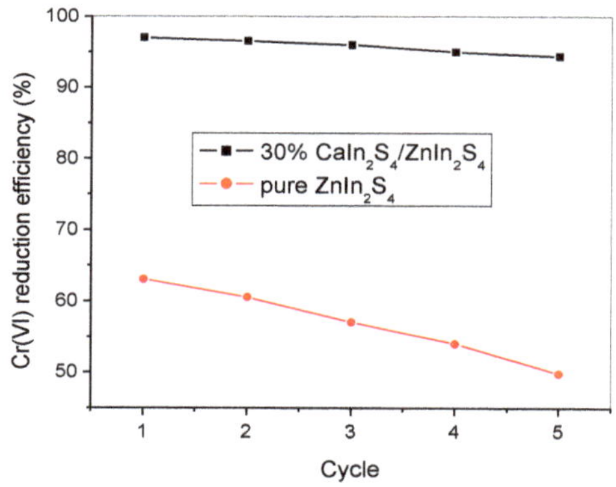

Figure 7. Photocatalytic stability tests of $ZnIn_2S_4$ and 30% $CaIn_2S_4/ZnIn_2S_4$ toward the Cr(VI) reduction.

3.6. Enhancement Mechanism of Photocatalytic Activity and Stability

On the basis of the above discussion and results, it is obvious that coupling a suitable amount of $CaIn_2S_4$ can dramatically improve the photocatalytic performance for the Cr(VI) reduction, including the reduction efficiency and cycling stability of the $CaIn_2S_4/ZnIn_2S_4$ composites. The remarkably enhanced photocatalytic activity of the $CaIn_2S_4/ZnIn_2S_4$ samples, compared with the pure $CaIn_2S_4$ and $ZnIn_2S_4$, can be ascribed to the effective separation of the photogenerated electron/hole pairs due to the forming of flower-like heterostructures between $ZnIn_2S_4$ and $CaIn_2S_4$. Generally, the photoluminescence spectrum (PLs) is considered as a vital technology to study the migration and fate of photoinduced charge carriers [39,41]. Higher intensities of PL signals usually represent higher recombination rates of photogenerated charge carriers, thus resulting in a lower photocatalytic performance. The comparison of the PL spectrum for the pristine $ZnIn_2S_4$ and 30% $CaIn_2S_4/ZnIn_2S_4$ with an excitation wavelength of 330 nm is presented in Figure 8. The pristine $ZnIn_2S_4$ showed emissions at 520, 571, and 656 nm. The strong emission peak centered at 520 nm for the pristine $ZnIn_2S_4$ was attributed to the intrinsic luminescence of $ZnIn_2S_4$. The relatively weak PL peaks at 571, and 656 nm can be ascribed to the surface state emissions, which were mainly caused by the surface defects in the $ZnIn_2S_4$ structure. In comparison with the pristine $ZnIn_2S_4$, there was no new emission signal in the PL spectrum of 30% $CaIn_2S_4/ZnIn_2S_4$, but the intensities of the PL peaks decreased obviously. The results imply that the photogenerated electrons and holes can transfer effectively between $CaIn_2S_4$ and $ZnIn_2S_4$, thus suppressing the recombination of charge carriers. This can be a primary reason for the $CaIn_2S_4/ZnIn_2S_4$ composites possessing excellent photocatalytic reduction performance under simulated sunlight irradiation.

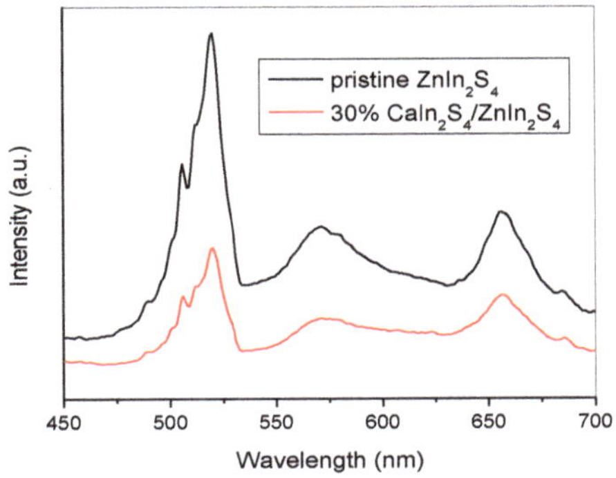

Figure 8. Room temperature photoluminescence spectra (PLs) of the pure $ZnIn_2S_4$ and 30% $CaIn_2S_4/ZnIn_2S_4$ under the excitation wavelength of 330 nm.

Most of the heterostructured nanocomposites follow the bidirectional charge transfer mechanism [42,43]. For instance, Kumar and his colleagues have reported on the photoinduced electrons enriched on the conduction band of Ag_3PO_4 and holes on the valence band of g-C_3N_4, which was conducted through a bidirectional charge transfer process between Ag_3PO_4 and g-C_3N_4 [42]. As for the as-prepared $CaIn_2S_4/ZnIn_2S_4$ composites, the aforementioned bidirectional charge transfer was also the primary charge migration process. Based on the bandgap energies of $CaIn_2S_4$ and $ZnIn_2S_4$ estimated from Figure 5 and Equations (2) and (3), the energy band structure diagram of $CaIn_2S_4$ and $ZnIn_2S_4$ can be schematically illustrated, as shown in Figure 9. Under light illumination, the components of $CaIn_2S_4$ and $ZnIn_2S_4$ were simultaneously excited, generating photoinduced electron/hole pairs. Due to the more negative conduction-band edge of $CaIn_2S_4$ (−1.12 eV) than that of $ZnIn_2S_4$ (−0.85 eV), the photogenerated electrons prefer to transfer from the CB of $CaIn_2S_4$ to $ZnIn_2S_4$, whereas the photogenerated holes on the more positive VB of $ZnIn_2S_4$ (+1.58 eV) would migrate to that of $CaIn_2S_4$ (+0.91 eV). This bidirectional charge transfer process results in efficient separation of photogenerated electron/hole pairs in the synthesized $CaIn_2S_4/ZnIn_2S_4$ composites.

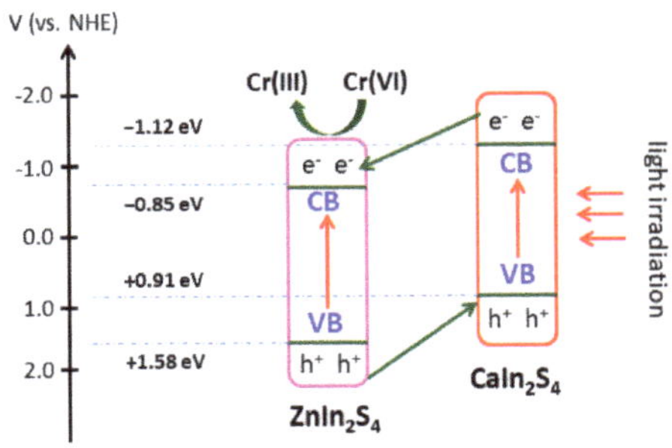

Figure 9. Schematic diagram of the transfer and separation of photogenerated charges in the $CaIn_2S_4/ZnIn_2S_4$ composites under simulated sunlight irradiation.

To further investigate the important function of heterostructures in enhancing the separation of charge carriers, the transient photocurrent responses were measured over working electrodes made of the pure $CaIn_2S_4$, $ZnIn_2S_4$, and 30% $CaIn_2S_4/ZnIn_2S_4$ composite. As depicted in Figure 10, the fast and steady photocurrent response can be detected for each light-on and light-off cycle over the pure $CaIn_2S_4$, $ZnIn_2S_4$, and 30% $CaIn_2S_4/ZnIn_2S_4$ composite. The pristine $CaIn_2S_4$ exhibited a very low photocurrent density, whereas the pure $ZnIn_2S_4$ showed a relatively higher photocurrent than that of

$CaIn_2S_4$ under simulated sunlight irradiation. This could be due to the fact that $CaIn_2S_4$ possesses a narrower bandgap than $ZnIn_2S_4$, which is unfavorable for the effective separation of photoinduced electron/hole pairs. This would result in the short survival time of photogenerated electrons and weak photocurrent density. However, the composite sample of 30% $CaIn_2S_4/ZnIn_2S_4$ exhibits a dramatically enhanced photocurrent density compared with that of the pure $ZnIn_2S_4$ and $CaIn_2S_4$, which further substantiates the efficient separation of the photoinduced electron/hole pairs in the obtained $CaIn_2S_4/ZnIn_2S_4$ composites.

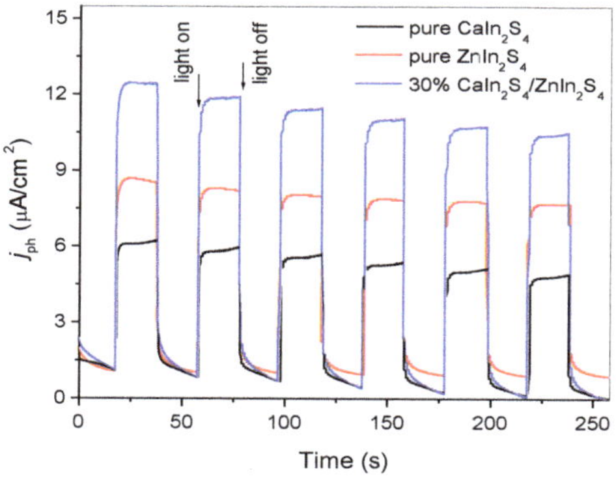

Figure 10. Photocurrent spectra of the as-synthesized pure $CaIn_2S_4$, $ZnIn_2S_4$, and 30% $CaIn_2S_4/ZnIn_2S_4$ samples under simulated sunlight irradiation with 20 s light on/off cycles.

Based on the above experimental results and the charge transfer process depicted in Figure 9, a possible enhancement mechanism of photocatalytic activity and stability for the Cr(VI) reduction can be proposed. Under simulated sunlight irradiation, the components of $CaIn_2S_4$ and $ZnIn_2S_4$ are excited to produce photoinduced holes and electrons. Owing to the relatively high recombination rate of photogenerated electron/hole pairs, the pristine $CaIn_2S_4$ and $ZnIn_2S_4$ exhibited low photocatalytic performance. Due to the well-matched energy band structures and the intimate interfacial contact between $CaIn_2S_4$ and $ZnIn_2S_4$ in the as-synthesized composites, the photoinduced electrons located on the CB of $CaIn_2S_4$ can easily migrate to that of $ZnIn_2S_4$, and on the contrary, the photoinduced holes on the VB of $ZnIn_2S_4$ spontaneously transfer to that of $CaIn_2S_4$ (Figure 9). These photoelectrons (e^-) accumulated on the CB of $ZnIn_2S_4$ possess a strong reduction ability (−0.85 eV vs. NHE (normal hydrogen electrode)), which can reduce toxic Cr(VI) to Cr(III) ($E_{Cr(VI)/Cr(III)}$ = +0.55 eV vs. NHE) [44,45]. The photoholes located on the VB of $CaIn_2S_4$ react with the sacrificial reagents immediately. The heterostructured composites formed between $CaIn_2S_4$ and $ZnIn_2S_4$ effectively prevent the recombination of photogenerated electrons and holes, and thus, the photocatalytic activities of the Cr(VI) reduction are enhanced greatly. Meanwhile, the effective transfer of photogenerated holes from the VB of $ZnIn_2S_4$ to that of $CaIn_2S_4$ is beneficial for preventing the oxidation of S^{2-} by holes, which significantly improves the photostability of $ZnIn_2S_4$ in the composite catalysts.

4. Conclusions

In short, the $CaIn_2S_4/ZnIn_2S_4$ composite photocatalysts were successfully prepared through a one-step hydrothermal process. XRD patterns show that the as-synthesized flower-like composites consist of hexagonal phase $ZnIn_2S_4$ and cubic phase $CaIn_2S_4$. Compared with the pristine $ZnIn_2S_4$, the heterostructured composites $CaIn_2S_4/ZnIn_2S_4$ show significantly improved photocatalytic activity and stability for the Cr(VI) reduction under simulated sunlight illumination. The molar content of $CaIn_2S_4$ has a great influence on the photocatalytic activity of the $CaIn_2S_4/ZnIn_2S_4$ composites, and 30% $CaIn_2S_4/ZnIn_2S_4$ exhibits the optimal photocatalytic performance for the Cr(VI) reduction.

A possible mechanism of the photogenerated charge transfer was proposed to illustrate the superior photocatalytic performance and photostability of the $CaIn_2S_4/ZnIn_2S_4$ composite catalysts. This study is of great importance in the design and synthesis of heterostructured sulfide composites with excellent photocatalytic performance and consistent stability toward the elimination of toxic metal ions in water.

Supplementary Materials: The following are available online at http://www.mdpi.com/2079-4991/8/7/472/s1, Figure S1: Concentrations of Cr(VI) and total Cr ions in the photocatalytic reaction solution over 30% CaIn2S4/ZnIn2S4 catalyst under simulated sunlight irradiation., Figure S2: XPS survey spectra (a), high-resolution XPS spectra of Zn 2p (b), In 3d (c), Ca 2p (d), and S 2p (e) of 30% CaIn2S4/ZnIn2S4 composite sample before and after the photocatalytic reaction, respectively.

Author Contributions: J.Y. (Juan Yang) proposed and planned the research and supervised the experiments. J.D. designed the synthesis of the photocatalysts. S.X. and J.Y. (Jun You) prepared the photocatalysts and performed the characterization and photocatalytic measurements. J.D. and J.H. improved the data analysis. All the authors collaborated and participated in writing of the manuscript.

Acknowledgments: This works has been supported by the National Nature Science Foundation of China (21307027), the Scientific and Technological Project of Henan Province (172102310725), the Funding Scheme for the Young Backbone Teachers of Higher Education Institutions in Henan Province (2015GGJS-071), the Fundamental Research Funds for the Universities of Henan Province (NSFRF170303), the Foundation for Distinguished Young Scientists (J2016-4), and the Innovative Research Team (T2018-2) of Henan Polytechnic University.

Conflicts of Interest: The authors declare no conflict of interest.

References

1. Yang, L.X.; Luo, S.L.; Li, Y.; Xiao, Y.; Kang, Q.; Cai, Q.Y. High efficient photocatalytic degradation of p-nitrophenol on a unique Cu_2O/TiO_2 p-n heterojunction network catalyst. *Environ. Sci. Technol.* **2010**, *44*, 7641–7646. [CrossRef] [PubMed]
2. Liu, Y.M.; Hou, C.L.; Jiao, T.F.; Song, J.W.; Zhang, X.; Xing, R.R.; Zhou, J.X.; Zhang, L.X.; Peng, Q.M. Self-assembled AgNP-containing nanocomposites constructed by electrospinning as efficient dye photocatalyst materials for wastewater treatment. *Nanomaterials* **2018**, *8*, 35. [CrossRef] [PubMed]
3. Wang, X.L.; Pehkonen, S.O.; Ray, A.K. Removal of aqueous Cr(VI) by a combination of photocatalytic reduction and coprecipitation. *Ind. Eng. Chem. Res.* **2004**, *43*, 1665–1672. [CrossRef]
4. Yin, H.B.; Wada, Y.; Kitamura, T.; Yanagida, S. Photoreductive dehalogenation of halogenated benzene derivatives using ZnS or CdS nanocrystallites as photocatalysts. *Environ. Sci. Technol.* **2001**, *35*, 227–231. [CrossRef] [PubMed]
5. Zhou, R.X.; Guzman, M.I. CO_2 reduction under periodic illumination of ZnS. *J. Phys. Chem. C* **2014**, *118*, 11649–11656. [CrossRef]
6. Zhou, R.X.; Guzman, M.I. Photocatalytic reduction of fumarate to succinate on ZnS mineral surfaces. *J. Phys. Chem. C* **2016**, *120*, 7349–7357. [CrossRef]
7. Baran, T.; Wojtyla, S.; Dibenedetto, A.; Aresta, M.; Macyk, W. Zinc sulfide functionalized with ruthenium nanoparticles for photocatalytic reduction of CO_2. *Appl. Catal. B Environ.* **2015**, *178*, 170–176. [CrossRef]
8. Wojtyla, S.; Baran, T. Insight on doped ZnS and its activity towards photocatalytic removing of Cr(VI) from wastewater in the presence of organic pollutants. *Mater. Chem. Phys.* **2018**, *212*, 103–112. [CrossRef]
9. Wang, W.J.; Ng, T.W.; Ho, W.K.; Huang, J.H.; Liang, S.J.; An, T.C.; Li, G.Y.; Yu, J.C.; Wong, P.K. $CdIn_2S_4$ microsphere as an efficient visible-light-driven photocatalyst for bacterial inactivation: Synthesis, characterizations and photocatalytic inactivation mechanisms. *Appl. Catal. B Environ.* **2013**, *129*, 482–490. [CrossRef]
10. Tu, X.L.; Lu, J.; Li, M.; Su, Y.J.; Yin, G.L.; He, D.N. Hierarchically $ZnIn_2S_4$ nanosheet-constructed microwire arrays: Template-free synthesis and excellent photocatalytic performances. *Nanoscale* **2018**, *10*, 4735–4744. [CrossRef] [PubMed]
11. Gou, X.L.; Cheng, F.Y.; Shi, Y.H.; Zhang, L.; Peng, S.J.; Chen, J.; Shen, P.W. Shape-controlled synthesis of ternary chalcogenide $ZnIn_2S_4$ and CuIn(S,Se)(2) nano-/microstructures via facile solution route. *J. Am. Chem. Soc.* **2006**, *128*, 7222–7229. [CrossRef] [PubMed]
12. Li, L.L.; Peng, S.J.; Wang, N.; Srinivasan, M.; Mhaisalkar, S.G.; Yan, Q.Y.; Ramakrishna, S. A general strategy toward carbon cloth-based hierarchical films constructed by porous nanosheets for superior photocatalytic activity. *Small* **2015**, *11*, 2429–2436. [CrossRef] [PubMed]

13. Yu, X.; Shavel, A.; An, X.Q.; Luo, Z.S.; Ibáñez, M.; Cabot, A. Cu_2ZnSnS_4-Pt and Cu_2ZnSnS_4-Au heterostructured nanoparticles for photocatalytic water splitting and pollutant degradation. *J. Am. Chem. Soc.* **2014**, *136*, 9236–9239. [CrossRef] [PubMed]
14. Gao, B.; Liu, L.F.; Liu, J.D.; Yang, F.L. Photocatalytic degradation of 2,4,6-tribromophenol over Fe-doped $ZnIn_2S_4$: Stable activity and enhanced debromination. *Appl. Catal. B Environ.* **2013**, *129*, 89–97. [CrossRef]
15. Guan, Z.J.; Xu, Z.Q.; Li, Q.Y.; Wang, P.; Li, G.Q.; Yang, J.J. $AgIn_5S_8$ nanoparticles anchored on 2D layered $ZnIn_2S_4$ to form OD/2D heterojunction for enhanced visible-light photocatalytic hydrogen evolution. *Appl. Catal. B Environ.* **2018**, *227*, 512–518. [CrossRef]
16. Yuan, Y.J.; Chen, D.Q.; Zhong, J.S.; Yang, L.X.; Wang, J.J.; Liu, M.J.; Tu, W.G.; Yu, Z.T.; Zou, Z.G. Interface engineering of a noble-metal-free 2D-2D MoS_2/Cu-$ZnIn_2S_4$ photocatalyst for enhanced photocatalytic H_2 production. *J. Mater. Chem. A* **2017**, *5*, 15771–15779. [CrossRef]
17. Tan, C.W.; Zhu, G.Q.; Hojamberdiev, M.; Lokesh, K.S.; Luo, X.C.; Jin, L.; Zhou, J.P.; Liu, P. Adsorption and enhanced photocatalytic activity of the {0001} faceted Sm-doped $ZnIn_2S_4$ microspheres. *J. Hazard. Mater.* **2014**, *278*, 572–583. [CrossRef] [PubMed]
18. Xu, F.Y.; Zhang, J.J.; Zhu, B.C.; Yu, J.G.; Xu, J.S. $CuInS_2$ sensitized TiO_2 hybrid nanofibers for improved photocatalytic CO_2 reduction. *Appl. Catal. B Environ.* **2018**, *230*, 194–202. [CrossRef]
19. Aguirre, M.E.; Zhou, R.X.; Eugene, A.J.; Guzman, M.I.; Grela, M.A. Cu_2O/TiO_2 heterostructures for CO_2 reduction through a direct Z-scheme: Protecting Cu_2O from photocorrosion. *Appl. Catal. B Environ.* **2017**, *217*, 485–493. [CrossRef]
20. Wu, Z.; Gong, C.; Yu, J.; Sun, L.; Xiao, W.; Lin, C. Enhanced visible light photoelectrocatalytic activity over $Cu_xZn_{1-x}In_2S_4$@TiO_2 nanotube array hetero-structures. *J. Mater. Chem. A* **2017**, *5*, 1292–1299. [CrossRef]
21. Hou, J.; Yang, C.; Cheng, H.; Wang, Z.; Jiao, S.; Zhu, H. Ternary 3D architectures of CdS QDs/graphene/$ZnIn_2S_4$ heterostructures for efficient photocatalytic H_2 production. *Phys. Chem. Chem. Phys.* **2013**, *15*, 15660–15668. [CrossRef] [PubMed]
22. Zhang, Z.; Liu, K.; Feng, Z.; Bao, Y.; Dong, B. Hierarchical sheet-on-sheet $ZnIn_2S_4$/g-C_3N_4 heterostructure with highly efficient photocatalytic H_2 production based on photoinduced interfacial charge transfer. *Sci. Rep.* **2016**, *6*, 19221. [CrossRef] [PubMed]
23. Huang, S.; Lin, Y.; Yang, J.H.; Yu, Y. CdS-based semiconductor photocatalysts for hydrogen production from water splitting under solar light. *ACS Symp. Ser.* **2013**, *1140*, 219–241. [CrossRef]
24. Chen, J.Y.; Zhang, H.M.; Liu, P.R.; Li, Y.B.; Liu, X.L.; Li, G.Y.; Wong, P.K.; An, T.C.; Zhao, H.J. Cross-linked $ZnIn_2S_4$/rGO composite photocatalyst for sunlight-driven photocatalytic degradation of 4-nitrophenol. *Appl. Catal. B Environ.* **2015**, *168–169*, 266–273. [CrossRef]
25. Yuan, W.H.; Yang, S.; Li, L. Synthesis of g-C_3N_4/$CaIn_2S_4$ composites with enhanced photocatalytic activity under visible light irradiation. *Dalton Trans.* **2015**, *44*, 16091–16098. [CrossRef] [PubMed]
26. Ding, J.J.; Yan, W.H.; Sun, S.; Bao, J.; Gao, C. Hydrothermal synthesis of $CaIn_2S_4$-reduced graphene oxide nanocomposites with increased photocatalytic performance. *ACS Appl. Mater. Interfance* **2014**, *6*, 12877–12884. [CrossRef] [PubMed]
27. Jo, W.K.; Natarajan, T.S. Facile synthesis of novel redox-mediator-free direct Z-Scheme $Caln_2S_4$ marigold-flower-like/TiO_2 photocatalysts with superior photocatalytic efficiency. *ACS Appl. Mater. Interfance* **2015**, *7*, 17138–17154. [CrossRef] [PubMed]
28. Testa, J.J.; Grela, M.A.; Litter, M.I. Heterogeneous photocatalytic reduction of chromium(VI) over TiO_2 particles in the presence of oxalate: Involvement of Cr(V) species. *Environ. Sci. Technol.* **2004**, *38*, 1589–1594. [CrossRef] [PubMed]
29. Fu, F.L.; Wang, Q. Removal of heavy metal ions from wastewaters: A review. *J. Environ. Manag.* **2011**, *92*, 407–418. [CrossRef] [PubMed]
30. Liu, F.H.; Yu, J.; Tu, G.Y.; Qu, L.; Xiao, J.C.; Liu, Y.D.; Wang, L.Z.; Lei, J.Y.; Zhang, J.L. Carbon nitride coupled Ti-SBA15 catalyst for visible-light-drivenphotocatalytic reduction of Cr (VI) and the synergistic oxidation of phenol. *Appl. Catal. B Environ.* **2017**, *201*, 1–11. [CrossRef]
31. Nayak, A.K.; Lee, S.; Choi, Y.I.; Yoon, H.J.; Sohn, Y.; Pradhan, D. Crystal phase and size-controlled synthesis of tungsten trioxide hydrate nanoplates at room temperature: Enhanced Cr(VI) photoreduction and methylene blue adsorption properties. *ACS Sustain. Chem. Eng.* **2017**, *5*, 2741–2750. [CrossRef]

32. Yang, J.; You, J.; Dai, J.; Chen, Y.M.; Li, Y. In-situ synthesis of hydrogen-titanate nanotubes/graphene composites with chemically bonded interface and enhanced visible photocatalytic activity. *Nanomaterials* **2018**, *8*, 229. [CrossRef] [PubMed]
33. Wang, L.; Zhang, C.B.; Gao, F.; Mailhot, G.; Pan, G. Algae decorated TiO_2/Ag hybrid nanofiber membrane with enhanced photocatalytic activity for Cr(VI) removal under visible light. *Chem. Eng. J.* **2017**, *314*, 622–630. [CrossRef]
34. Ye, L.; Fu, J.L.; Xu, Z.; Yuan, R.S.; Li, Z.H. Facile one-pot solvothermal method to synthesize sheet-on-sheet reduced graphene oxide (RGO)/$ZnIn_2S_4$ nanocomposites with superior photocatalytic performance. *ACS Appl. Mater. Interfaces* **2014**, *6*, 3483–3490. [CrossRef] [PubMed]
35. Ding, J.; Sun, S.; Yan, W.; Bao, J.; Gao, C. Photocatalytic H_2 evolution on a novel $CaIn_2S_4$ photocatalyst under visible light irradiation. *Int. J. Hydrogen Energy* **2013**, *38*, 13153–13158. [CrossRef]
36. Zhang, X.; Zhang, J.; Yu, J.Q.; Zhang, Y.; Cui, Z.X.; Sun, Y.; Hou, B.R. Fabrication of $InVO_4$/$AgVO_3$ heterojunctions with enhanced photocatalytic antifouling efficiency under visible-light. *Appl. Catal. B Environ.* **2018**, *220*, 57–66. [CrossRef]
37. Shi, L.; Yin, P.Q.; Dai, Y.M. Synthesis and photocatalytic performance of $ZnIn_2S_4$ nanotubes and nanowires. *Langmuir* **2013**, *29*, 12818–12822. [CrossRef] [PubMed]
38. Deng, Y.C.; Tang, L.; Zeng, G.M.; Zhu, Z.J.; Yan, M.; Zhou, Y.Y.; Wang, J.J.; Liu, Y.N.; Wang, J.J. Insight into highly efficient simultaneous photocatalytic removal of Cr(VI) and 2,4-diclorophenol under visible light irradiation by phosphorus doped porous ultrathin g-C_3N_4 nanosheets from aqueous media: Performance and reaction mechanism. *Appl. Catal. B Environ.* **2017**, *203*, 343–354. [CrossRef]
39. Liu, S.Q.; Yang, M.Q.; Tang, Z.R.; Xu, Y.J. A nanotree-like CdS/ZnO nanocomposite with spatially branched hierarchical structure for photocatalytic fine-chemical synthesis. *Nanoscale* **2014**, *6*, 7193–7198. [CrossRef] [PubMed]
40. Gilja, V.; Novakovic, K.; Travas-Sejdic, J.; Hrnjak-Murgic, Z.; Rokovic, M.K.; Zic, M. Stability and synergistic effect of polyaniline/TiO_2 photocatalysts in degradation of Azo dye in wastewater. *Nanomaterials* **2017**, *7*, 412. [CrossRef] [PubMed]
41. Wu, N.; Wang, J.; Tafen, D.N.; Wang, H.; Zheng, J.-G.; Lewis, J.P.; Liu, X.; Leonard, S.S.; Manivannan, A. Shape-enhanced photocatalytic activity of single-crystalline anatase TiO_2 (101) nanobelts. *J. Am. Chem. Soc.* **2010**, *132*, 6679–6685. [CrossRef] [PubMed]
42. Kumar, S.; Surendar, T.; Baruah, A.; Shanker, V. Synthesis of a novel and stable g-C_3N_4-Ag_3PO_4 hybrid nanocomposite photocatalyst and study of the photocatalytic activity under visible light irradiation. *J. Mater. Chem. A* **2013**, *1*, 5333–5340. [CrossRef]
43. Xiong, P.; Zhu, J.W.; Wang, X. Cadmium sulfide-ferrite nanocomposite as a magnetically recyclable photocatalyst with enhanced visible-light-driven photocatalytic activity and photostability. *Ind. Eng. Chem. Res.* **2013**, *52*, 17126–17133. [CrossRef]
44. Qiu, J.H.; Zhang, X.F.; Zhang, X.G.; Feng, Y.; Li, Y.X.; Yang, L.Y.; Lu, H.Q.; Yao, J.F. Constructing $Cd_{0.5}Zn_{0.5}S$@ZIF-8 nanocomposites through self-assembly strategy to enhance Cr(VI) photocatalytic reduction. *J. Hazard. Mater.* **2018**, *349*, 234–241. [CrossRef] [PubMed]
45. Yuan, Y.J.; Chen, D.Q.; Shi, X.F.; Tu, J.R.; Hu, B.; Yang, L.X.; Yu, Z.T.; Zou, Z.G. Facile fabrication of "green" SnS_2 quantum dots/reduced graphene oxide composites with enhanced photocatalytic performance. *Chem. Eng. J.* **2017**, *313*, 1438–1446. [CrossRef]

© 2018 by the authors. Licensee MDPI, Basel, Switzerland. This article is an open access article distributed under the terms and conditions of the Creative Commons Attribution (CC BY) license (http://creativecommons.org/licenses/by/4.0/).

nanomaterials

MDPI

Article

Morphology, Optical Properties and Photocatalytic Activity of Photo- and Plasma-Deposited Au and Au/Ag Core/Shell Nanoparticles on Titania Layers

Alexander Müller [1,2,†,‡], Sandra Peglow [3,†], Michael Karnahl [4,§], Angela Kruth [3], Henrik Junge [4], Volker Brüser [3] and Christina Scheu [1,*]

1 Max-Planck-Institut für Eisenforschung GmbH (MPIE), Max-Planck-Straße 1, 40237 Düsseldorf, Germany; amueller@lbl.gov
2 Department of Chemistry and Center for NanoScience (CeNS), University of Munich (LMU), Butenandtstrasse 5-13, 81377 Munich, Germany
3 Leibniz Institute for Plasma Science and Technology (INP), Felix-Hausdorff-Straße 2, 17489 Greifswald, Germany; sandra.peglow@inp-greifswald.de (S.P.); angela.kruth@inp-greifswald.de (A.K.); brueser@inp-greifswald.de (V.B.)
4 Leibniz Institute for Catalysis at the University of Rostock (LIKAT), Albert-Einstein-Straße 29a, 18059 Rostock, Germany; michael.karnahl@oc.uni-stuttgart.de (M.K.); henrik.junge@catalysis.de (H.J.)
* Correspondence: scheu@mpie.de; Tel.: +49-211-6792-720
† These authors contributed equally to this work.
‡ Current Address: National Center for Electron Microscopy, Molecular Foundry, Lawrence Berkeley National Laboratory, 1 Cyclotron Rd, Berkeley, CA 94720, USA.

Received: 4 June 2018; Accepted: 4 July 2018; Published: 6 July 2018

Abstract: Titania is a promising material for numerous photocatalytic reactions such as water splitting and the degradation of organic compounds (e.g., methanol, phenol). Its catalytic performance can be significantly increased by the addition of co-catalysts. In this study, Au and Au/Ag nanoparticles were deposited onto mesoporous titania thin films using photo-deposition (Au) and magnetron-sputtering (Au and Au/Ag). All samples underwent comprehensive structural characterization by grazing incidence X-ray diffraction (XRD), scanning electron microscopy (SEM), and transmission electron microscopy (TEM). Nanoparticle distributions and nanoparticle size distributions were correlated to the deposition methods. Light absorption measurements showed features related to diffuse scattering, the band gap of titania and the local surface plasmon resonance of the noble metal nanoparticles. Further, the photocatalytic activities were measured using methanol as a hole scavenger. All nanoparticle-decorated thin films showed significant performance increases in hydrogen evolution under UV illumination compared to pure titania, with an evolution rate of up to 372 $\mu L\ H_2\ h^{-1}\ cm^{-2}$ representing a promising approximately 12-fold increase compared to pure titania.

Keywords: noble metal nanoparticles; core-shell structures; photodeposition; magnetron sputtering; photocatalysis; hydrogen production; localized surface plasmon resonance; structure-property relationships

1. Introduction

For decades, mankind has relied heavily on unsustainable energy sources such as coal, oil, gas or nuclear power [1–3]. In recent years, concerns over climate change, environmental pollution, resource depletion, and safety issues have led to global efforts towards the development of sustainable and clean energy sources [1–5], with some of the most promising approaches making use of the almost unlimited energy of the sun [2,6–12]. One of these approaches is photocatalytic water splitting, where solar energy is used to photogenerate charge carriers in a semiconductor. The excited electrons then reduce protons, producing "green" hydrogen, which can be stored, transported, and finally converted back to energy using fuel cells. This approach therefore opens a promising route towards a carbon-neutral energy landscape.

Titania in the rutile phase was the first material found to have conduction and valence band positions suitable for photocatalytic water splitting [13–15]. It also has several other desirable properties such as high corrosion-resistance, abundance, low price, and non-toxicity [16]. Therefore, titania, both in the rutile and in the low-temperature anatase modification, is still one of the most studied materials for water splitting and other photocatalytic applications like the degradation of organic compounds [5,15,17–19]. Its biggest drawback is the comparatively large band gaps of 3.2 eV for anatase and of 3.0 eV for rutile, corresponding to absorption edges at wavelengths of 386 and 416 nm, respectively [13,16,17,20]. Accordingly, significant light absorption and photogeneration of charge carriers can only occur within or near the UV region, which accounts for merely 5% of the total energy of the solar spectrum [21,22]. This limits the theoretical maximum efficiency to 1.3% for anatase and 2.2% for rutile [17]. As the performance strongly depends on the band gap, it can be enhanced by either introducing additional electronic states into the band gap via doping and/or by depositing a second light-absorbing material that absorbs within the visible region and acts as a photosensitizer for titania [21,23,24]. Popular photosensitizers, that are also used in solar cells, are (metal) organic dyes [25–30]. They are, however, often unstable under UV radiation and in the chemical environment present during water splitting [25].

More suitable for this application are nanoparticles based on noble metals such as Au, Ag or Cu. In these metals, light can induce a localized collective electron oscillation, a so-called "plasmon", near the nanoparticle surface [22,23,31,32]. The energy of this surface plasmon strongly depends on the size, shape and dielectric constant of the nanoparticle as well as of the surrounding medium [22,23,32,33]. By optimizing these parameters, the wavelength of the localized surface plasmon resonance (LSPR) can be shifted to the desired range, allowing for visible or even near-infrared light absorption [22,23,32–35]. These plasmon-induced electrons can transfer to an electron-acceptor such as titania [24], increasing visible light-induced charge separation and enhancing the efficiency of photoreactions. If the LSPR frequency is at or near the semiconductor band gap, the electromagnetic field stemming from the plasmon excitation can lead to charge carrier formation in the supporting semiconductor [36–38]. Under these conditions, the large scattering cross section of the plasmon oscillation in noble metal nanoparticles enhances the optical pathway of the incident photons leading to increased light absorption [39]. In addition to their function as photosensitizer, noble metal nanoparticles can enhance the performance of the semiconductor as co-catalysts by providing chemically active sites with low activation barriers [40], prolonging charge carrier lifetime [40], and serve as a reservoir for electrons generated in the titania by UV light [24,41–43].

Several studies on photosensitizing titania using noble metal nanoparticles were carried out on suspended powders [40,42,44–48]. In contrast, we deposited Au-based nanoparticles onto anatase thin films. Compared to powdered samples, immobilizing the catalyst allows for precise control over morphology, light absorption and catalytic behavior, and therefore for high reproducibility of the measured photocatalytic performance [49,50]. The samples are also easier to handle, and agglomeration of particles is not an issue. The performance typically depends greatly on the morphology of the sample, and studying the correlation of synthesis conditions, morphology and properties is of paramount importance. In this study, two different methods were used to deposit metallic nanoparticles. First, gold nanoparticles were prepared by in situ photodeposition, which is a widely established approach that works well at the laboratory scale [51]. Second, both gold and silver/gold nanoparticles, which often show significantly increased catalytic activity compared to the respective monometallic nanoparticles [52,53], were deposited by a radio frequency (RF) magnetron sputtering process followed by an annealing step [54,55]. In comparison to photodeposition, physical deposition processes such as this one can often be scaled up to an industrial scale. Care was taken that the noble-metal loading of the two Au-TiO_2 samples was comparable. The crystal structures of all phases were confirmed via grazing incidence X-ray diffraction, and the morphologies of the samples were analyzed using scanning and transmission electron microscopy. As expected, both synthesis methods resulted in considerably different particle distributions, particle size distributions and defect structures. We further measured the UV-vis spectra and the photocatalytic evolution of hydrogen from water using the sacrificial electron donor methanol. Methanol was chosen as past studies had shown it to be an excellent hole scavenger, allowing us to neglect surface kinetics in the discussion and simplifying the complex behavior of this system [56–58]. All results were correlated to each other, and we hope that our combined findings contribute to an in-depth understanding of the interplay between synthesis, structure, and properties.

2. Materials and Methods

Titania films were deposited onto fluorine-doped tin oxide substrates (fluorine doped tin oxide (FTO), TCO 22-7, Solaronix, 25 × 25 mm, Aubonne, Switzerland) by a direct current (DC) reactive magnetron sputtering process previously described by Kruth et al. [54]. The cylindrical Ti target (Ti-133, Bekaert Advanced Coatings NV, Deinze, Belgium, 135 mm diameter, 58.5 mm length) was sputter-cleaned in an Ar atmosphere at 8 kW for 5 min. After stabilizing the process conditions in an $O_2/N_2/Ar$ atmosphere (6 standard cm^3/min (sccm) O_2, 3 sccm N_2, and 60 sccm Ar) at 3 Pa for 8 min, TiO_2 was plasma-deposited at a magnetron power of about 5.3 kW and a magnetron voltage of 450 V. To transform the resulting amorphous TiO_2 into anatase, the samples were annealed for 1 h at 400 °C with a heating rate of 10 °C/min in an oxygen atmosphere at a flow rate of 0.05 standard L/min (slm).

Au and Au/Ag core-shell nanoparticles were deposited onto the titania films described above using a RF-magnetron sputtering process previously published by Peglow et al. [55]. Au and Ag sputtering targets (both of 3 mm thickness, two inch diameter and purity of 99.999%, MaTeck, Juelich, Germany, were placed at respective distances of 9.5 and 5.5 cm from the substrate. Small sputtering rates were achieved by shielding the magnetic field with a 1 mm thick iron disk (99.95%, MaTeck) placed between the magnetron and the two targets. The deposition was performed at a magnetron power of 50 W at a working pressure of 5 Pa in an argon atmosphere (15 sccm gas flow). After each deposition, the samples were annealed by placing them in a quartz tube that was inserted into a tube furnace (Zirox GmbH, Greifswald, Germany, kept at 400 °C by a thermal controller (Eurotherm 2416, Limburg an der Lahn, Germany) for 30 min. The O_2 atmosphere (0.05 slm) was regulated using a gas flow controller (MKS Instruments Multi Gas Controller 647B, Andover, MA, USA). Au nanoparticles were synthesized by depositing gold over a period of 300 s, resulting in a nominal layer thickness of (6.6 ± 0.7) nm. Deposition was followed by an annealing step, a second, 300 s long deposition step and a final annealing step leading to an estimated total layer thickness of (13.2 ± 1.5) nm. To obtain Au/Ag core-shell nanoparticles, Au-deposition for 188 s resulted in a nominal layer thickness of (4.1 ± 0.5) nm and was followed by Ag-deposition over 36 s, resulting in a nominal layer thickness of (2.5 ± 0.1) nm, and one final annealing step.

A second series of Au nanoparticles was prepared by in situ photodeposition onto titania films following a synthesis procedure described by Gärtner et al. [40]. The temperature of a double-walled reaction vessel was adjusted to 25 °C by a thermostat. The titania film (25 × 25 mm) was inserted into the reactor with a glass holder. Subsequently, the gold precursor ($NaAuCl_4 \cdot 2\,H_2O$, 3.1 mg) was added. The whole system was evacuated and flushed with argon to remove any other gases. Then, 40 mL freshly distilled water and 40 mL methanol were added under argon counter flow, resulting in a final concentration of the gold precursor of about 0.1 mmol/L. The photodeposition was initiated with a Hg-lamp (7.2 W output, Lumatec Superlite 400, Deisenhofen, Germany) equipped with a 320–500 nm filter [40]. A swift color change from light yellow to dark red occurred, with the formed hydrogen escaping by a bubbler. The reaction was stopped after 3 h and the sample was washed with deionized water and ethanol prior to drying in air.

Phase identity and average crystallite sizes were determined by grazing incidence X-ray diffractometry (GIXRD). Diffractograms were obtained using a Bruker D8 Advance (Billerica, MA, USA) with a Cu-Kα source. The measurement was carried out in a 2θ-range of 20°–80° at an incident angle of 0.5°, with a step width of 0.02° and a measurement time of 5 s per step. Crystallite sizes were calculated from the (200) reflection of Au as well as the (200) and (101) reflections of anatase using a combination of the Stokes–Wilson and the Variance model and fitting the correlated integral widths by a Pearson VII function [59].

Scanning electron microscopy (SEM) was performed on a JEOL JSM 7500F (Tokyo, Japan) with a field emission gun, a semi-in-lens conical objective lens, and a secondary electron in-lens detector. At an acceleration voltage of 15 keV, a resolution of 1.0 nm was achieved.

A comprehensive structural analysis was carried out using a FEI Titan 80-300 transmission electron microscope (TEM, Hillosboro, OR, USA). Bright-field (BF) and high-resolution TEM (HRTEM) images were recorded on a Gatan UltraScan 1000 CCD (Pleasanton, CA, USA), scanning TEM (STEM) images with a Fischione Model 3000 high angle annular dark-field (HAADF) detector (Export, PA, USA) and energy-dispersive X-ray (EDX) spectra with an EDAX detector (Mahwah, NJ, USA). Samples were prepared by either scratching material of the substrate and depositing it onto a TEM grid with a holey carbon film or by preparing a cross-section according to a procedure adapted from Strecker et al. so that the sample was prepared at room temperature [60].

The optical properties of the different samples were investigated using a PerkinElmer Lambda UV-vis 850 spectrophotometer with a L6020322 150 mm integrating sphere and a Spectralon Reflectance Standard (>99% R, USRS-99-020, PerkinElmer Inc., Waltham, MA, USA). The UV-vis spectra were recorded by measuring the diffuse transmission at wavelengths from 250 nm to 850 nm. Calculation of the absorbance A was carried out under the assumption that no reflection occurs at the sample using Equation (1) [54,61],

$$A = -\log 10(I_T/I_0), \quad (1)$$

where A is the absorbance in arbitrary units, I_T is the measured transmission intensity in percent and I_0 is the incident light intensity, which equals 100%.

Photocatalytic hydrogen evolution experiments were performed under argon atmosphere and the strict exclusion of oxygen using freshly distilled and degassed solvents. The sample was introduced into a double-walled, thermostatically-controlled reaction vessel by a glass holder and aligned in parallel to the planar optical window. This setup allowed for a reproducible experimental arrangement and a direct illumination of the sample without blocking by the cooling water. Furthermore, a complete irradiation of the 25 × 25 mm thin film layer was ensured. Subsequently, the photoreactor was connected to an automatic gas burette and repeatedly evacuated and filled with argon in order to exclude any oxygen. Then, the solvent mixture (80 mL), composed of water and methanol in a ratio of 1/1 (v/v), was added, fully covering the layer. The temperature of the whole system was maintained at 25 °C by a thermostat. After stirring for at least 10 min at 300 rounds per minute to reach thermal equilibrium, the reaction was started by switching on a Hg-lamp (Lumatec Superlite 400, Deisenhofen, Germany) equipped with either a 320–500 nm or a 400–700 nm filter. In both cases, the light intensity was set to 7.2 W. The amount of evolved gases was continuously monitored by the automatic gas burette, while the gas composition was analyzed by gas chromatography. A more detailed description of the experimental setup can be found in the literature [40].

3. Results and Discussion

3.1. Structural and Morphological Characterization

GIXR diffractograms (Figure 1a) indicate the phase identities of all samples [62–64]. Annealing of the titania films led to crystallization of the material in the anatase modification, with an average crystallite size of approximately 25 nm. This is in agreement with earlier results [27,54]. Monometallic Au and bimetallic Au/Ag nanoparticles crystallize in the face-centered cubic Fm-3m symmetry (space group 225), with the lattice parameters of Ag and Au being too similar to differentiate between the two phases. The (200) peaks of Au and Au/Ag did not show any overlap with those of other phases and were used for further analysis (Figure 1b). The intensities of the (200) peaks of Au in photodeposited and plasma-deposited Au-TiO_2 are similar, indicating similar metal loading. In contrast, a weaker peak is observed for plasma-deposited Au/Ag-TiO_2, explicable by the lower metal loading. Calculated average crystallite sizes of all three samples were comparable at around 6–7 nm, with no drop for the bimetallic sample with significantly lower metal loading. It should, however, be considered that crystallite size calculations from XRD are limited to average values for crystalline domains and only assume the presence of ideal, spherical crystallites.

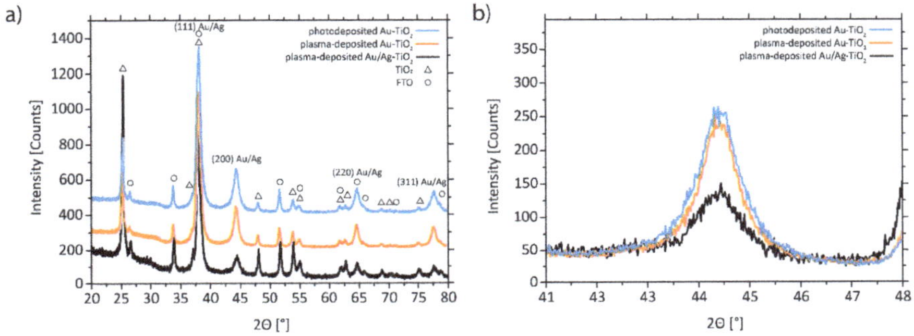

Figure 1. (**a**) Grazing incidence X-ray (GIXR) diffractograms of the three noble metal-decorated samples and (**b**) close-up of the Au/Ag (200) peaks. The intensities of the curves in (**a**) were shifted vertically by constant factors, as they would overlap otherwise.

In Figure 2, electron microscopy images of the three samples are shown. Top-view images were acquired by scanning electron microscope (SEM) (Figure 2a–c), cross-sections by HAADF-STEM (high angle annular dark-field cross-section scanning transmission electron microscopy) (Figure 2d–f).

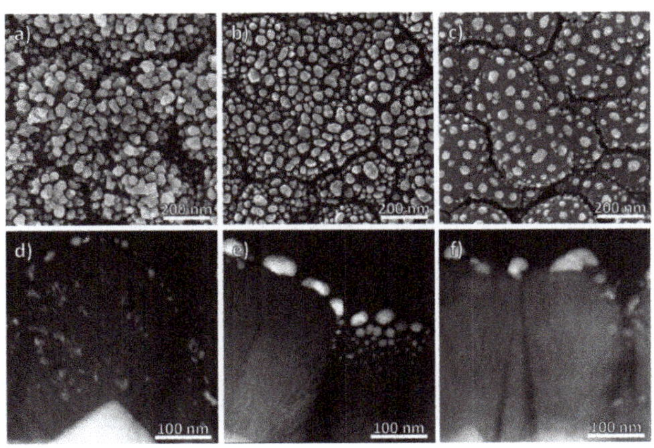

Figure 2. Top-view SEM (scanning electron microscopy) and cross-section scanning transmission electron microscopy (STEM) images of photodeposited Au-TiO_2 (**a**,**d**), plasma-deposited Au-TiO_2 (**b**,**e**) and plasma-deposited Au/Ag-TiO_2 films (**c**,**f**).

The titania layer is polycrystalline, approximately 300 nm thick and composed of individual pillars, each of them grown on top of a FTO pyramid. This microstructure is typical for ZI thin film growth [65,66]. The fibrous titania pillars are in the anatase modification, with 3–5 nm wide pores elongated in the direction perpendicular to the TiO_2/FTO interface. The porosity of the titania was quantified from HAADF-STEM images. The signal intensity I in such images scales with the mean atomic number Z raised by an exponent y (Equation (2)) [67]. As the FTO and the underlying SiO_2 substrate are compact layers, y can be calculated. This calculation, as well as those following, were done using several hundred acquisition points for each material and working with average values. The mean atomic number of the titania layer is given by:

$$\frac{I_{Titania}}{I_{FTO}} = \left(\frac{Z_{Titania}}{Z_{FTO}}\right)^{y} \rightarrow Z_{Titania} = Z_{FTO} \cdot \sqrt[y]{\frac{I_{Titania}}{I_{FTO}}}, \quad (2)$$

The porosity is then equal to the ratio of the mean atomic numbers of the measured, porous and the theoretical, compact layer and was determined as ~10%, indicating low porosity.

The photodeposited Au nanoparticles are found both on top of the columns and incorporated into pores of the TiO_2 layer. The latter indicates that some of the pores are open at the surface and can be filled with the gold precursor solution during the photodeposition process. In contrast, RF-sputtered Au and Au/Ag nanoparticles are found on top of the titania layer which is typical for such sputter deposition processes. The particles located in the cavities are significantly smaller than the grains formed on top of the columns (Figure 2). Compared to plasma-deposited Au-TiO_2, approximately half of the nominal layer thickness was deposited during the synthesis of plasma-deposited Au/Ag-TiO_2 (Figure 2b,c). This reduction leads to a sparser distribution of nanoparticles of roughly the same size. The half-as-high loading was also confirmed by EDX measurements (Table 1). The noble metal content could be determined by calculating the mass of the TiO_2 layer from the thickness and the density and comparing it with the ratio of noble metals to Ti. In contrast to Au/Ag-TiO_2, the Au-loading in photodeposited Au-TiO_2 and plasma-deposited Au-TiO_2 is very similar.

Table 1. Noble metal content of the three samples.

Sample	Au-Content (μg/cm²)	Ag-Content (μg/cm²)
photodeposited Au	17.5 ± 3.7	-
plasma-deposited Au	19.6 ± 6.5	-
plasma-deposited Au/Ag	7.8 ± 3.4	1.4 ± 0.8

In Figure 3, the size distributions of all three samples are shown. As already mentioned, photodeposited Au nanoparticles grew both inside and on top of the titania layer, which is reflected by the two different log-normal size distributions used to describe the experimentally determined size distribution. Photodeposited Au nanoparticles inside the titania layer, which account for 66% of all Au nanoparticles, have a different size distribution than those found on top of the titania layer (Figure 3a). However, a log-normal distribution, which has previously been applied to the size distributions of Au nanoparticles synthesized by several, solution-based synthesis procedures, could be used to describe both [68]. The size distribution of Au nanoparticles inside the titania is shifted towards smaller diameters, indicating that the growth is slowed down or stopped within the pores of the TiO_2 layer.

To interpret the size distributions of plasma-deposited particles (Figure 3b,c), two underlying processes, deposition and annealing, have to be considered. Previous studies have shown that sputter deposition produces thin films, which dewet during annealing [55,69,70]. Some of these isolated particles then grow via a coarsening mechanism. This coarsening step is expected to depend strongly on a low surface roughness to prevent particle pinning and facilitate particle diffusion. We used these assumptions to split each size distribution in two by considering large nanoparticles on top of smooth TiO_2 surfaces as resulting from a coarsening mechanism. With this assumption, the non-coarsened particles, which account for 86% in plasma-deposited Au-TiO_2 and 83% in plasma-deposited Au/Ag-TiO_2, can be fit very well to a log-normal distribution. Attempts to model the other particle fraction with a size distribution failed due to their relative scarcity. Compared to pure plasma-deposited Au nanoparticles, the maximum of the log-normal distribution of the smaller, non-coarsened Au/Ag nanoparticles is shifted from 4 to 10 nm (Figure 3b,c). In spite of the reduced nominal layer thickness, the increase in size indicates that the initial deposition of Au directs their size, and not the subsequent Ag deposition or the annealing step. This hints at the observed shift resulting from Ag being added to a pre-existing Au nanostructure. Of course, this argument only applies to non-coarsened, small nanoparticles.

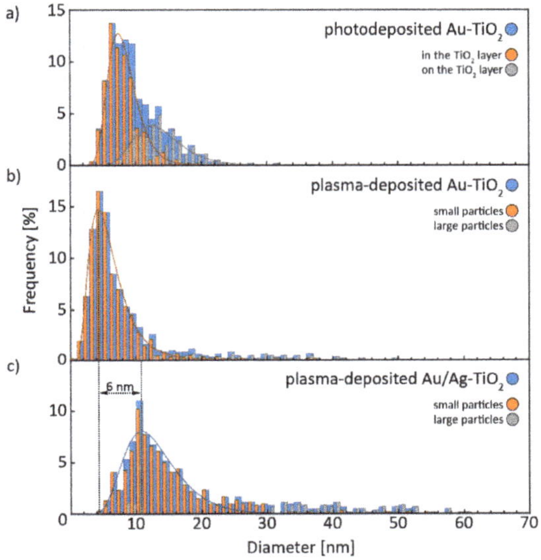

Figure 3. Size distributions of the noble metal nanoparticles in (**a**) photodeposited Au-TiO_2, (**b**) plasma-deposited Au-TiO_2 and (**c**) plasma-deposited Ag/Au-TiO_2. All size distributions were split into two sub-distributions each. As explained in the main text, "large particles" refers to particles we believe have coarsened, "small particles" to those that have not. Please note that the frequency values only apply to the size distribution of the whole sample, but not to the sub-distributions.

The nanoparticles can possess several different defect structures (Figure 4). The photodeposited Au nanoparticles can be inside and outside of the titania layer, with each fraction having its own predominant defect structure. Photodeposited nanoparticles on top of the titania are predominantly five-fold twinned (Figure 4a), with few occurrences of other defect structures such as grain boundaries. Such a twinning is energetically favorable for small nanoparticles and therefore very common [71–75]. In contrast, all particles observed within the titania layer were monocrystalline (Figure 4b). However, a definite correlation of nanoparticle size and defect structure could not be concluded. The existence of defects not inherent to the metal or of an oxide surface layer large enough to form a defined crystal structure could be excluded from HRTEM images.

For the plasma-deposited nanoparticles, single-crystallinity, five-fold twinning, stacking faults, and grain boundaries were observed (Figure 4d,e). As with photodeposited nanoparticles, we could not conclude a correlation of size and defect structure, with the exception of grain boundaries, which were very common in big nanoparticles. We tentatively ascribe these to the coarsening process. These particles also often have little protrusions that fill nooks in the titania substrate. Once again, other defects can be excluded from HRTEM images.

Bimetallic Au/Ag nanoparticles could potentially be alloyed or form core-shell nanoparticles [76,77]. The melting temperatures of Ag and Au decreases with the nanoparticle size, but are always high compared to the highest temperature reached during synthesis (400 °C) [78–80]. Alloying therefore seems unlikely. In accordance, EDX maps confirmed the formation of a uniform, 2–3 nm thick Ag shell around the Au core. As both metals crystallize in the face-centered cubic structure and their lattice parameters differ by only 0.2%, we observe defect-free continuation of the crystal structure of Au by Ag without any phase boundary (Figure 5c) [81]. Previous studies have shown inhomogeneous deposition of gold and silver, and the core-shell nanoparticles presumably result from nanoparticle attachment during the annealing treatment [82,83]. Interestingly, former experiments by one of the co-authors, in which the Ag/Au deposition order was reversed, also yielded Au/Ag-core-shell nanoparticles [55]. The deposition order can therefore not be the decisive factor when determining which metal becomes the core and which the shell. Unfortunately, growth mechanism studies so far mostly focus on wet-chemical synthesis methods and do not apply to our synthesis method [82,83]. Looking at the thermodynamics of the two possible core-shell configurations, four different energies contribute to the total energy: those of bulk Au and bulk Ag, the interface energy of the interface between Ag and Au, and the surface energy of the shell material. Assuming the amount of bulk material is the same in both possible configuration, the volumes and therefore the enthalpies associated with the interface and the two bulk phases are identical for both and only the contribution of the surface changes when exchanging core and shell material. As the surface energy of Au is approximately 40% higher than that of Ag, the total enthalpy is reduced by forming a silver instead of a gold shell. We assume this to be the dominant driving force for the preferred creation of Au/Ag core-shell nanoparticles over Ag/Au core-shell nanoparticles [84].

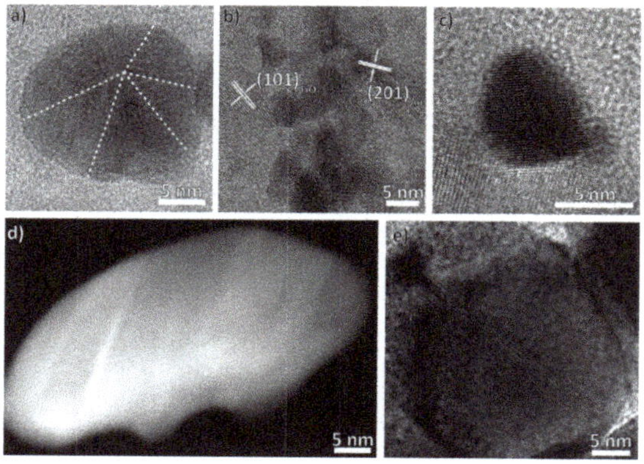

Figure 4. Representative images of different defect structures of the noble metal nanoparticles. In (**a**) and (**b**) high-resolution TEM (HRTEM) images of the photodeposited Au nanoparticles are shown: (**a**) is a five-fold twinned particle on top of the titania and (**b**) defect-free nanoparticles inside the titania. (**c–e**) show plasma-deposited nanoparticles: (**c**) is representative for small, defect free nanoparticles, (**d**) of those with stacking faults and (**e**) of those with grain boundaries.

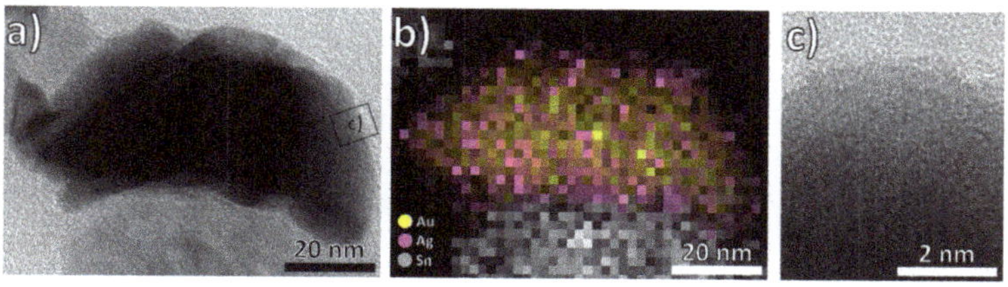

Figure 5. (**a**) BF (Bright-field) images of a representative Au/Ag nanoparticle; (**b**) energy dispersive X-ray (EDX) map of the same particle, showing an accumulation of Ag at the surface; (**c**) HRTEM image of the nanoparticle surface.

3.2. Optical Properties

Figure 6 shows the absorbance spectra of the Au, Au/Ag and pure TiO_2 samples calculated from diffuse transmission measurements using Equation (1). The strong absorbance at wavelengths shorter than approximately 400 nm corresponds to the anatase band gap of 3.2 eV [13,16,17,20]. Throughout the measurement range, the absorbance of the samples never reach an intensity of zero, which is attributed to Rayleigh scattering at crystalline domains as well as pores of the TiO_2 layer, with the noble metal nanoparticles acting as additional scattering sites [85–88]. Consequently, the intensities of this background absorption can be correlated to nanoparticle loading and distribution of the different samples. Nanoparticle-decoration leads to stronger absorbance than pure TiO_2, with photodeposited Au-TiO_2 reaching the highest background absorbance, followed by plasma deposited Au-TiO_2 and then plasma-deposited Au/Ag-TiO_2. The total noble metal mass in Au/Ag-TiO_2 is roughly half that of both Au-TiO_2 samples (Table 1), and it consequently absorbs less than those two. The high absorbance of photodeposited Au-TiO_2 could be attributed to the dense coverage of the titania surface (Figure 2a) and the additional presence of nanoparticles embedded in TiO_2 pores (Figure 2d).

Both plasma- and photodeposited Au-TiO_2 have additional absorption bands at 550–800 nm and 480–580 nm, respectively, caused by the excitation of localized surface plasmon resonances (LSPR). The positions and shapes of these bands are determined by the particle shape [89], the contact area with the titania [90], the size [91], and the size distribution [92]. An overlap of these factors hinders the interpretation of polydispersed particle ensembles. The center of the absorption band of photodeposited Au-TiO_2 matches the LSPR frequency of 520 nm previously described for isolated spherical gold nanoparticles [93,94]. TEM images confirm the existence of such particles (Figure 4a), however, many irregularly shaped nanoparticles with diameters more than 100 nm are found in top view SEM micrographs (Figure 2a). The intense band of plasma-deposited Au-TiO_2, however, is red-shifted. This could be explained by the non-spherical shape of the large Au nanoparticles [39,94–96], plasmon-coupling [22,94,95,97,98], and/or a large contact area with the TiO_2 [97] as suggested by SEM and TEM micrographs (Figures 2 and 4). Although embedded nanoparticles were not found in plasma-deposited samples, the particles adapt to the titania surface (Figures 4 and 5). The band broadening could result from large nanoparticle size and/or shape distributions. No distinct bands were observed for plasma-deposited Au/Ag-TiO_2, even though two would be expected: one stemming from the outer shell surface, and one from the Au/Ag interface [99–101]. A possible explanation is the significantly lower metal loading than the two Au-TiO_2 samples, which should lead to weaker plasmon bands. Furthermore, the peak related to the outer silver shell (which would be expected at around 400 nm) could overlap with the absorbance band of the TiO_2 substrate [99].

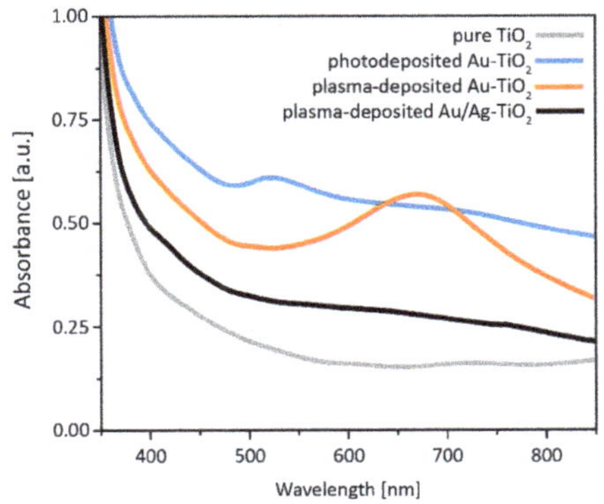

Figure 6. UV-vis spectra showing the absorbance of all samples.

3.3. Catalytic Properties

The photocatalytic performance of the different layers to reduce protons to molecular hydrogen was studied under visible light illumination (400–700 nm) and under UV-vis illumination (320–500 nm) with methanol as a hole scavenger (Figure 7, Table 2).

All hydrogen evolution curves (Figure 7) show a strong initial increase during the first minutes, followed by a smaller, but constant hydrogen evolution rate. The steep rise at the beginning of each measurement is caused by an increase in pressure in the automatic gas burette due to heating of the photoreactor upon irradiation with UV-vis light. Within an hour, thermal equilibrium is reached by external cooling with a thermostat (25 °C). Therefore, the later, constant region is more representative of the catalytic activity and the first hour of each measurement was disregarded when discussing the curves (Figure 7) or determining the hydrogen evolution rates presented in Table 2. No further loss in activity could be observed for up to 18 h, indicating stable operation.

Under visible light illumination, the hydrogen evolution rate was zero and no hydrogen was measured apart from the initial rise (Figure 7). In this regime, light is absorbed in the noble metal nanoparticles via surface plasmon resonance, and then electrons are injected into TiO_2 [42,44]. Given this mechanism and the UV-vis spectra (Figure 6), which show only weak bands attributed to plasmon formation, the weak hydrogen evolution can be explained.

Under UV illumination, only TiO_2 absorbs light and the noble metal nanoparticles act as co-catalysts (Figure 6) [40–42,44,45] by providing chemically active sites with low activation barriers [40], prolonging charge carrier lifetime [40], and by serving as an electron reservoir [24,41–43]. Previous photocatalytic experiments under similar conditions revealed a drastic increase in H_2 evolution upon deposition of noble metal nanoparticles [40–42,44,45]. This is confirmed in our experiments, with all samples having a significantly increased hydrogen evolution rate compared to pure TiO_2.

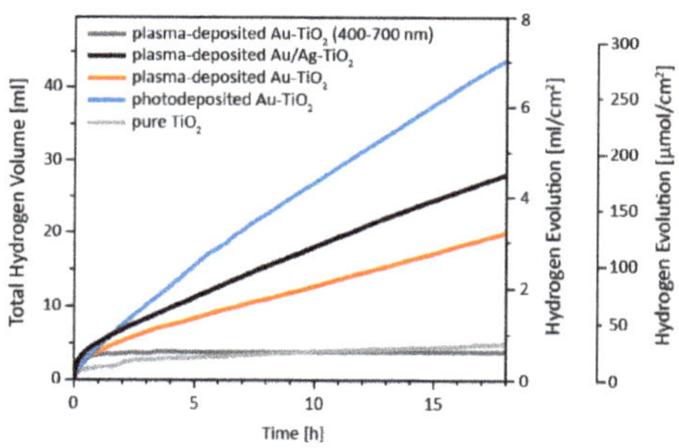

Figure 7. Hydrogen evolution of all samples from a methanol/water mixture. Conditions: 80 mL $MeOH/H_2O$ (1/1, *v*/*v*), Lumatec Hg-light source equipped with a 320–500 nm filter, 7.2 W output, 25 °C. In addition, the H_2 evolution of plasma-deposited Au-TiO_2, acquired using a 400–700 nm filter, is plotted exemplarily (dark grey). Measurements of all samples under the same visible light illumination yielded perfectly overlapping curves and we therefore chose to present only the measurement of the sample with the strongest localized surface plasmon resonances (LSPR) band.

Table 2. Comparison of the produced amount of hydrogen and the respective photocatalytic activities normalized to the evolution of hydrogen per area and time.

Sample	Total H_2 Volume (mL)	H_2 evolution Rate [a] ($\mu L\ h^{-1}\ cm^{-2}$)	H_2 evolution Rate [b] ($\mu mol\ h^{-1}\ cm^{-2}$)
pure TiO_2	5.0	31	1.3
photodeposited Au-TiO_2	44.1	372	15.2
plasma-deposited Au-TiO_2	20.0	152	6.2
plasma-deposited Ag/Au-TiO_2	28.2	217	8.9

[a] calculated as an average over 17 h, starting after the first hour. [b] calculated by applying the van der Waals equation at 25 °C, which yields a molar volume of 24.48 µL/µmol.

As mentioned above, all samples onto which noble metal nanoparticles were deposited outperform pure TiO_2. The largest hydrogen evolution rate of 372 μL h^{-1} cm^{-2} was measured for photodeposited Au-TiO_2, representing a 12-fold increase. This was followed by plasma-deposited Au/Ag-TiO_2 (217 μL h^{-1} cm^{-2}, 7-fold) and plasma-deposited Au-TiO_2 (152 μL h^{-1} cm^{-2}, 5-fold).

To interpret this order, we first consider the influence of the hole scavenger methanol. Two reactions compete at the surface of the noble metal nanoparticle co-catalysts: the transfer of holes to the electrolyte and the recombination with photogenerated electrons. Adding hole scavengers leads to the transfer reaction being favored over recombination, and the latter is suppressed. Methanol is a very efficient hole scavenger, near-perfect transfer can be assumed [56–58], and the kinetics of surface reactions can be neglected. In contrast, light absorption plays an important role and the order in which the samples perform is indeed closely correlated to the amount of light being absorbed in the UV-region. Only plasma-deposited Au/Ag-TiO_2 is more efficient than expected from light absorption measurements. We attribute this behavior to a reduction of bulk recombination caused by the electric field gradient at the Au/Ag interface, which is known to positively affect the efficiencies of catalytic reactions [52,53].

4. Conclusions

Au nanoparticles were deposited onto a mesoporous anatase thin film using two different deposition methods, photo- and plasma-deposition. The second method was further used to prepare bimetallic Au/Ag nanoparticles. Both methods resulted in different particle distributions and particle size distributions. Photodeposited nanoparticles both infiltrated the titania itself and were deposited on top of it, with the latter group of particles being larger on average. In contrast, plasma-deposited nanoparticles were found only at the surface of the titania.

These differences in the morphology could then be correlated to several materials properties. Strong light absorption up to about 400 nm is due to the band gap of titania. Above 400 nm, Rayleigh scattering at crystalline domain boundaries and pores within the TiO_2 layer, as well as at noble metal nanoparticles, leads to a strong background. Both samples with pure Au nanoparticles further show peaks attributed to LSPR.

The light absorption properties, in turn, strongly influence the photocatalytic performance of the hydrogen evolution reaction from a methanol/water solution under UV illumination. In this wavelength range, Au only acts as a co-catalyst and the performance of all samples but one, plasma-deposited Au/Ag-TiO_2, was correlated to the light absorption efficiency. The higher-than-expected performance of plasma-deposited Au/Ag-TiO_2 can be explained by an electric field gradient at the Au/Ag interface, concluding a comprehensive correlation of structure and properties. Consequently, this study shows that a systematic investigation of the interplay of synthesis method, structure and catalytic activity plays an important role in furthering our understanding of such complex systems. We believe that similar studies, along with those on different hole scavenger, surface reaction kinetics, thermal stability etc., will eventually lead to commercial applications and thereby contribute to a sustainable energy mix.

Author Contributions: Conceptualization, All authors; Methodology, All authors; Investigation, A.M., S.P., M.K., A.K.; Writing-Original Draft Preparation, A.M., S.P.; Writing-Review & Editing, All authors; Visualization, A.M., S.P.; Supervision, C.S., V.B., H.J.; Funding Acquisition, C.S., V.B., H.J.

Funding: This research was funded by the German Research Foundation (Deutsche Forschungsgesellschaft, DFG) via the Schwepunktsprogramm (SPP) 1613.

Acknowledgments: We are grateful to Markus Döblinger and Steffen Schmidt for support with TEM measurements, Christoph Heinzl for additional SEM measurements, Harm Wulff for in-depth analysis of the GIXRD measurements and Petra Bartels for performing photocatalytic experiments.

Conflicts of Interest: The authors declare no conflict of interest. The funders had no role in the design of the study; in the collection, analyses, or interpretation of data; in the writing of the manuscript, and in the decision to publish the results.

References

1. Armaroli, N.; Balzani, V. The future of energy supply: Challenges and opportunities. *Angew. Chem. Int. Ed.* **2007**, *46*, 52–66. [CrossRef] [PubMed]
2. Schiermeier, Q.; Tollefson, J.; Scully, T.; Witze, A.; Morton, O. Electricity without carbon. *Nature* **2008**, *5*, 10–15. [CrossRef]
3. Lewis, N.S.; Nocera, D.G. Powering the planet: Chemical challenges in solar energy utilization. *Proc. Natl. Acad. Sci. USA* **2006**, *103*, 15729–15735. [CrossRef] [PubMed]
4. Styring, S. Artificial photosynthesis for solar fuels. *Faraday Discuss.* **2012**, *155*, 357–376. [CrossRef] [PubMed]
5. Concepcion, J.J.; House, R.L.; Papanikolas, J.M.; Meyer, T.J. Chemical approaches to artificial photosynthesis. *Proc. Natl. Acad. Sci. USA* **2012**, *109*, 15560–15564. [CrossRef] [PubMed]
6. Grätzel, M. Photoelectrochemical cells. *Nature* **2001**, *414*, 338–344. [CrossRef] [PubMed]
7. Choi, J.; Amaranatha Reddy, D.; Han, N.S.; Jeong, S.; Hong, S.; Praveen Kumar, D.; Song, J.K.; Kim, T.K. Modulation of charge carrier pathways in CdS nanospheres by integrating MoS_2 and Ni_2P for improved migration and separation toward enhanced photocatalytic hydrogen evolution. *Catal. Sci. Technol.* **2017**, *7*, 641–649. [CrossRef]
8. Reddy, D.A.; Park, H.; Gopannagari, M.; Kim, E.H.; Lee, S.; Kumar, D.P.; Kim, T.K. Designing CdS Mesoporous Networks on Co-C@Co_9S_8 Double-Shelled Nanocages as Redox-Mediator-Free Z-Scheme Photocatalyst. *ChemSusChem* **2018**, *11*, 245–253. [CrossRef] [PubMed]
9. Reddy, D.A.; Park, H.; Ma, R.; Kumar, D.P.; Lim, M.; Kim, T.K. Heterostructured WS_2-MoS_2 Ultrathin Nanosheets Integrated on CdS Nanorods to Promote Charge Separation and Migration and Improve Solar-Driven Photocatalytic Hydrogen Evolution. *ChemSusChem* **2017**, *10*, 1563–1570. [CrossRef] [PubMed]
10. Reddy, D.A.; Choi, J.; Lee, S.; Kim, Y.; Hong, S.; Kumar, D.P.; Kim, T.K. Hierarchical dandelion-flower-like cobalt-phosphide modified CdS/reduced graphene oxide-MoS_2 nanocomposites as a noble-metal-free catalyst for efficient hydrogen evolution from water. *Catal. Sci. Technol.* **2016**, *6*, 6197–6206. [CrossRef]
11. Reddy, D.A.; Park, H.; Hong, S.; Kumar, D.P.; Kim, T.K. Hydrazine-assisted formation of ultrathin MoS_2 nanosheets for enhancing their co-catalytic activity in photocatalytic hydrogen evolution. *J. Mater. Chem. A* **2017**, *5*, 6981–6991. [CrossRef]
12. Reddy, D.A.; Kim, H.K.; Kim, Y.; Lee, S.; Choi, J.; Islam, M.J.; Kumar, D.P.; Kim, T.K. Multicomponent transition metal phosphides derived from layered double hydroxide double-shelled nanocages as an efficient non-precious co-catalyst for hydrogen production. *J. Mater. Chem. A* **2016**, *4*, 13890–13898. [CrossRef]
13. Scanlon, D.O.; Dunnill, C.W.; Buckeridge, J.; Shevlin, S.A.; Logsdail, A.J.; Woodley, S.M.; Catlow, C.R.A.; Powell, M.J.; Palgrave, R.G.; Parkin, I.P.; et al. Band alignment of rutile and anatase TiO_2. *Nat. Mater.* **2013**, *12*, 798–801. [CrossRef] [PubMed]
14. Reddy, K.M.; Manorama, S.V.; Reddy, A.R. Bandgap studies on anatase titanium dioxide nanoparticles. *Mater. Chem. Phys.* **2003**, *78*, 239–245. [CrossRef]
15. Fujishima, A.; Honda, K. Electrochemical photolysis of water at a semiconductor electrode. *Nature* **1972**, *238*, 37–38. [CrossRef] [PubMed]
16. Zhang, J.; Zhou, P.; Liu, J.; Yu, J. New understanding of the difference of photocatalytic activity among anatase, rutile and brookite TiO_2. *Phys. Chem. Chem. Phys.* **2014**, *16*, 20382–20386. [CrossRef] [PubMed]
17. Chen, X.; Shen, S.; Guo, L.; Mao, S.S. Semiconductor-based Photocatalytic Hydrogen Generation. *Chem. Rev.* **2010**, *110*, 6503–6570. [CrossRef] [PubMed]
18. Primo, A.; Corma, A.; García, H. Titania supported gold nanoparticles as photocatalyst. *Phys. Chem. Chem. Phys.* **2011**, *13*, 886–910. [CrossRef] [PubMed]
19. Ahmed, S.; Rasul, M.G.; Brown, R.; Hashib, M.A. Influence of parameters on the heterogeneous photocatalytic degradation of pesticides and phenolic contaminants in wastewater: A short review. *J. Environ. Manag.* **2011**, *92*, 311–330. [CrossRef] [PubMed]
20. Lan, Y.; Lu, Y.; Ren, Z. Mini Review on photocatalysis of titanium dioxide nanoparticles and their solar applications. *Nano Energy* **2013**, *2*, 1031–1045. [CrossRef]
21. Pu, Y.C.; Wang, G.; Chang, K.D.; Ling, Y.; Lin, Y.K.; Fitzmorris, B.C.; Liu, C.M.; Lu, X.; Tong, Y.; Zhang, J.Z.; et al. Au nanostructure-decorated TiO_2 nanowires exhibiting photoactivity across entire UV-visible region for photoelectrochemical water splitting. *Nano Lett.* **2013**, *13*, 3817–3823. [CrossRef] [PubMed]

22. Linic, S.; Christopher, P.; Ingram, D.B. Plasmonic-metal nanostructures for efficient conversion of solar to chemical energy. *Nat. Mater.* **2011**, *10*, 911–921. [CrossRef] [PubMed]
23. Wang, C.; Astruc, D. Nanogold plasmonic photocatalysis for organic synthesis and clean energy conversion. *Chem. Soc. Rev.* **2014**, *43*, 7188–7216. [CrossRef] [PubMed]
24. Furube, A.; Du, L.; Hara, K.; Katoh, R.; Tachiya, M. Ultrafast plasmon-induced electron transfer from gold nanodots into TiO_2 nanoparticles. *J. Am. Chem. Soc.* **2007**, *129*, 14852–14853. [CrossRef] [PubMed]
25. Youngblood, W.J.; Lee, S.A.; Maeda, K.; Mallouk, T.E. Visible Light Water Splitting Using Dye- Sensitized Oxide Semiconductors. *Acc. Chem. Res.* **2009**, *42*, 1966–1973. [CrossRef] [PubMed]
26. Yum, J.-H.; Chen, P.; Grätzel, M.; Nazeeruddin, M.K. Recent developments in solid-state dye-sensitized solar cells. *ChemSusChem* **2008**, *1*, 699–707. [CrossRef] [PubMed]
27. Kruth, A.; Quade, A.; Brüser, V.; Weltmann, K.-D. Plasma-Enhanced Synthesis of Poly (allylamine)-Encapsulated Ruthenium Dye-Sensitized Titania Photocatalysts. *J. Phys. Chem. C* **2013**, *117*, 3804–3811. [CrossRef]
28. Kruth, A.; Peglow, S.; Rockstroh, N.; Junge, H.; Brüser, V.; Weltmann, K.D. Enhancement of photocatalyic activity of dye sensitised anatase layers by application of a plasma-polymerized allylamine encapsulation. *J. Photochem. Photobiol. A Chem.* **2014**, *290*, 31–37. [CrossRef]
29. Kruth, A.; Hansen, S.; Beweries, T.; Brüser, V.; Weltmann, K.D. Plasma synthesis of polymer-capped dye-sensitised anatase nanopowders for visible-light-driven hydrogen evolution. *ChemSusChem* **2013**, *6*, 152–159. [CrossRef] [PubMed]
30. Karnahl, M.; Mejía, E.; Rockstroh, N.; Tschierlei, S.; Luo, S.P.; Grabow, K.; Kruth, A.; Brüser, V.; Junge, H.; Lochbrunner, S.; et al. Photocatalytic hydrogen production with copper photosensitizer-titanium dioxide composites. *ChemCatChem* **2014**, *6*, 82–86. [CrossRef]
31. Warren, S.C.; Thimsen, E. Plasmonic solar water splitting. *Energy Environ. Sci.* **2012**, *5*, 5133–5146. [CrossRef]
32. Petryayeva, E.; Krull, U.J. Localized surface plasmon resonance: Nanostructures, bioassays and biosensing—A review. *Anal. Chim. Acta* **2011**, *706*, 8–24. [CrossRef] [PubMed]
33. Noguez, C. Surface Plasmons on Metal Nanoparticles: The Influence of Shape and Physical Environment. *J. Phys. Chem. C* **2007**, *111*, 3806–3819. [CrossRef]
34. Link, S.; El-Sayed, M.A. Shape and size dependence of radiative, non-radiative and photothermal properties of gold nanocrystals. *Int. Rev. Phys. Chem.* **2000**, *19*, 409–453. [CrossRef]
35. Jain, P.K.; Lee, K.S.; El-Sayed, I.H.; El-Sayed, M.A. Calculated absorption and scattering properties of gold nanoparticles of different size, shape, and composition: Applications in biological imaging and biomedicine. *J. Phys. Chem. B* **2006**, *110*, 7238–7248. [CrossRef] [PubMed]
36. Pu, Y.; Zhang, J.Z. Mechanisms Behind Plasmonic Enhancement of Photocurrent in Metal Oxides. *Austin J. Nanomed. Nanotechnol.* **2014**, *2*, 1–4.
37. Cushing, S.K.; Li, J.; Meng, F.; Senty, T.R.; Suri, S.; Zhi, M.; Li, M.; Bristow, A.D.; Wu, N. Photocatalytic Activity Enhanced by Plasmonic Resonant Energy Transfer Transfer from Metal to Semiconductor. *J. Am. Chem. Soc.* **2012**, *134*, 105033–105041. [CrossRef] [PubMed]
38. Ma, X.C.; Dai, Y.; Yu, L.; Huang, B.B. Energy transfer in plasmonic photocatalytic composites. *Light Sci. Appl.* **2016**, *5*, 1–13. [CrossRef]
39. Garcia, M.A. Surface plasmons in metallic nanoparticles: Fundamentals and applications. *J. Phys. D Appl. Phys.* **2012**, *45*, 389501. [CrossRef]
40. Gärtner, F.; Losse, S.; Boddien, A.; Pohl, M.-M.; Denurra, S.; Junge, H.; Beller, M. Hydrogen evolution from water/Alcohol mixtures: Effective in situ generation of an active Au/TiO_2 catalyst. *ChemSusChem* **2012**, *5*, 530–533. [CrossRef] [PubMed]
41. Yang, J.; Wang, D.; Han, H.; Li, C. Roles of cocatalysts in photocatalysis and photoelectrocatalysis. *Acc. Chem. Res.* **2013**, *46*, 1900–1909. [CrossRef] [PubMed]
42. Priebe, J.B.; Karnahl, M.; Junge, H.; Beller, M.; Hollmann, D.; Brückner, A. Water Reduction with Visible Light: Synergy between Optical Transitions and Electron Transfer in Au-TiO_2 Catalysts Visualized by In situ EPR Spectroscopy. *Angew. Chem. Int. Ed.* **2013**, *52*, 11420–11424. [CrossRef] [PubMed]
43. Tian, Y.; Tatsuma, T. Plasmon-induced photoelectrochemistry at metal nanoparticles supported on nanoporous TiO_2. *Chem. Commun.* **2004**, 1810–1811. [CrossRef] [PubMed]

44. Priebe, J.B.; Radnik, J.; Lennox, A.J.J.; Pohl, M.-M.; Karnahl, M.; Hollmann, D.; Grabow, K.; Bentrup, U.; Junge, H.; Beller, M.; et al. Solar Hydrogen Production by Plasmonic Au-TiO_2 Catalysts: Impact of Synthesis Protocol and TiO_2 Phase on Charge Transfer Efficiency and H_2 Evolution Rates. *ACS Catal.* **2015**, *5*, 2137–2148. [CrossRef]
45. Silva, C.G.; Juárez, R.; Marino, T.; Molinari, R.; García, H. Influence of Excitation Wavelength (UV or Visible Light) on the Photocatalytic Activity of Titania Containing Gold Nanoparticles for the Generation of Hydrogen or Oxygen from Water. *J. Am. Chem. Soc.* **2011**, *133*, 595–602. [CrossRef] [PubMed]
46. Kowalska, E.; Rau, S.; Ohtani, B. Plasmonic titania photocatalysts active under UV and visible-light irradiation: Influence of gold amount, size, and shape. *J. Nanotechnol.* **2012**, *2012*, 1–11. [CrossRef]
47. Yan, J.; Wu, G.; Guan, N.; Li, L. Synergetic promotion of the photocatalytic activity of TiO_2 by gold deposition under UV-visible light irradiation. *Chem. Commun.* **2013**, *49*, 11767–11769. [CrossRef] [PubMed]
48. Su, R.; Tiruvalam, R.; Logsdail, A.J.; He, Q.; Downing, C.A.; Jensen, M.T.; Dimitratos, N.; Kesavan, L.; Wells, P.P.; Bechstein, R.; et al. Designer titania-supported Au-Pd nanoparticles for efficient photocatalytic hydrogen production. *ACS Nano* **2014**, *8*, 3490–3497. [CrossRef] [PubMed]
49. Osterloh, F.E. Inorganic nanostructures for photoelectrochemical and photocatalytic water splitting. *Chem. Soc. Rev.* **2013**, *42*, 2294–2320. [CrossRef] [PubMed]
50. Zhang, P.; Gao, L.; Song, X.; Sun, J. Micro- and Nanostructures of Photoelectrodes for Solar-Driven Water Splitting. *Adv. Mater.* **2015**, *27*, 562–568. [CrossRef] [PubMed]
51. Wenderich, K.; Mul, G. Methods, Mechanism, and Applications of Photodeposition in Photocatalysis: A Review. *Chem. Rev.* **2016**, *116*, 14587–14619. [CrossRef] [PubMed]
52. Zhang, H.; Haba, M.; Okumura, M.; Akita, T.; Hashimoto, S.; Toshima, N. Novel formation of Ag/Au bimetallic nanoparticles by physical mixture of monometallic nanoparticles in dispersions and their application to catalysts for aerobic glucose oxidation. *Langmuir* **2013**, *29*, 10330–10339. [CrossRef] [PubMed]
53. Jiang, H.; Akita, T.; Ishida, T. Synergistic Catalysis of Au@Ag Core−Shell Nanoparticles Stabilized on Metal−Organic Framework. *J. Am. Chem. Soc.* **2011**, *133*, 1304–1306. [CrossRef] [PubMed]
54. Kruth, A.; Peglow, S.; Quade, A.; Pohl, M.-M.; Weltmann, K.-D. Structural and Photoelectrochemical Properties of DC Magnetron- Sputtered TiO_2 Layers on FTO. *J. Phys. Chem. C* **2014**, *118*, 25234–25244. [CrossRef]
55. Peglow, S.; Pohl, M.-M.; Kruth, A.; Brüser, V. Plasma Based Synthesis, Electron Microscopy, and Optical Characterization of Au-, Ag-, and Ag/Au-Core−Shell Nanoparticles. *J. Phys. Chem. C* **2015**, *119*, 563–572. [CrossRef]
56. Schneider, J.; Bahnemann, D.W. Undesired Role of Sacrificial Reagents in Photocatalysis. *J. Phys. Chem. Lett.* **2013**, *4*, 3479–3483. [CrossRef]
57. Nada, A.A.; Barakat, M.H.; Hamed, H.A.; Mohamed, N.R.; Veziroglu, T.N. Studies on the photocatalytic hydrogen production using suspended modified TiO_2 photocatalysts. *Int. J. Hydrogen Energy* **2005**, *30*, 687–691. [CrossRef]
58. Kawai, T.; Sakata, T. Conversion of carbohydrate into hydrogen fuel by a photocatalytic process. *Nature* **1980**, *286*, 474–476. [CrossRef]
59. Wulff, H. *Introduction to Complex Plasmas*; Bonitz, M., Horing, N., Ludwig, P., Eds.; Springer: Berlin/Heidelberg, Germany, 2010.
60. Strecker, A.; Salzberger, U.; Mayer, J. Specimen Preparation for Transmission Electron Microscopy: Reliable Method for Cross-Sections and Brittle Materials. *Prakt. Metallogr.* **1993**, *30*, 482–495.
61. Gottwald, W.; Heinrich, K.H. *UV/VIS-Spektroskopie für Anwender*; Wiley VCH: Weinheim, Germany, 1999.
62. Djerdj, I.; Tonejc, A.M. Structural investigations of nanocrystalline TiO_2 samples. *J. Alloys Compd.* **2006**, *413*, 159–174. [CrossRef]
63. Suh, I.; Ohta, H.; Waseda, Y. High-temperature thermal expansion of six metallic elements measured by dilatation method and X-ray diffraction. *J. Mater. Sci.* **1988**, *23*, 757–760. [CrossRef]
64. Yamanaka, T.; Kurashima, R.; Mimaki, J. X-ray diffraction study of bond character of rutile-type SiO_2, GeO_2 and SnO_2. *Z. Kristallogr.* **2000**, *215*, 424–428. [CrossRef]
65. Petrov, I.; Barna, P.B.; Hultman, L.; Greene, J.E. Microstructural evolution during film growth. *J. Vac. Sci. Technol. A* **2003**, *21*, 117–128. [CrossRef]
66. Thornton, J.A. The microstructure of sputter-deposited coatings. *J. Vac. Sci. Technol. A* **1986**, *4*, 3059–3065. [CrossRef]

67. Williams, D.; Carter, C. *Transmission Electron Microscopy—A Textbook for Materials Science*; Springer: Boston, MA, USA, 1996; ISBN 9780387765006.
68. Claus, P.; Brückner, A.; Mohr, C.; Hofmeister, H. Supported Gold Nanoparticles from Quantum Dot to Mesoscopic Size Scale: Effect of Electronic and Structural Properties on Catalytic Hydrogenation of Conjugated Functional Groups. *J. Am. Chem. Soc.* **2000**, *122*, 11430–11439. [CrossRef]
69. Söderlund, J.; Kiss, L.; Niklasson, G.; Granqvist, C. Lognormal Size Distributions in Particle Growth Processes without Coagulation. *Phys. Rev. Lett.* **1998**, *80*, 2386–2388. [CrossRef]
70. Robson, J.D. Modelling the evolution of particle size distribution during nucleation, growth and coarsening. *Mater. Sci. Technol.* **2004**, *20*, 441–448. [CrossRef]
71. Cleveland, C.; Luedtke, W.; Landman, U. Melting of Gold Clusters: Icosahedral Precursors. *Phys. Rev. Lett.* **1998**, *81*, 2036–2039. [CrossRef]
72. Marks, L.D. Experimental studies of small particle structures. *Rep. Prog. Phys.* **1999**, *57*, 603–649. [CrossRef]
73. Nam, H.-S.; Hwang, N.; Yu, B.; Yoon, J.-K. Formation of an Icosahedral Structure during the Freezing of Gold Nanoclusters: Surface-Induced Mechanism. *Phys. Rev. Lett.* **2002**, *89*, 1–4. [CrossRef] [PubMed]
74. Ascencio, J.A.; Gutiérrez-Wing, C.; Espinosa, M.E.; Marín, M.; Tehuacanero, S.; Zorrilla, C.; José-Yacamán, M. Structure determination of small particles by HREM imaging: Theory and experiment. *Surf. Sci.* **1998**, *396*, 349–368. [CrossRef]
75. Barnard, A.S. A thermodynamic model for the shape and stability of twinned nanostructures. *J. Phys. Chem. B* **2006**, *110*, 24498–24504. [CrossRef] [PubMed]
76. Hubenthal, F.; Ziegler, T.; Hendrich, C.; Alschinger, M.; Träger, F. Tuning the surface plasmon resonance by preparation of gold-core/silver-shell and alloy nanoparticles. *Eur. Phys. J. D* **2005**, *34*, 165–168. [CrossRef]
77. Douglas, F.; Yañez, R.; Ros, J.; Marín, S.; De La Escosura-Muñiz, A.; Alegret, S.; Merkoçi, A. Silver, gold and the corresponding core shell nanoparticles: Synthesis and characterization. *J. Nanopart. Res.* **2008**, *10*, 97–106. [CrossRef]
78. Shyjumon, I.; Gopinadhan, M.; Ivanova, O.; Quaas, M.; Wulff, H.; Helm, C.A.; Hippler, R. Structural deformation, melting point and lattice parameter studies of size selected silver clusters. *Eur. Phys. J. D* **2006**, *37*, 409–415. [CrossRef]
79. Buffat, P.; Borel, J.-P. Size Effect on the Melting Temperature of Gold Particles. *Phys. Rev. A* **1976**, *13*, 2287–2298. [CrossRef]
80. Jiang, Q.; Zhang, S.; Zhao, M. Size-dependent melting point of noble metals. *Mater. Chem. Phys.* **2003**, *82*, 225–227. [CrossRef]
81. Jette, E.R.; Foote, F. Precision Determination of Lattice Constants. *J. Chem. Phys.* **1935**, *3*, 605–616. [CrossRef]
82. Tsuji, M.; Yamaguchi, D.; Matsunaga, M.; Alam, M.J. Epitaxial growth of Au@Cu core-shell nanocrystals prepared using the PVP-assisted polyol reduction method. *Cryst. Growth Des.* **2010**, *10*, 5129–5135. [CrossRef]
83. Fan, F.R.; Liu, D.Y.; Wu,'Y.F.; Duan, S.; Xie, Z.X.; Jiang, Z.Y.; Tian, Z.Q. Epitaxial growth of heterogeneous metal nanocrystals: From gold nano-octahedra to palladium and silver nanocubes. *J. Am. Chem. Soc.* **2008**, *130*, 6949–6951. [CrossRef] [PubMed]
84. Skriver, H.L.; Rosengaard, N.M. Surface energy and work function of elemental metals. *Phys. Rev. B* **1992**, *46*, 7157–7168. [CrossRef]
85. Koo, J.B.; Hong, K.J.; Park, J.S.; Shin, D.C. Effect of grain size on transmittance and mechanical strength of sintered alumina. *Mater. Sci. Eng. A* **2004**, *374*, 191–195. [CrossRef]
86. Apetz, R.; Bruggen, M.P.B. Van Transparent Alumina: A Light-Scattering Model. *J. Am. Ceram. Soc.* **2003**, *86*, 480–486. [CrossRef]
87. Dericioglu, A.F.; Kagawa, Y. Effect of grain boundary microcracking on the light transmittance of sintered transparent $MgAl_2O_4$. *J. Eur. Ceram. Soc.* **2003**, *23*, 951–959. [CrossRef]
88. Barbe, C.J.; Arendse, F.; Comte, P.; Jirousek, M.; Lenzmann, F.; Shklover, V.; Gra, M. Nanocrystalline Titanium Oxide Electrodes for Photovoltaic Applications. *J. Am. Ceram. Soc.* **1997**, *71*, 3157–3171. [CrossRef]
89. Nehl, C.L.; Hafner, J.H. Shape-dependent plasmon resonances of gold nanoparticles. *J. Mater. Chem.* **2008**, 2415–2419. [CrossRef]
90. Rubio, A.; Serra, L. Dielectric screening effects on the photoabsorption cross section of embedded metallic clusters. *Phys. Rev. B* **1993**, *48*, 18222–18229. [CrossRef]
91. Amendola, V.; Meneghetti, M. Size Evaluation of Gold Nanoparticles by UV-vis Spectroscopy. *J. Phys. Chem. C* **2009**, 4277–4285. [CrossRef]

92. Zook, J.M.; Rastogi, V.; Maccuspie, R.I.; Keene, A.M.; Fagan, J. Measuring Agglomerate Size Distribution and Dependence of Localized Surface Plasmon Resonance Absorbance on Gold Nanoparticle Agglomerate Size Using Analytical Ultracentrifugation. *ACS Nano* **2011**, *5*, 8070–8079. [CrossRef] [PubMed]
93. Huang, X.; El-Sayed, M.A. Gold nanoparticles: Optical properties and implementations in cancer diagnosis and photothermal therapy. *J. Adv. Res.* **2010**, *1*, 13–28. [CrossRef]
94. Jain, P.K.; El-Sayed, M.A. Plasmonic coupling in noble metal nanostructures. *Chem. Phys. Lett.* **2010**, *487*, 153–164. [CrossRef]
95. Hayashi, S.; Okamoto, T. Plasmonics: Visit the past to know the future. *J. Phys. D. Appl. Phys.* **2012**, *45*, 433001. [CrossRef]
96. Hartland, G. V Optical Studies of Dynamics in Noble Metal Nanostructures. *Chem. Rev.* **2011**, *111*, 3858–3887. [CrossRef] [PubMed]
97. Stockman, M.I. Nanoplasmonics: The physics behind the applications. *Phys. Today* **2011**, 39–44. [CrossRef]
98. Romero, I.; Aizpurua, J.; Bryant, G.W.; Abajo, F.J.G. De Plasmons in nearly touching metallic nanoparticles: Singular response in the limit of touching dimers. *Opt. Express* **2006**, *14*, 1627–1631. [CrossRef]
99. Zhu, J. Surface Plasmon Resonance from Bimetallic Interface in Au-Ag Core-Shell Structure Nanowires. *Nanoscale Res. Lett.* **2009**, 977–981. [CrossRef] [PubMed]
100. Chen, Y.; Wu, H.; Li, Z.; Wang, P. The Study of Surface Plasmon in Au/Ag Core/Shell Compound Nanoparticles. *Plasmonics* **2012**, 509–513. [CrossRef]
101. Zhang, C.; Chen, B.-Q.; Li, Z.-Y.; Xia, Y.; Chen, Y.-G. Surface Plasmon Resonance in Bimetallic Core–Shell Nanoparticles. *J. Phys. Chem. C* **2015**, *119*, 16836–16845. [CrossRef]

© 2018 by the authors. Licensee MDPI, Basel, Switzerland. This article is an open access article distributed under the terms and conditions of the Creative Commons Attribution (CC BY) license (http://creativecommons.org/licenses/by/4.0/).

nanomaterials

MDPI

Article

Band Gap Implications on Nano-TiO_2 Surface Modification with Ascorbic Acid for Visible Light-Active Polypropylene Coated Photocatalyst

Chiara Anna D'Amato [1,*], Rita Giovannetti [1,*], Marco Zannotti [1,*], Elena Rommozzi [1], Marco Minicucci [2], Roberto Gunnella [2] and Andrea Di Cicco [2]

1 School of Science and Technology, Chemistry Division, University of Camerino, 62032 Camerino, Italy; elena.rommozzi@unicam.it

2 School of Science and Technology, Physics Division, University of Camerino, 62032 Camerino, Italy; marco.minicucci@unicam.it (M.M.); roberto.gunnella@unicam.it (R.G.); andrea.dicicco@unicam.it (A.D.C.)

* Correspondence: chiaraanna.damato@unicam.it (C.A.D.); rita.giovannetti@unicam.it (R.G.); marco.zannotti@unicam.it (M.Z.); Tel.: +39-0737402272 (C.A.D. & R.G. & M.Z.); Fax: +39-0737404508 (C.A.D. & R.G. & M.Z.)

Received: 16 July 2018; Accepted: 4 August 2018; Published: 7 August 2018

Abstract: The effect of surface modification using ascorbic acid as a surface modifier of nano-TiO_2 heterogeneous photocatalyst was studied. The preparation of supported photocatalyst was made by a specific paste containing ascorbic acid modified TiO_2 nanoparticles used to cover Polypropylene as a support material. The obtained heterogeneous photocatalyst was thoroughly characterized (scanning electron microscope (SEM), RAMAN, X-ray diffraction (XRD), X-ray photoelectron spectroscopy (XPS), photoluminescence (PL), and Diffuse Reflectance Spectra (DRS) and successfully applied in the visible light photodegradation of Alizarin Red S in water solutions. In particular, this new supported TiO_2 photocatalyst showed a change in the adsorption mechanism of dye with respect to that of only TiO_2 due to the surface properties. In addition, an improvement of photocatalytic performances in the visible light photodegration was obtained, showing a strict correlation between efficiency and energy band gap values, evidencing the favorable surface modification of TiO_2 nanoparticles.

Keywords: heterogeneous photocatalysis; TiO_2; ascorbic acid; surface modification; band gap energy; Alizarin Red S

---

1. Introduction

In recent years, progress in industrialized society has caused serious environmental problems due, for example, to the discharge of a wide variety of environmental contaminants from residential, commercial, and industrial sources [1]. Azo-dyes and contaminants released from the textile industry are mostly non-biodegradable pollutants, toxic, and also resistant to degradation using the traditional treatment methods. For these reasons they represent an important source of environmental contamination. Color removal from wastewater is an important issue because only small amounts of dyes present high effects both on the color and water quality. Consequently, it is necessary to find an effective method of wastewater treatment to remove dye pollutants and their colors from textile effluents [2]. Nowadays, a greater challenge in the environmental field relies on the treatment of contaminants and advanced oxidation processes (AOPs) have been considered as alternatives to traditional water treatment technologies [1,3]. Among several AOPs, TiO_2 based photocatalysis has received huge attention as one of the most viable environmental clean-up technologies [1]; TiO_2 has been considered as among the most promising materials due to its high chemical stability, low cost, chemical inertness, commercial availability, and outstanding photocatalytic activity [4–7]. TiO_2 has been extensively used in many industrially relevant processes ranging from environmental applications to clean energy, from paints to cosmetics and medicine [7]. The semiconductor materials find application in medicine as a photosensitizer for photodynamic and photothermal therapy of cancer, as well as for drug delivery [7–9]. In particular, TiO_2 photocatalysis is used to destroy hazardous compounds in water or air [2,10] through a process that require low energy, operates at ambient conditions and is able to mineralize organic pollutants using only atmospheric oxygen as the additional chemical species [11]. Unlike the bulk counterpart, nanosized TiO_2 demonstrated improved performance, thanks to its high surface-to-volume ratio that greatly increases the density of active surface sites available for adsorption and catalysis. In addition, the size-dependent band gap of nanosized semiconductors allows to adjust the redox potentials of photogenerated electron–hole pairs to selectively control the photochemical reactions. Therefore, the reduced dimensions of the nanocatalyst allow the photo-generated charges to reach the catalyst surface, thus reducing the probability of undesired bulk recombination [4,12,13]. As a drawback, large bandgap semiconductors like TiO_2 (3.2 eV for Anatase) respond only to UV light, thus resulting in low efficiency for the visible spectrum. The band gap excitation of semiconductor causes charge separation followed by the scavenging of electrons and holes by surface adsorbed species [14]. Visible-light-driven photocatalytic processes can thus be realized by doping TiO_2 with non-metal [15–18], noble metal [19–21] or reduced and defect TiO_2 engineering [22] etc., which represent different methods widely employed to narrow the wide band gap of TiO_2. By controlling the surface treatment and medium conditions, it is possible to fine-tune photocatalytic properties of TiO_2 to desired applications [14]. It is also well-known that, due to large curvature, TiO_2 particles with sizes smaller than 20 nm have under-coordinated surface structure with square pyramidal geometry instead of an octahedral one [7,23]. Therefore, Ti atoms surface are very reactive, leading to the formation of charge transfer (CT) complexes with a red absorption shift, due to their binding with electron-donating ligands. The visible light activation of TiO_2 has been observed upon the surface modification of colloidal TiO_2 with L (+)-Ascorbic Acid (AA) [2,24,25]. AA, known as Vitamin C, is an important natural compound in the biology and chemistry fields, and particular interest is tuned towards its complex with metals. Complexes of AA with titanium (IV) should be relative strong and not prone to the metal catalyzed ligand oxidation that renders many metal ascorbate complexes so reactive [26]. The surface modifiers tend to enhance the surface coverage of the pollutant molecules on TiO_2, inhibit the recombination process by separating the charge pairs, and extend the wavelength response [24]. It was found that compounds such as AA modify the surface of particles through the formation of π-π donor-acceptor complexes [2,24,27,28].

Our previous studies regarded the preparation of Polypropylene (PP) coated with only TiO_2, TiO_2 in combination with graphene and with gold nanoparticles and their application in the visible light photodegradation of Alizarin Red S (1,2-dihydroxy-9,10-anthraquinonesulfonic acid sodium salt or ARS) obtaining highly efficient dye degradation with an easy separation of the photocatalyst from the solution [29–32]. ARS is a widely used synthetic water soluble dye considered a refractory pollutant because of the difficulty in removing it through general treatments [29–32].

In this study, we want to continue our efforts regarding the extension of TiO_2 light absorption into the visible range. For this purpose, AA modified TiO_2 NPs supported on PP were prepared and investigated for the first time as visible light photocatalysts in water. We focused on three main points: first in the establishing a procedure for the preparation of specific new paste of AA modified TiO_2 nanoparticles (NPs), second, to use this to obtain supported photocatalyst of defined qualities and third, to demonstrate the increasing photocatalytic ability in the degradation of ARS as target pollutants under visible light irradiation. This new photocatalyst showed a change in the adsorption mechanism with respect to that of pure TiO_2 NPs and an improvement of the photocatalytic efficiency. The main advantages of the present approach were the easy preparation of the photocatalyst together with the use of the green compound AA. A comprehensive and in-depth characterization of the obtained photocatalyst permitted us to understand the reasons and types of surface modifications and the correlation between all the results.

2. Materials and Methods

2.1. Photocatalyst Preparation

Different types of new modified TiO_2 pastes, named $[AA\text{-}TiO_2]_A$ were prepared by the addition of 6 g of Titanium (IV) dioxide Anatase nano-powdered (<25 nm), into 10 mL of distilled water containing different amounts of AA from 0.5 to 3.4 wt %, acetyl acetone 10% *v*/*v* and few drops of Triton X-100 with continuous grinding for 3 min. All the used chemicals were Sigma Aldrich products (Sigma Aldrich, St. Louis, MO, USA). Five different heterogeneous photocatalysts, named $[PP@AA\text{-}TiO_2]_A$, and containing different amount wt % of AA were prepared through the dip-coating technique on 20 cm^2 surface of Polypropylene (PP) strips (3 M 2500 material). After the preparation, the photocatalyst was thermally dried in the oven at 110 °C.

2.2. Photocatalyst Characterization

A morphological study on the modified $[AA\text{-}TiO_2]_A$ and pure $[TiO_2]_A$ photocatalyst was made using Field Emission Scanning Electron Microscopy (FE-SEM, Sigma Family, Zeiss, Oberkochen, Germany) operated at 5–7 KV. All samples have been carefully prepared by removing the TiO_2 paste from the plastic support material without a change of its properties. Then the obtained powder was deposited on aluminum stabs using self-adhesive carbon conductive tabs. The samples were sputtered with chromium (5 nm) by Quorum QT150 (Quorum, Laughton, UK) to prevent charging during the analysis. To study the structural variations of two compounds, pure $[TiO_2]_A$ and modified $[AA\text{-}TiO_2]_A$ were removed from the PP support and characterized by using the X-ray diffraction (XRD) technique. A customized horizontal Debye-Scherrer diffractometer was used for XRD measurements; this instrument is equipped with an INEL CPS 180 (INEL, Artenay, France) curved position sensitive detector in order to reduce drastically the acquisition time for each pattern. In order to optimize the efficiency this detector is filled with a Kr/CO_2 gas mixture while, the absence of moving parts eliminates the need for mechanical scanning devices such as complex scanning goniometers used in conventional XRD instruments. A Mo K-alpha (lambda = 0.7093 Å) X-ray source is used generated by a Philips sealed X-ray tube and monochromatized through a graphite crystal along the 002 plane. The samples were positioned on the beam into glass capillaries (diameter 100 microns).

X-ray photoelectron spectroscopy (XPS) analysis has been obtained by means of an unmonochromatized X-ray source (Al Kα) and CLAM IV hemispherical spectrometer (VG Scientific Ltd., East Grinstead, UK) a constant passing energy (50 eV) for an overall lower than 1 eV half width at half-maximum (HWHM). The Raman analysis was performed using a micro-Raman spectrometer iHR320 (Horiba, Kyoto, Japan) in which the photocatalysts were excited with a green laser emitting at λ = 532 nm, at room temperature and the objective outlet was 100×. The photoluminescence (PL) measurements were achieved using a Perkin Elmer LS 45 luminescence spectrometer (Perkin Elmer, Waltham, MA, USA) equipped with a pulsed Xe flash lamp and, in particular, the PL spectra were collected at room temperature using an excitation wavelength of 290 nm in the range from 300–900 nm. The Diffuse Reflectance Spectra (DRS) were collected using an UV-Vis Spectrometer Lambda35 (Perkin Elmer, Waltham, MA, USA) with an integration sphere (P/N C6951014) in a range of wavelength from 200–1100 nm.

2.3. Adsorption and Photodegradation Processes

The adsorption isotherms of $[PP@AA\text{-}TiO_2]_A$ in the ARS adsorption under dark condition were analyzed by using four different concentration of ARS from 2.92×10^{-5} to 7.30×10^{-5} mol L^{-1}.

The photocatalytic performance of $[PP@AA\text{-}TiO_2]_A$ photocatalyst was evaluated in the degradation of ARS 5.843×10^{-5} mol L^{-1} by using nine equal strips of $[PP@AA\text{-}TiO_2]_A$ inserted in a typical thermostated photoreactor system [29] connected with Cary 8454 Diode Array System spectrophotometer (Agilent Technologies, Santa Clara, CA, USA) with a continuous flux quartz cuvette (178.710-QS, light path 10 mm, Hellma Analytics, Müllheim, Germany) allowing a real-time analysis; the photoreactor was irradiated with visible light by using a tubular lamp (100 W, 1800 Lumen,

LYVIA, (Arteleta International S.p.A., Milano, Italy); the spectral features are reported in Figure S1. All the spectrophotometric data were collected monitoring the decrease of ARS absorbance at fixed wavelength of 424 nm. The adsorption kinetics was evaluated in the same way under dark conditions.

3. Results and Discussion

3.1. Morphological and Structure Characterization

The overall procedure of the photocatalyst is schematically presented in Figure 1 where the addition of AA in the TiO_2 pastes preparation, give the formation of yellow-brown color paste of $[AA\text{-}TiO_2]_A$ at pH equal to 5 with an increase color intensity as a function of AA amount wt %.

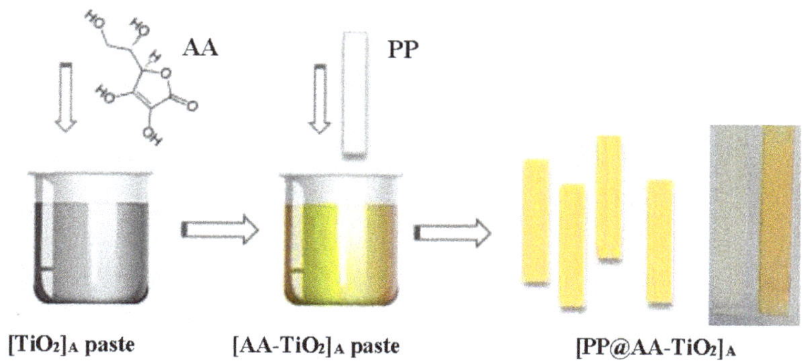

Figure 1. Schematic representation of the operative steps for the preparation of the new modified heterogeneous $[PP@AA\text{-}TiO_2]_A$ photocatalyst and photograph of $[PP@AA\text{-}TiO_2]_A$ prepared with two different AA amount wt %.

SEM analysis was performed on the photocatalyst removed from the PP support; Figure 2 shows the SEM micrographs of pure $[TiO_2]_A$ and modified $[AA\text{-}TiO_2]_A$ samples at the same magnification revealing that the presence of AA as a surface modifier changes the morphological aspect of $[AA\text{-}TiO_2]_A$ photocatalyst.

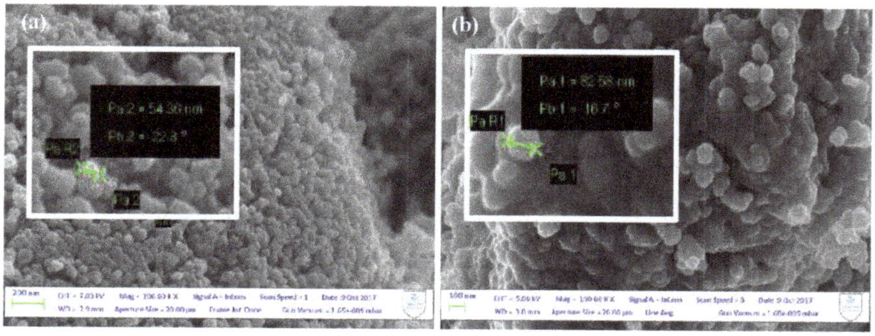

Figure 2. Scanning electron microscope (SEM) micrographs of (**a**) pure $[TiO_2]_A$; (**b**) modified $[AA\text{-}TiO_2]_A$ containing 2.5 wt % of AA.

In particular, the images reveal that the particle sizes change as a consequence of the addition of AA to TiO_2 from around 55 nm for the pure $[TiO_2]_A$ photocatalyst (Figure 2a) to around 80 nm for the modified $[AA\text{-}TiO_2]_A$ (Figure 2b).

To investigate the effect of AA as a surface modifier, the structural features of pure $[TiO_2]_A$ were characterized by XRD measurements and compared to that of $[AA\text{-}TiO_2]_A$ samples. Figure 3a showed the diffraction patterns of pure $[TiO_2]_A$ (black circle) and the modified $[AA\text{-}TiO_2]_A$ samples containing 2.5 wt % of AA (blue circle), the unit cell refinement (red and green lines), and the theoretical pattern of TiO_2-anatase (orange line). From the analysis of these spectra, it is clearly visible that both samples exhibited a series of well-defined diffraction peaks attributable to the Anatase TiO_2 crystal structure and no extra peaks have been observed in the XRD patterns. The values for the structural parameters of the cells obtained by the data refinement are shown in the Table 1. We found a close agreement between TiO_2 [33] and $[TiO_2]_A$ values, while there is evidence of a parameter cell expansion of the $[AA\text{-}TiO_2]_A$ especially along the c-axis: this effect is clearly visible in Figure 3b where the data, in the range of 20–23 deg, show the shift of the 004 reflection directly connected with the vertical axis. It is known that the pH has an important role on the change of average crystallite size and the results obtained at about five pH derive from an increase of the average crystallite size as a consequence of the tensile strain [33]. The obtained results may be attributed to the lattice expansion with consequent incorporation of AA inside the crystalline lattice of the semiconductor material.

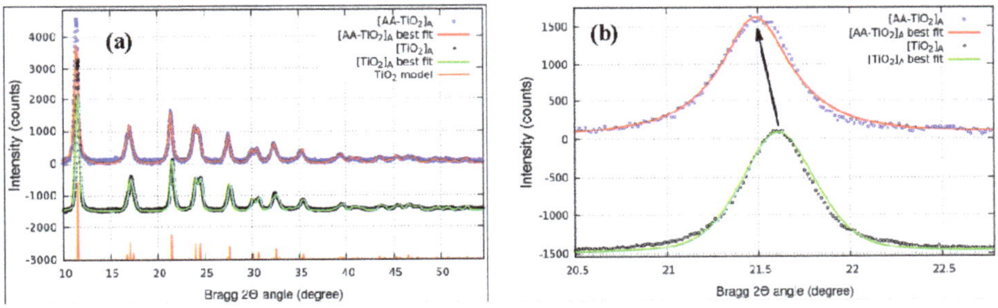

Figure 3. X-ray diffraction (XRD) patterns of (**a**) $[TiO_2]_A$ and $[AA\text{-}TiO_2]_A$ containing 2.5 wt % of AA; (**b**) Magnification of the 004 peak in the range in the range 20.5–23.0 deg.

Table 1. Parameter cell values for TiO_2 [33], $[TiO_2]_A$ and $[AA\text{-}TiO_2]_A$.

Sample	a (Å)	c (Å)
TiO_2 [33]	3.785	9.514
$[TiO_2]_A$	3.787	9.526
$[AA\text{-}TiO_2]_A$	3.808	9.565

The high resolution XPS spectra shown in Figure 4 for commercial Anatase, pure $[TiO_2]_A$ paste and modified $[AA\text{-}TiO_2]_A$ samples were realized in order to analyze the surface modification due to the presence of AA through the formation of the bidentate binuclear binding-bridging of AA-TiO_2. All the obtained spectra were calibrated to the C 1s electron peak at 284.6 eV. Figure 4a,b show the peaks deconvolution of XPS spectra for Ti 2p and O 1s respectively. In Figure 4a (top), for commercial Anatase, two binding energy peaks at 458 and 463.5 eV are observed and are assigned to the Ti^{4+} $2p_{3/2}$ and $2p_{1/2}$ core levels, respectively [34–37]. Figure 4a (in the middle) for the pure $[TiO_2]_A$ heterogeneous photocatalyst shows additional strong peaks at 456.2 and 461.6 eV that are attributed to the Ti^{3+} $2p_{3/2}$ and $2p_{1/2}$ respectively formed on the TiO_2 surface [38]. These peaks derived from a change of the TiO_2 surface because of the paste preparation, in which the presence of water and acetylacetone obtain a partial complexation of monomeric Ti precursor [39] and the concomitant presence of Ti–OH. Figure 4a (bottom) shows strong binding energy peaks at 458.4 and 464 eV that are ascribed to the $2p_{3/2}$ and $2p_{1/2}$ core levels of Ti^{4+} and assigned to the chemical interaction between TiO_2 and AA molecules [34–37]. The O 1s spectrum of commercial Anatase showed in Figure 4b (top) present two binding energy peaks at 528.6 and 532.1 eV that are attributed to Ti–O bond of lattice oxygen of TiO_2 and non-lattice oxygen respectively as the Ti–OH terminal groups [40,41]. The pure $[TiO_2]_A$ sample is shown in Figure 4 (b, in the middle) in which two binding energy peaks at 531.2 and 533.6 eV are visible due to the presence of the OH group with oxygen at the bridging oxygen site (Ti-OH_b) [42] and to the physiosorbed H_2O which is present as a consequence of the influence of water molecules on the sample surface due to the paste preparation [41,42]. Figure 4b (bottom) for the modified $[AA\text{-}TiO_2]_A$ sample shows strong binding energy peaks at 529.8 and 532.4 eV that are assigned to the Ti–O surface bulk oxide lattice of TiO_2 and OH as a terminal group with oxygen attached to the five-coordinated Ti^{4+} with an O–Ti^{4+} covalent bond [42] or C–OH due to the interaction of TiO_2 with the AA [43]. It is known that almost 40% of the TiO_2 surface consist of Ti atoms with incomplete coordination [24] that are four-fold coordinated to oxygen with two unfilled orbitals; consequently, they can accept two lone pair from electron donors to complete the octahedral coordination. The most possible conformation that leads to chelate ring structure that also offers higher stability derived by AA binding as a bidentate ligand through the two ene-diolate oxygen atoms with the function of the electron donor [27]. The five member AA ring structure is favorable for the Ti surface atoms, showing little distortion of bond angles and distances, while no evidence shows the involvement of glycolic side chain in the complex formation [27].

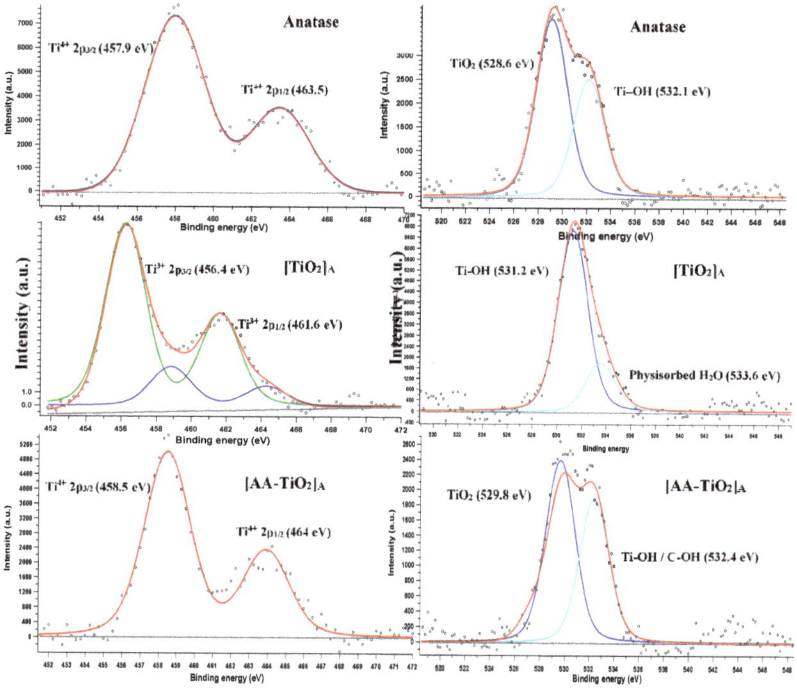

(**a**) Ti 2p (**b**) O 1s

Figure 4. X-ray photoelectron spectroscopy (XPS) spectra of commercial Anatase (top), pure $[TiO_2]_A$ (in the middle) and modified $[AA\text{-}TiO_2]_A$ (bottom) containing 2.5 wt % of AA for (**a**) Ti 2p and (**b**) for O 1s.

Raman spectra collected at room temperature did not show significant change with respect to the same in the absence of AA (Figure S2), the Raman active modes typical of TiO_2 anatase remain unchanged, only a broadened band at 2800 cm^{-1} is detected that can be attributed to the fluorescence of AA. These results are in accordance also with the same sample modified by AuNPs, evidencing that AA not influences the crystallinity, crystallite size, and defects of TiO_2 [32].

PL spectra (Figure 5) for pure $[TiO_2]_A$ and modified $[AA\text{-}TiO_2]_A$ sample with AA amount of 2.5 wt %, were also achieved in order to monitor the electron-hole pair recombination in response to the photon irradiation occurring on TiO_2 surface mediated by the presence of AA as a surface modifier.

The PL spectra shows the peaks attributed to TiO_2 [32], which decrease in intensity in the case of modified $[AA\text{-}TiO_2]_A$, indicated that the recombination process has been suppressed, resulting in higher photocatalytic activity. A de-convoluted PL emission spectra of both samples were reported in Figure S3a,b respectively.

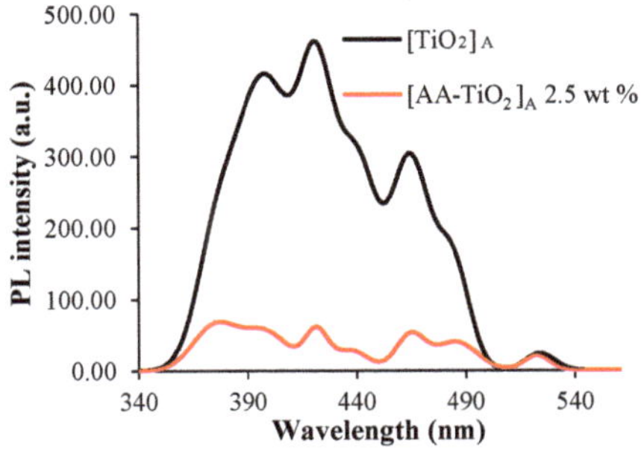

Figure 5. Photoluminescence (PL) spectra of $[TiO_2]_A$ (black line) and $[AA\text{-}TiO_2]_A$ with AA amount of 2.5 wt % (red line) excited at 290 nm in the wavelength range of 300–900 nm.

3.2. Optical Characterization

Figure 6a shows the UV-Vis spectral change due to different AA amounts wt % of $[PP@AA\text{-}TiO_2]_A$ photocatalysts, demonstrating an increase in the absorption in the range of 370–570 nm. The appearance of a yellow-brown color on the modified $[AA\text{-}TiO_2]_A$ paste can be explained as a result of an intense ligand to metal charge transfer (LMCT) transition [27]; this is also clearly evidenced in the DRS spectra reported in Figure 6b that shows the change of optical properties of modified $[PP@AA\text{-}TiO_2]_A$ as a function of the AA amount.

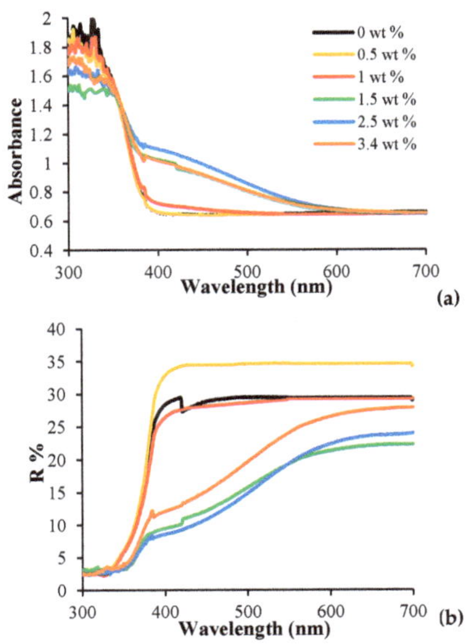

Figure 6. *Cont.*

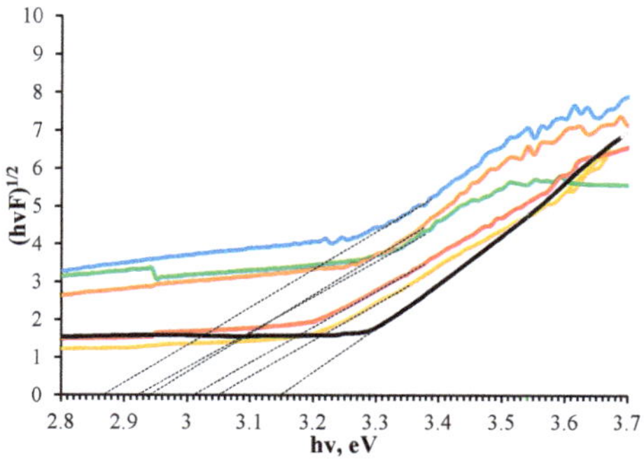

Figure 6. (**a**) UV-Vis Light Diffuse Reflectance spectra (DRS); (**b**) DRS spectra; (**c**) E_g values, calculated with Kubelka–Munk method of [PP@AA-TiO_2]$_A$ photocatalysts containing different AA from 0 to 3.4 wt %.

The DRS spectrum of pure [PP@TiO_2]$_A$ presents a sharp adsorption edge around 390 nm attributed to the electron's excitation from the VB to CB (band gap 3.2 for Anatase [27] while AA does not absorb any light above 300 nm [27]). Generally, activated samples show a shift of absorption peak in the visible part of spectrum and, in particular, contain an extended absorption edge above 400 nm and a broad absorption peak between 550 and 900 nm [44]. The formation of a Ti^{IV}-AA surface complex results from a change of the absorption threshold [27] of the modified [PP@AA-TiO_2]$_A$ photocatalyst that shifted towards the visible part of the spectrum. In fact, in this case, the modified sample presents a long tail extending up to ca. 600 nm as a consequence of the charge transfer complex formation between Ti^{IV} atoms. AA introduces electronic states that are spread across the band gap, resulting in a diffused absorption spectrum [24]. These results confirm the formation of the AA-TiO_2 charge-transfer complex into TiO_2 paste that could narrow the energy band gap (E_g) of the modified [PP@AA-TiO_2]$_A$ photocatalyst. E_g values of pure and modified samples were calculated by applying the Kubelka-Munk method [32] by the linear fit of the curves of Figure 6c, where F represents the Kubelka-Munk function, obtaining the respective values (Table 2) by the intercept in the x-axis.

The results of Table 2 clearly show that the surface modification with AA positively influences the E_g, improving the photocatalytic activity. In fact, the E_g value is around 3.15 eV for [PP@TiO_2]$_A$ [32], while the minimum E_g value of 2.87 eV is obtained for [PP@AA-TiO_2]$_A$ in the presence of AA 2.5 wt %.

Table 2. E_g , k_{ads} and k_{photo} values for [PP@AA-TiO_2]$_A$ containing different AA wt %.

AA wt %	E_g	10^2 k_{ads} (min^{-1})	10^2 k_{photo} (min^{-1})
0	3.15	3.75	1.99
0.5	3.05	3.55	3.50
1	3.01	3.47	3.72
1.5	2.92	3.46	4.04
2.5	2.87	3.18	4.15
3.4	2.94	3.09	3.85

3.3. Equilibrium and Kinetic Studies of ARS Adsorption

In order to study the presence of surface modifiers and how they influence ARS adsorption on [PP@AA-TiO_2]$_A$ photocatalyst, the adsorption data has been analyzed by the application of the adsorption isotherm models of Freundlich and Langmuir [29]. While ARS adsorption on [PP@TiO_2]$_A$ photocatalyst occurred according to the Freundlich isotherm model [29], instead the results obtained with [PP@AA-TiO_2]$_A$ (Figure 7) showed that the dye adsorption fitted well the Langmuir isotherm model $C_e/Q_e = 1/K_L + a_L C_e/K_L$, where C_e is the ARS solution concentration (mol L^{-1}), Q_e is the adsorbed ARS amount at equilibrium (mol L^{-1}). The results showed Langmuir constants K_L and

a_L of 8.16 and 1.24 × 10^5, respectively, a theoretical saturation capacity of the TiO_2 surface Q_0 of 6.56 × 10^{-5} indicating therefore a change in the adsorption mechanism due to the presence of AA. [PP@AA-TiO_2]$_A$ permitted therefore the adsorption of AA molecules in monolayer mode and with the same adsorption energy.

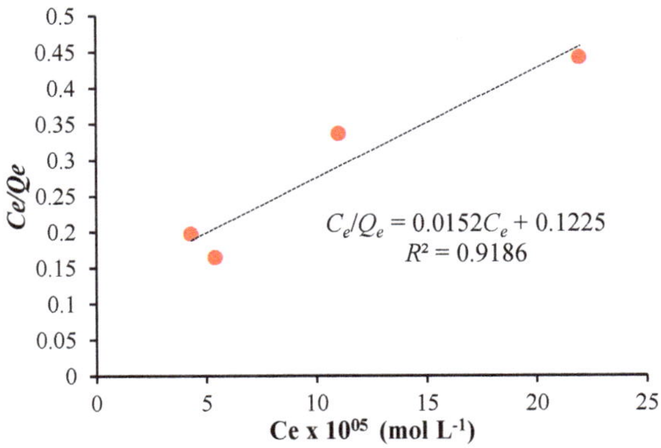

Figure 7. Langmuir isotherm graph for ARS adsorption on [PP@AA-TiO_2]$_A$.

3.4. *Visible Light Photoactivity of [PP@AA-TiO_2]$_A$*

ARS adsorption under dark condition and photodegradation under visible light were monitored by the decrease of the ARS absorption spectra at 424 nm in acidic conditions.

As reported in Figure 8a, the adsorption of ARS on the modified [PP@AA-TiO_2]$_A$ photocatalyst follows a pseudo first order kinetic in which the adsorption kinetic constant k_{ads} is expressed by the equation $\ln[(q_e - q_t)]/q_e = k_{ads}t$, where q_t and q_e are the amount of adsorbed dye at time t and its equilibrium concentration, respectively [29–32].

In addition, in the presence of [PP@AA-TiO_2]$_A$ the photodegradation rate becomes proportional to the ARS concentration during time, in accordance with zero order kinetic, $[ARS]_t = -k_{photo}t$ [45], where k_{photo} is the photodegradation kinetic constant, while $[ARS]_t$ is ARS concentration at time t. In particular, in this case, the simultaneous presence of two components as AA and ARS onto TiO_2, influenced the kinetic order of the photodegradation process with respect to that of first-order kinetics with only TiO_2, probably due to AA distribution on the TiO_2 surface. Table 2 reports the results obtained for k_{ads} and k_{photo} values relative to the process of ARS 5.84 × 10^{-5} mol L^{-1} on [PP@AA-TiO_2]$_A$ prepared with different AA wt %, while in Figure 8 shows the linear graphs about the kinetics of adsorption (Figure 8a) and photodegradation (Figure 8b).

According to Table 2, as shown in Figure 8c, a correlation among the k_{ads} and the AA amount used for the preparation of the modified photocatalyst is observed demonstrating that by increasing the concentration of AA in the TiO_2 paste, a decrease of the k_{ads} was found. In addition, an increase of the concentration of AA, corresponds to a decrease of E_g and an increase of k_{photo} until the sample with AA concentration of 2.5 wt % after which both show a reverse trend (Figure 8d).

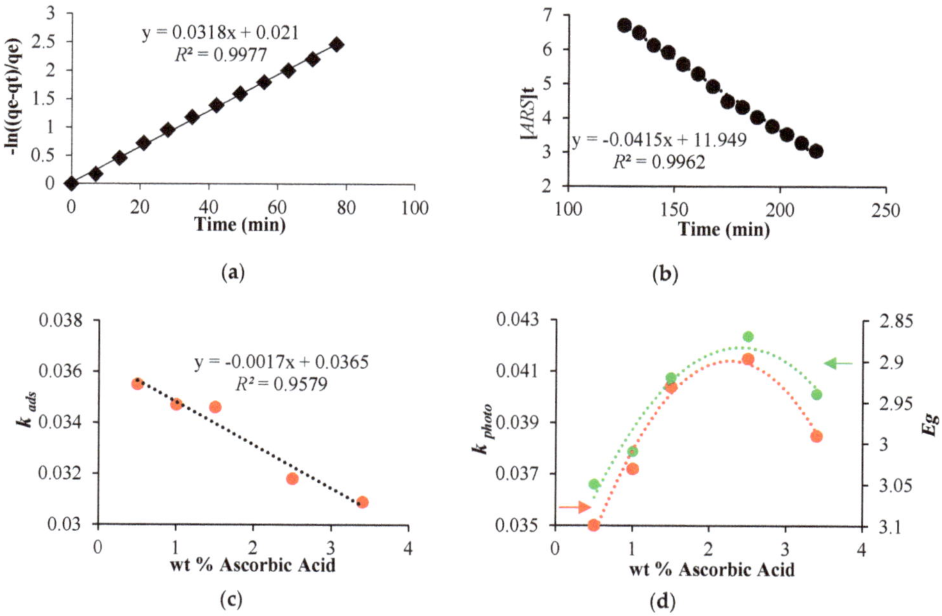

Figure 8. (**a**) Pseudo first order adsorption kinetic; (**b**) zero order photodegradation kinetic by using $[PP@AA\text{-}TiO_2]_A$ containing 2.5 wt %; (**c**) correlation between k_{ads} and AA wt %; (**d**) correlations between the k_{photo} and E_g vs. AA wt %.

In Figure 9 is reported the ARS photodegradation time for different $[PP@AA\text{-}TiO_2]_A$, $[PP@\text{-}TiO_2]_A$ photocatalysts, and without photocatalyst under visible light, these results show the positive effects of the presence of AA; in particular the best condition is obtained with AA 2.5 wt %.

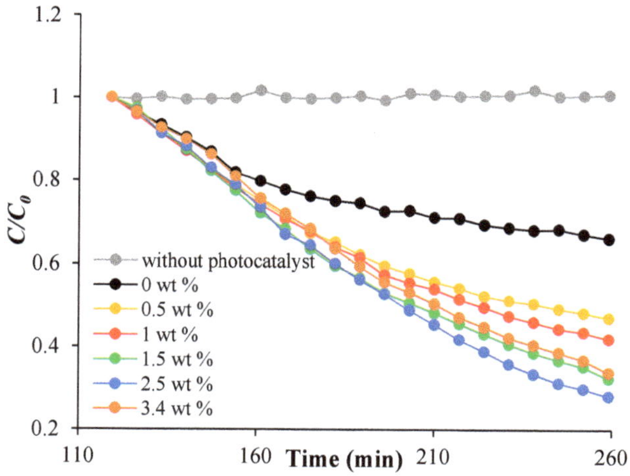

Figure 9. ARS photodegradation versus time for $[PP@\text{-}TiO_2]_A$, different $[PP@AA\text{-}TiO_2]_A$ photocatalysts and without photocatalyst under visible light.

The mechanism of the photocatalytic process under visible light of the modified [PP@AA-TiO_2]$_A$ photocatalyst can be resumed as reported in Figure 10. When [PP@AA-TiO_2]$_A$ is irradiated by visible light, an electron transfer from the AA to CB semiconductor occurred and superoxide molecular ions are formed by the presence of molecular oxygen. Then, the formed radicals drive the photocatalytic degradation of ARS adsorbed on a modified photocatalyst [46].

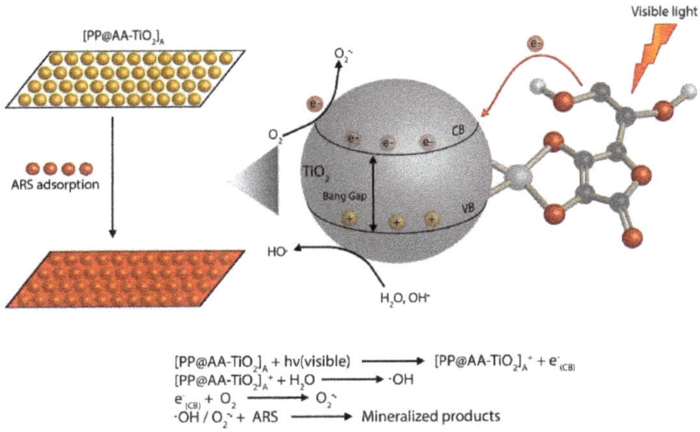

Figure 10. Mechanism of photocatalytic process by using [PP@AA-TiO_2]$_A$ for the degradation of ARS solution.

4. Conclusions

A [PP@AA-TiO_2]$_A$ yellow-brown photocatalyst with high visible-light photocatalytic activity in the ARS dye degradation was successfully and for the first time realized by cover PP material with a specific TiO_2 paste modified with AA.

The new modified photocatalyst has been widely characterized by using SEM, XRD, XPS, PL and Raman techniques. SEM images reveal that the particle sizes changed as a consequence of the addition of AA to TiO_2, XRD measurements demonstrate a lattice expansion with consequent incorporation of AA inside the crystalline lattice of TiO_2 material, while XPS measurements showed a superficial change of Ti oxidation state that change from Ti(III) to Ti(IV).

In addition, due to the interaction of AA with TiO_2, a lower PL emission intensity has been obtained, demonstrating a lower charge recombination that enhances the photo-produced electron transition to AA with an improved electron-hole separation.

The experimentally calculated E_g values, by using [PP@AA-TiO_2]$_A$ photocatalysts, decreased with the increase of surface modifier concentration according to the increase of performances. The best E_g value of 2.87 eV obtained with 2.5 wt % of AA corresponds to a k_{photo} of 0.0415. The obtained results have demonstrated that this new photocatalyst has proved to be 2.08 times more effective of only TiO_2 prepared in absence of AA.

In addition, a change in the adsorption mechanism with respect to that of pure [PP@TiO_2]$_A$ has been observed, while kinetic studies on the photocatalytic performance of [PP@AA-TiO_2]$_A$ in the visible light photodegradation of ARS showed an improvement of the photocatalytic efficiency that is strictly correlated with E_g values.

Supplementary Materials: The following are available online at http://www.mdpi.com/2079-4991/8/8/599/s1, Figure S1: Raman spectra of [PP@TiO_2]$_A$ and [PP@AA-TiO_2]$_A$ containing 2.5 wt % of AA. Figure S2. Raman spectra of [PP@TiO_2]$_A$ and [PP@AA-TiO_2]$_A$ containing 2.5 wt % of AA. Figure S3. Gaussian fitted PL spectra of [TiO_2]$_A$ (a) and [AA-TiO_2]$_A$ (b), dotted line: fitting of deconvolution study.

Author Contributions: C.A.D. and R.G. proposed and designed the experiments; C.A.D., M.Z. performed the experiments; M.Z. and E.R. analyzed the data; M.M., R.G. and A.D.C. physically characterized the materials. R.G. contributed to reagents, materials, analysis tools; C.A.D., R.G. and M.Z. wrote the paper. All the authors participated in discussions of the research.

Funding: Financial support derived from the FAR Project 2015–2017 of Camerino University.

Conflicts of Interest: The authors declare no conflict of interest.

References

1. Park, H.; Park, Y.; Kim, W.; Choi, W. Surface modification of TiO_2 photocatalyst for environmental applications. *J. Photochem. Photobiol. C* **2013**, *15*, 1–20. [CrossRef]
2. Ou, Y.; Lin, J.D.; Zou, H.M.; Liao, D.W. Effects of surface modification of TiO_2 with ascorbic acid on photocatalytic decolorization of an azo dye reactions and mechanisms. *J. Mol. Catal. A Chem.* **2005**, *241*, 59–64. [CrossRef]
3. Daghrir, R.; Drogui, P.; Robert, D. Modified TiO_2 for environmental photocatalytic applications: A review. *Ind. Eng. Chem. Res.* **2013**, *52*, 3581–3599. [CrossRef]
4. Truppi, A.; Petronella, F.; Placido, T.; Striccoli, M.; Agostiano, A.; Curri, M.L.; Comparelli, R. Visible-light-active TiO_2-based hybrid nanocatalysts for environmental applications. *Catalysts* **2017**, *7*, 100. [CrossRef]
5. Petronella, F.; Truppi, A.; Ingrosso, C.; Placido, T.; Striccoli, M.; Curri, M.L.; Agostiano, A.; Comparelli, R. Nanocomposite materials for photocatalytic degradation of pollutants. *Catal. Today* **2017**, *281*, 85–100. [CrossRef]
6. Schneider, J.; Matsuoka, M.; Takeuchi, M.; Zhang, J.; Horiuchi, Y.; Anpo, M.; Bahnemann, D.W. Understanding TiO_2 Photocatalysis: Mechanisms and Materials. *Chem. Rev.* **2014**, *114*, 9919–9986. [CrossRef] [PubMed]
7. Bajic, V.; Spremo-Potparevic, B.; Zivkovic, L.; Cabarkapa, A.; Kotur-Stevuljevic, J.; Isenovic, E.; Sredojevic, D.; Vukojea, I.; Lazic, V.; Ahrenkiel, S.P.; et al. Surface-modified TiO_2 nanoparticles with ascorbic acid: Antioxidant properties and efficiency against DNA damage in vitro. *Colloids Surf. B* **2017**, *155*, 323–331. [CrossRef] [PubMed]
8. Shetty, P.K.; Venuvanka, V.; Jagani, H.V.; Chethan, G.H.; Ligade, V.S.; Musmade, P.M.; Nayak, U.Y.; Reddy, M.S.; Kalthur, G.; Udupa, N.; et al. Development and evaluation of sunscreen creams containing morin-encapsulated nanoparticles for enhanced UV radiation protection and antioxidant activity. *Int. J. Nanomed.* **2015**, *10*, 6477–6491. [CrossRef]
9. Wang, Q.; Huang, J.Y.; Li, H.Q.; Zhao, A.Z.; Wang, Y.; Zhang, K.Q.; Sun, H.T.; Lai, Y.K. Recent advances on smart TiO_2 nanotube platforms for sustainable drug delivery applications. *Int. J. Nanomed.* **2017**, *12*, 151–165. [CrossRef] [PubMed]
10. Schneider, J.; Bahnemann, D.; Ye, J.; Li Puma, G.; Dionysios, D.D. *Photocatalysis: Fundamentals and Perspectives*; Royal Society of Chemistry: London, UK, 2016; ISBN 978-1-78262-041-9.
11. Cinar, Z. The role of molecular modeling in TiO_2 photocatalysis. *Molecules* **2017**, *22*, 556. [CrossRef] [PubMed]
12. Panniello, A.; Curri, M.L.; Diso, D.; Licciulli, A.; Locaputo, V.; Agostiano, A.; Comparelli, R.; Mascolo, G. Nanocrystalline TiO_2 based films onto fibers for photocatalytic degradation of organic dye in aqueous solution. *Appl. Catal. B Environ.* **2012**, *121–122*, 190–197. [CrossRef]
13. Stolarczyk, J.K.; Bhattacharyya, S.; Polavarapu, L.; Feldmann, J. Challenges and Prospects in Solar Water Splitting and CO_2 Reduction with Inorganic and Hybrid Nanostructures. *ACS Catal.* **2018**, *8*, 3602–3635. [CrossRef]
14. Ibhadon, A.O.; Fitzpatrick, P. Heterogeneous Photocatalysis: Recent Advances and Applications. *Catalysts* **2013**, *3*, 189–218. [CrossRef]
15. Ma, Y.; Wang, X.; Jia, Y.; Chen, X.; Han, H.; Li, C. Titanium dioxide-based nanomaterials for photocatalytic fuel generations. *Chem. Rev.* **2014**, *114*, 9987–10043. [CrossRef] [PubMed]
16. Zaleska, A. Doped-TiO_2: A Review. *Recent Patents Eng.* **2008**, *2*, 157–164. [CrossRef]
17. Di Valentin, C.; Pacchioni, G. Trends in non-metal doping of anatase TiO_2: B, C, N and F. *Catal. Today* **2013**, *206*, 12–18. [CrossRef]
18. Zhuang, H.; Zhang, Y.; Chu, Z.; Long, J.; An, X.; Zhang, H.; Lin, H.; Zhang, Z.; Wang, X. Synergy of metal and nonmetal dopants for visible-light photocatalysis: A case-study of Sn and N co-doped TiO_2. *Phys. Chem. Chem. Phys.* **2016**, *18*, 9636–9644. [CrossRef] [PubMed]

19. Roy, N.; Sohn, Y.; Leung, K.T.; Pradhan, D. Engineered electronic states of transition metal doped TiO_2 nanocrystals for low overpotential oxygen evolution reaction. *J. Phys. Chem. C* **2014**, *118*, 29499–29506. [CrossRef]
20. Ola, O.; Maroto-Valer, M.M. Transition metal oxide based TiO_2 nanoparticles for visible light induced CO_2 photoreduction. *Appl. Catal. A Gen.* **2015**, *502*, 114–121. [CrossRef]
21. Jang, D.M.; Kwak, I.H.; Kwon, E.L.; Jung, C.S.; Im, H.S.; Park, K.; Park, J. Transition-metal doping of oxide nanocrystals for enhanced catalytic oxygen evolution. *J. Phys. Chem. C* **2015**, *119*, 1921–1927. [CrossRef]
22. Das, T.K.; Ilaiyaraja, P.; Mocherl, P.S.V.; Bhalerao, G.M.; Sudakar, C. Influence of surface disorder, oxygen defects and bandgap in TiO_2 nanostructures on the photovoltaic properties of dye sensitized solar cells. *Sol. Energy Mater. Sol. Cells C* **2016**, *144*, 194–209. [CrossRef]
23. Chen, L.X.; Rajh, T.; Jäger, W.; Nedeljkovic, J.; Thurnauer, M.C. X-ray absorption reveals surface structure of titanium dioxide nanoparticles. *J. Synchrotron Rad.* **1999**, *6*, 445–447. [CrossRef] [PubMed]
24. Mert, E.H.; Yalçın, Y.; Kılıç, M.; San, N.; Çınar, Z. Surface modification of TiO_2 with ascorbic acid for heterogeneous photocatalysis: Theory and experiment. *J. Adv. Oxid. Technol.* **2008**, *11*, 199–207. [CrossRef]
25. Shah, M.W.; Zhu, Y.; Fan, X.; Zhao, J.; Li, Y.; Asim, S.; Wang, C. Facile Synthesis of Defective TiO_{2-x} Nanocrystals with High Surface Area and Tailoring Bandgap for Visible-light Photocatalysis. *Sci. Rep.* **2015**, *5*, 15804. [CrossRef] [PubMed]
26. Buettner, K.M.; Collins, J.M.; Valentine, A.M. Titanium(IV) and Vitamin C: Aqueous complexes of a bioactive form of Ti(IV). *Inorg. Chem.* **2012**, *51*, 11030–11039. [CrossRef] [PubMed]
27. Xagas, A.P.; Bernard, M.C.; Hugot-Le Goff, A.; Spyrellis, N.; Loizos, Z.; Falaras, P. Surface modification and photosensitisation of TiO_2 nanocrystalline films with ascorbic acid. *Photochem. Photobiol. A* **2000**, *132*, 115–120. [CrossRef]
28. Pichat, P. *Photon-Involving Purification of Water and Air*; MDPI: Basel, Switzerland, 2018; ISBN 978-3-03842-700-1.
29. Giovannetti, R.; D'Amato, C.A.; Zannotti, M.; Rommozzi, E.; Gunnella, R.; Minicucci, M.; Di Cicco, A. Visible light photoactivity of polypropylene coated nano-TiO_2 for dyes degradation in water. *Sci. Rep.* **2015**, *5*, 17801. [CrossRef] [PubMed]
30. Giovannetti, R.; Rommozzi, E.; D'Amato, C.A.; Zannotti, M. Kinetic model for simultaneous adsorption/photodegradation process of Alizarin Red S in water solution by nano-TiO_2 under visible light. *Catalysts* **2016**, *6*, 84. [CrossRef]
31. Giovannetti, R.; Rommozzi, E.; Zannotti, M.; D'Amato, C.A.; Ferraro, S.; Cespi, M.; Bonacucina, G.; Minicucci, M.; Di Cicco, A. Exfoliation of graphite into graphene in aqueous solution: An application as graphene/TiO_2 nanocomposite to improve visible light photocatalytic activity. *RSC Adv.* **2016**, *6*, 93048–93055. [CrossRef]
32. D'Amato, C.A.; Giovannetti, R.; Zannotti, M.; Rommozzi, E.; Ferraro, S.; Seghetti, C.; Minicucci, M.; Gunnella, R.; Di Cicco, A. Enhancement of visible-light photoactivity by polypropylene coated plasmonic Au/TiO_2 for dye degradation in water solution. *Appl. Surf. Sci.* **2018**, *441*, 575–587. [CrossRef]
33. Tsega, M.; Dejene, F.B. Influence of acidic pH on the formulation of TiO_2 nanocrystalline powders with enhanced photoluminescence property. *Heliyon* **2017**, *3*, e00246. [CrossRef] [PubMed]
34. Li, L.; Yan, J.; Wang, T.; Zhao, Z.J.; Zhang, J.; Gong, J.; Guan, N. Sub-10 nm rutile titanium dioxide nanoparticles for efficient visible-light-driven photocatalytic hydrogen production. *Nat. Commun.* **2015**, *6*, 5881. [CrossRef] [PubMed]
35. Lin, Z.; Liu, P.; Yana, J.; Yang, G. Matching energy levels between TiO_2 and α-Fe_2O_3 in a core–shell nanoparticle for visible-light photocatalysis. *J. Mater. Chem. A* **2015**, *3*, 14853–14863. [CrossRef]
36. Pan, S.S.; Lu, W.; Zhao, Y.H.; Tong, W.; Li, M.; Jin, L.M.; Choi, J.Y.; Qi, F.; Chen, S.G.; Fei, L.F.; et al. Self-doped rutile titania with high performance for direct and ultrafast assay of H_2O_2. *Appl. Mater. Interfaces* **2013**, *5*, 12784–12788. [CrossRef] [PubMed]
37. An, H.R.; Park, S.Y.; Kim, H.; Lee, C.Y.; Choi, S.; Lee, S.C.; Seo, S.; Park, E.C.; Oh, Y.K.; Song, C.G.; et al. Advanced nanoporous TiO_2 photocatalysts by hydrogen plasma for efficient solar-light photocatalytic application. *Sci. Rep.* **2016**, *6*, 29683. [CrossRef] [PubMed]
38. Mehta, M.; Kodan, N.; Kumar, S.; Kaushal, A.; Mayrhofer, L.; Walter, M.; Moseler, M.; Dey, A.; Krishnamurthy, S.; Basu, S.; et al. Hydrogen treated anatase TiO_2: A new experimental approach and further insights from theory. *J. Mater. Chem. A* **2016**, *4*, 2670–2681. [CrossRef]

39. Fernández-González, R.; Julián-López, B.; Cordoncillo, E.; Escribano, P. New insights on the structural and optical properties of Ce–Ti mixed oxide nanoparticles doped with praseodymium. *J. Mater. Chem.* **2011**, *21*, 497–504. [CrossRef]
40. Bharti, B.; Kumar, S.; Lee, H.N.; Kuma, R. Formation of oxygen vacancies and Ti^{3+} state in TiO_2 thin film and enhanced optical properties by air plasma treatment. *Sci. Rep.* **2016**, *6*, 32355. [CrossRef] [PubMed]
41. Wu, C.Y.; Tu, K.J.; Deng, J.P.; Lo, Y.S.; Wu, C.H. Markedly enhanced surface hydroxyl groups of TiO_2 nanoparticles with superior water-dispersibility for photocatalysis. *Materials* **2017**, *10*, 566. [CrossRef] [PubMed]
42. Krishnan, P.; Liu, M.; Itty, P.A.; Liu, Z.; Rheinheimer, V.; Zhang, M.H.; Monteiro, P.J.M.; Yu, L.E. Characterization of photocatalytic TiO_2 powder under varied environments using near ambient pressure X-ray photoelectron spectroscopy. *Sci. Rep.* **2017**, *7*, 43298. [CrossRef] [PubMed]
43. Kim, H.B.; Jang, D.J. Morphological variation of anatase TiO_2 crystals via formation of titanium glycerolate precursors under microwave irradiation. *CrystEngComm* **2015**, *17*, 3325–3332. [CrossRef]
44. Choudhury, B.; Dey, M.; Choudhury, A. Defect generation, d-d transition, and band gap reduction in Cu-doped TiO_2 nanoparticles. *Int. Nano Lett.* **2013**, *3*, 25. [CrossRef]
45. Rumamurthy, V.; Schanze, K.S. *Semiconductor Photochemistry and Photophysics*; Marcel Dekker, Inc.: New York, NY, USA, 2003; Volume 10, ISBN 0824709586.
46. Zhou, N.; Lopez-Puente, V.; Wang, Q.; Polavarapu, L.; Pastoriza-Santos, I.; Xu, Q.-H. Plasmon-enhanced light harvesting: Applications in enhanced photocatalysis, photodynamic therapy and photovoltaics. *RSC Adv.* **2015**, *5*, 29076–29097. [CrossRef]

© 2018 by the authors. Licensee MDPI, Basel, Switzerland. This article is an open access article distributed under the terms and conditions of the Creative Commons Attribution (CC BY) license (http://creativecommons.org/licenses/by/4.0/).

nanomaterials

Article

Synergistic Effects of Active Sites' Nature and Hydrophilicity on the Oxygen Reduction Reaction Activity of Pt-Free Catalysts

Mariangela Longhi [1,*], Camilla Cova [1], Eleonora Pargoletti [1], Mauro Coduri [2], Saveria Santangelo [3], Salvatore Patanè [4], Nicoletta Ditaranto [5], Nicola Cioffi [5], Anna Facibeni [6] and Marco Scavini [1]

1 Dipartimento di Chimica, Università degli Studi di Milano, Via Golgi 19, 20133 Milano, Italy; camilla.cova@studenti.unimi.it (C.C.); eleonora.pargoletti@unimi.it (E.P.); marco.scavini@unimi.it (M.S.)
2 ESRF—The European Synchrotron, 71, Avenue des Martyrs, 38043 Grenoble, France; mauro.coduri@esrf.fr
3 Dipartimento di Ingegneria Civile, dell'Energia, dell'Ambiente e dei Materiali, Università "Mediterranea", Via Graziella, Loc. Feo di Vito, 89122 Reggio Calabria, Italy; saveria.santangelo@unirc.it
4 Dipartimento di Scienze Matematiche e Informatiche, Scienze Fisiche e Scienze della Terra, Università di Messina, Viale Stagno d'Alcontres 31, 98166 Messina, Italy; salvatore.patane@unime.it
5 Dipartimento di Chimica, Università degli Studi di Bari Aldo Moro, Via Orabona 4, 70125 Bari, Italy; nicola.cioffi@uniba.it (N.D.); nicoletta.ditaranto@uniba.it (N.C.)
6 Dipartimento di Energia and NEMAS—Centre for NanoEngineered MAterials and Surfaces, Politecnico di Milano, Via Ponzio 34/3, 20133 Milano, Italy; anna.facibeni@polimi.it
* Correspondence: mariangela.longhi@unimi.it; Tel.: +39-02-5031-4226

Received: 27 July 2018; Accepted: 21 August 2018; Published: 22 August 2018

Abstract: This work highlights the importance of the hydrophilicity of a catalyst's active sites on an oxygen reduction reaction (ORR) through an electrochemical and physico-chemical study on catalysts based on nitrogen-modified carbon doped with different metals (Fe, Cu, and a mixture of them). BET, X-ray Powder Diffraction (XRPD), micro-Raman, X-ray Photoelectron Spectroscopy (XPS), Scanning Electron Microscopy (SEM), Scanning Transmission Electron Microscopy (STEM), and hydrophilicity measurements were performed. All synthesized catalysts are characterized not only by a porous structure, with the porosity distribution centered in the mesoporosity range, but also by the presence of carbon nanostructures. In iron-doped materials, these nanostructures are bamboo-like structures typical of nitrogen carbon nanotubes, which are better organized, in a larger amount, and longer than those in the copper-doped material. Electrochemical ORR results highlight that the presence of iron and nitrogen carbon nanotubes is beneficial to the electroactivity of these materials, but also that the hydrophilicity of the active site is an important parameter affecting electrocatalytic properties. The most active material contains a mixture of Fe and Cu.

Keywords: oxygen reduction reaction; Pt-free catalysts; CNT N-doped carbons; active site hydrophilicity

1. Introduction

An oxygen reduction reaction (ORR) is a fundamental step in many electrochemical applications, e.g., fuel cells and zinc/air batteries. Being kinetically hindered, an ORR requires efficient catalysts typically based on Platinum Group Metals [1–3]. These materials, though very efficient, present some drawbacks, such as costs, availability, and technological problems. Finding cheaper substitutes with similar or better electrocatalytic properties and stability is a challenge for the scientific community [4–17]. Effective candidates are based on metal–nitrogen–carbon catalysts, in which the metal is usually iron or cobalt [4–11]. Many studies have been devoted to highlighting the complex nature and chemical environment of catalytic sites [12,13,18]. Principal results show that several types of active sites are present on the catalyst surface [14], and that iron coordination complexes with nitrogen promote ORRs to water [15]. Moreover, all these active centers are embedded in graphene planes [19]. Recently, by using N-doped carbon catalysts with supported and not embedded metal centers with different natures, specific surface areas, types, and amounts of surface nitrogen, it has been proposed that the hydrophilicity of the catalyst support has an important role on ORR activity [20]. In fact, a high hydrophilicity allows for a high dispersion of metal-containing active sites and a better accessibility of oxygen to catalytic centers [20]. However, a recent work has shown that, for nanoporous nitrogen-doped carbons, the adsorption of water molecules depends on the local density of nitrogen atoms rather than their nature [21]. Therefore, the simultaneous use of different hydrophilic supports, nitrogen amounts, and specific surface areas [20] could lead to cumulative effects, masking the importance of hydrophilicity either of the supporting carbon, as stated in [20], or of the active centers.

The aim of this work is to demonstrate the importance of the hydrophilicity of the active sites for ORRs. For this purpose, an electrochemical and physico-chemical study was carried out on catalysts based on nitrogen-modified carbon doped with different metals (Fe, Cu, and a mixture of them), having a comparable specific surface area and amount of nitrogen to avoid hiding cumulative effects.

The studied catalysts were synthesized by thermal decomposition of a mixture of a sugar, guanidine acetate, metal salts, and silica as templating agent. In the following, catalysts containing Fe, Cu, and Fe–Cu will be respectively labelled S_GA_Fe, S_GA_Cu, and S_GA_FeCu, whereas S_GA will indicate a metal-free (reference) sample.

2. Materials and Methods

2.1. Catalyst Preparation

All chemicals/solvents were purchased from Sigma Aldrich (Milano, Italy) and used as received without further purification. Gelling sugar (Südzucher AG, Mannheim, Germany) was purchased in a market (sucrose:pectin = 98%:2%, from NMR). A Pt-based commercial catalyst (EC20, 20% Pt/carbon) was tested and used as a reference material.

In the synthesis, gelling sugar (3 g), guanidine acetate (2 g), and metal (Me) acetate (1 wt.% total metal ion calculated on the total mass of sugar and guanidine acetate, Me = Fe, Cu, or a Fe:Cu = 0.5 wt.%:0.5 wt.% mixture) were added to silica powder (4.3 g) and mixed. This mixture underwent a first heating step (T = 600 °C, 1 h, N_2 flow rate = 100 $cm^3\ min^{-1}$) followed by lixiviation in boiling NaOH (3 mol dm^{-3}) to remove silica. After washing, the carbon was dried (T = 110 °C, 24 h, N_2) and, finally, pyrolysed (T = 900 °C, 3 h, N_2 flow rate = 100 $cm^3\ min^{-1}$) to activate catalytic sites.

2.2. Physical Characterizations

Specific surface area and porosity distribution were obtained from N_2 adsorption/desorption isotherms at 77 K using a Micromeritics Tristar II 3020 (Micromeritics, Milano, Italy) apparatus and the instrumental software (Version 1.03) and applying Brunauer–Emmett–Teller (BET) and Barrett–Joyner–Halenda analyses, respectively. Prior to measuring, sample powders were heat-treated (T = 150 °C, 4 h, N_2) to remove adsorbed foreign species.

Micro-Raman spectra were recorded in air at room temperature using a NTEGRA-Spectra SPM spectrometer (NT-MDT, Moscow, Russia) (350 mm monochromator, MS3504i, ANDOR Idus CCD cooled at T = −60 °C). Raman scattering was excited by a thermo-cooled solid-state laser (λ = 532 nm; 2.33 eV). The scattered light from the sample was collected using a 100x (0.75 Numerical Aperture) Mitutoyo objective. Spectra acquired from different random positions on each specimen were averaged to have a reliable picture of the sample bulk. The average spectra were analyzed by using a commercially available software package [22] to assess the graphitization degree of samples.

X-ray Photoelectron Spectroscopy (XPS) analyses were run on a PHI 5000 Versa Probe II Scanning XPS Microprobe spectrometer (ULVAC-PHI Inc., Kanagawa, Japan). Measurements were done with a monochromatic Al Kα source (X-ray spot 100 μm) at a power of 24.8 W. Wide scans and detailed spectra were acquired in Fixed Analyzer Transmission (FAT) mode with a pass energy of 117.40 eV and 29.35 eV, respectively. An electron gun was used for charge compensation (1.0 V 20.0 μA). Data processing was performed by using the MultiPak software v. 9.5.0.8 (ULVAC-PHI Inc., Kanagawa, Japan).

X-ray Powder Diffraction (XRPD) patterns were collected at room temperature at the ID22 beamline of the ESRF with a wavelength λ = 0.399946(4) Å up to 2θ = 40° and a ~0.5 h/pattern counting time. Data recorded continuously have been then merged using different step sizes spanning from $\Delta 2\theta$ = 0.005° to 0.050° to highlight different features in the patterns. Patterns merged with the former step size were analyzed via the Rietveld method using the General Structure Analysis System (GSAS) suite of the program [23].

Scanning Electron Microscopy (SEM) observations were carried out using a Leo 1430 SEM (Zeiss, Oberkochen, Germany).

Scanning Transmission Electron Microscopy (STEM) measurements were performed by using a Zeiss Supra 40 Field Emission Scanning Electron Microscope (FE-SEM) (Zeiss, Oberkochen, Germany) working in high vacuum conditions and equipped with the GEMINI column.

Wettability features of carbons were evaluated by a Krüss Easy Drop instrument (Krüss, Hamburg, Germany) using a water drop (V = 7 μL) gently placed on the surface of the samples compacted as a flat layer. By taking into account the complexity of the system and the relationship between the wetting properties and the physico-chemical features of the materials (the size of particles, the roughness of the samples, and their intrinsic surface free energy and packing degree) only the time necessary to achieve complete spreading was measured [20].

Electrochemical characterization was performed in 0.1 mol dm^{-3} KOH by the Thin Film Rotating Disk Electrode method using Cyclic Voltammetry (CV) as in [16]. ORR onset potential was evaluated at j = 1 mA cm^{-2} and the number of exchanged electrons was determined as described in Ref. [16].

The stability of materials was evaluated by comparing CV curves at 1600 rpm in O_2 at the beginning and at the end of electrochemical measurements (results not reported). In all cases, the stability was good.

3. Results and Discussion

Figure 1 shows experimental BET adsorption isotherms. All the samples exhibit a type IV isotherm with a H3-type hysteresis loop [24] characteristic of mesoporous materials and attributable to a non-fixed aggregation of plate-like particles giving rise to slit-shaped pores [25].

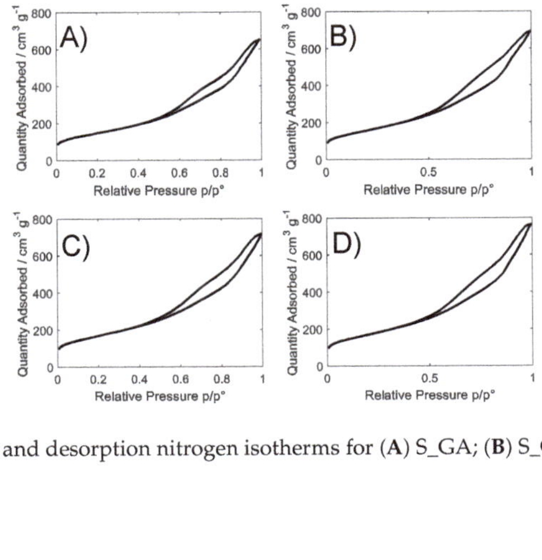

Figure 1. Adsorption and desorption nitrogen isotherms for (**A**) S_GA; (**B**) S_GA_Cu; (**C**) S_GA_Fe; and (**D**) S_GA_FeCu.

The specific surface area (Table 1) of samples is between ~520 and 600 m^2 g^{-1} without important differences among metal-doped samples (~560–600 m^2 g^{-1}, with a difference of 7%). A high mesoporosity (~55%), centered between 2 and 5 nm, characterizes all these materials (Table 1 and Figure S1). ORRs benefit from this pore range, since it favors the diffusion of reactants/products to/from active centers [26–28].

Table 1. Specific Surface Area (SSA) and Pore Area Distribution.

Sample Name	SSA/m^2 g^{-1}	Pores% $d < 2$ nm	Pores% $2 < d < 5$ nm	Pores% $5 < d < 20$ nm	Pores% $d > 20$ nm
S_GA	523	7.2	53.7	34.5	4.6
S_GA_Cu	556	3.9	56.3	35.6	4.2
S_GA_Fe	598	5.7	56.9	33.2	4.2
S_GA_FeCu	602	5.0	54.7	35.7	4.6

d = Pore diameter.

The influence of a metal precursor on the crystalline arrangement of the carbon phase in the catalysts was investigated by means of micro-Raman spectroscopy. Regardless of the presence and type of the metal, these materials exhibit the Raman profile peculiar to highly disordered graphitic carbons (Figure 2). Two very intense bands peaking at ~1360 cm^{-1} and ~1590 cm^{-1} dominate the lower frequency region (<2000 cm^{-1}) of the spectra. The band peaking at ~1360 cm^{-1} (D-band) is associated with the in-plane breathing mode of the C hexagonal rings [29]; the band peaking at ~1590 cm^{-1} (G-band) originates from the stretching of C=C pairs [29]. Moreover, a weaker and broader featureless second-order structure, typical of highly disordered carbon nanostructures, is detected at ~2700 cm^{-1}.

Since the D-band is generated by finite size effects and by lattice defects breaking the translational symmetry of graphitic layers [29], while the G-band is the fingerprint of the graphitic crystalline arrangement, the G/D integrated intensity ratio (IG/ID) is commonly regarded as a graphitization index [30] (the higher its value, the higher the graphitization degree of the sample).

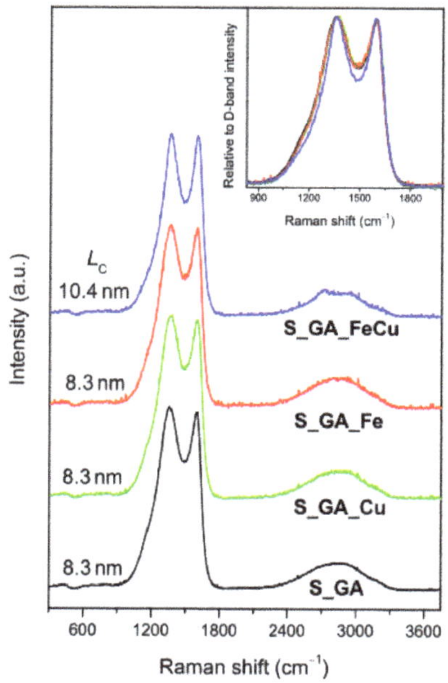

Figure 2. Micro-Raman spectra of the investigated catalysts (the average size of the graphitic crystallites (L_C), estimated from the G/D intensity ratio [31] is reported). Inset: comparison of the D- and G-band regions of the spectra.

No evident difference is observed in the lower frequency region of the spectra (inset of Figure 2) [22,29,30], but a slight D-band shrinking in the spectrum of S_GA_FeCu was observed. This hints at a little influence of both the presence and type of the metal on the crystalline arrangement of the carbon phase. The results of spectra decomposition, reported in detail in the Supplementary Material (Figure S2, Table S1), show that the highest graphitization degree (i.e., the greatest IG/ID value, 0.55 against 0.43–0.44) pertains to S_GA_FeCu. In this sample, the D-band is narrower than in the other metal-containing catalysts (183 against 211−215 cm^{-1}) and the average size of the graphitic crystallites, estimated from the relationship proposed by Cançado et al. [31] for nanographites, is larger (10.4 against 8.3 nm).

The results of XRPD analysis (Figure 3) fully agree with this picture. Despite the presence of different metals, all samples show broad signals at ~6.5, 11.1°, 13.3°, and 19.3° related to disordered graphitic carbon [16]. To highlight these peaks, in this figure patterns with a $\Delta 2\theta = 0.050°$ step size are shown with the corresponding Miller (hkl) indexes (Figure 3A). These signals are the only ones present in S_GA and S_GA_Fe patterns; conversely, in the other samples, Bragg peaks prove the presence of crystalline metal phases with a crystallite size of the order of ~400 nm. Rietveld refinements reveal that both S_GA_Cu and S_GA_FeCu patterns (Figure 3B,C) are suitably fitted by using one single ccp structure (space group Fm–3m) with sharp peaks and a carbon graphitic one (space group P63mc) where only a few broad peaks are apparent. In this case, a $\Delta 2\theta = 0.005°$ step size was adopted due to the sharpness of peaks of the metal phase.

The refined parameters are reported in Table S2. The lattice constants of the Fm–3m phase are 3.61711(1) and 3.61774(1) Å for S_GA_Cu and S_GA_FeCu, respectively. Though the difference is very small, the high resolution of the ID22 instrument makes it significant. Such a small expansion observed in S_GA_FeCu is consistent with a Cu-rich Cu1-xFex solid solution [32] as observed for xCu ≅ 0.03 at a T = 900 °C calcination temperature [33]. Biphasic Rietveld refinements show that the weight fraction of the metal phase is 2.6% and 1.6% in S_GA_Cu and S_GA_FeCu samples, respectively. We must warn the reader that, due to the noise in the broad peaks of the graphitic carbon peaks and to some arbitrariness in background subtraction, this phase analysis has to be considered as semi-quantitative. However, it points to an almost iron-free crystalline phase in the S_GA_FeCu sample. As in the case of S_GA_Fe one, most of the iron does not form extended crystalline phases.

Figures S3–S6 and Figure 4A,B report XPS results. The concentration of metal species on the surface ranges between 0.2 and 0.4%. Nitrogen and oxygen species are detected in all the samples (Figure S3, Table 2), but their concentration depends neither on the presence nor on the nature of the metal (Table 2), though, according to data in the literature, it would be expected to vary with the specific metal used in the synthesis [34]. This singular behavior, combined with the picture emerging from the micro-Raman and XRPD analyses, evidences the peculiarity of the present materials with respect to those generally described in the literature.

Table 2. Samples' Surface Atomic Composition *.

Sample Name	%C	%N	%O	%Fe	%Cu
S_GA	88.6 ± 0.7	8.3 ± 0.5	3.1 ± 0.6	-	-
S_GA_Cu	88.0 ± 0.5	6.9 ± 0.9	4.7 ± 1.3	-	0.4 ± 0.2
S_GA_Fe	88.6 ± 0.5	7.1 ± 0.5	4.0 ± 0.5	0.3 ± 0.2	-
S_GA_FeCu	89.4 ± 1.2	7.1 ± 0.5	3.0 ± 1.0	0.3 ± 0.2	0.2 ± 0.2

* The values are averaged out of three replicates. Error is expressed as the larger value between the error associated with a single quantification (0.2% for Cu and Fe, 0.5% for other elements) and one standard deviation.

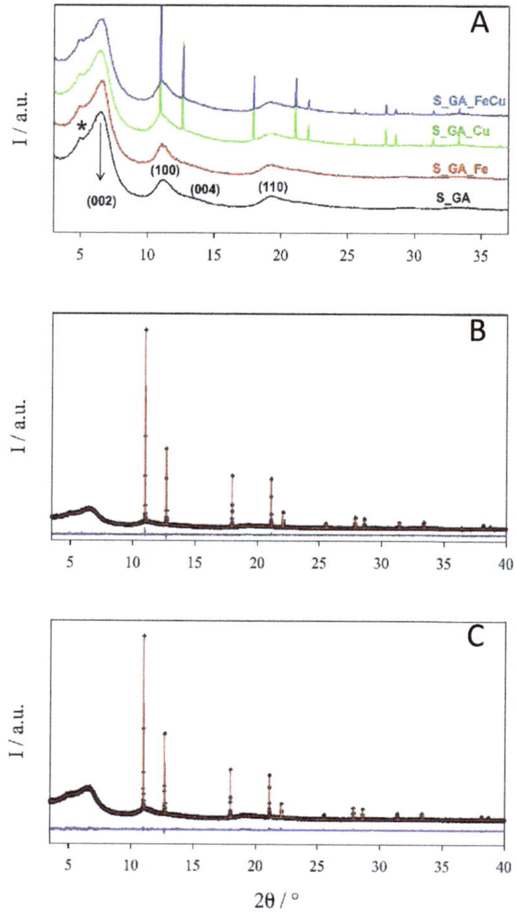

Figure 3. (**A**) XRPD patterns merged using a $\Delta 2\theta$ = 0.050° step size highlighting the graphitic contribution. Numbers in brackets are the Miller indexes of the hexagonal graphite phase. A star highlights a broad bump due to the kapton capillary. Rietveld refinements for samples (**B**) S_GA_Cu and (**C**) S_GA_FeCu. Observed (crosses) and calculated (continuous line) profiles and residuals (bottom).

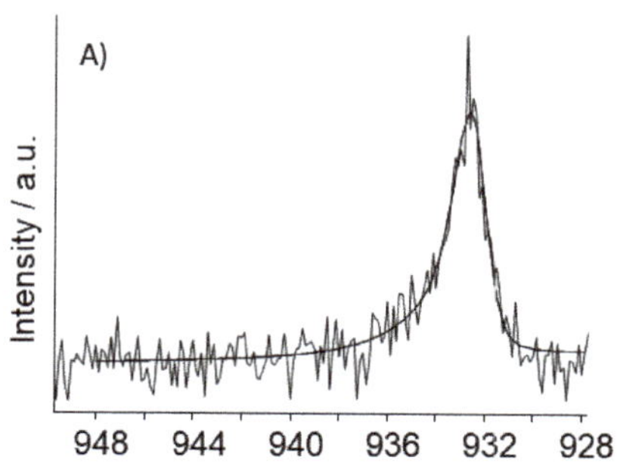

Figure 4. *Cont.*

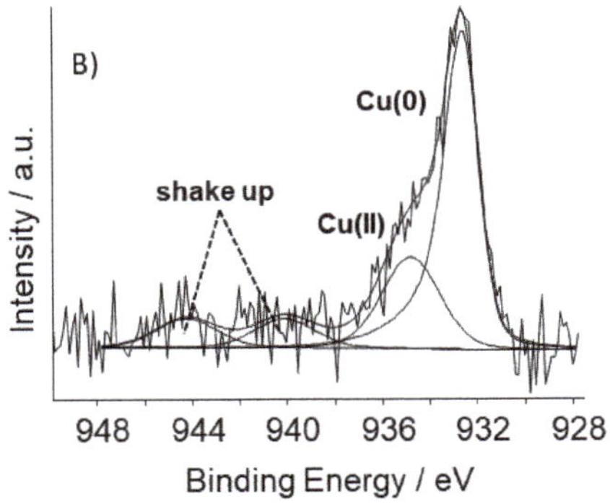

Figure 4. High-resolution photoelectron spectra of the $Cu2p^{3/2}$ core level of: (**A**) S_GA_FeCu; (**B**) S_GA_Cu.

The analysis of the high resolution photoelectron spectra of the N1s core level (Figure S4 and Table 3) evidences different types of nitrogen species on the surface of carbon [19,35]. Pyridinic, pyrrolic, and N_xMe nitrogen are the most important for ORR electroactivity [19]. Considering standard deviation, no appreciable differences among metal-doped samples are observed.

In Figure S5, high resolution spectra of the C1s core level for all synthesized samples are reported. It can be observed that in the range of the explored binding energies, all the spectra are overlapped. Therefore, the total amount of carbon-oxygen functionalities (C_xO_y) on the carbon surface, considered as a measure of defects or edge sites in the graphene-like network [19], and the sp^2/sp^3 ratio related to the graphitization degree [36], are the same for all materials. These data further confirm the results reported above.

The high resolution photoelectron spectrum of the $Cu2p^{3/2}$ core level in S_GA_FeCu (Figure 4A) can be fitted to a single peak attributable to Cu(0), while in S_GA_Cu, in addition to shake-up satellite peaks, two contributions assignable to Cu(0) and Cu(II) are found (Figure 4B) [37]. The absence of Cu(II) in S_GA_FeCu could be due to a Cu(II) surface concentration lower than the detection limit. On the other hand, the standard potentials of the redox couples Cu^{2+}/Cu and Fe^{2+}/Fe ($E°Cu^{2+}/Cu$ = +0.34 V/SHE and $E°Fe^{2+}/Fe$ = −0.44 V/SHE), suggest that the presence of iron during the synthesis lowers the stability of Cu(II) species.

The low signal-to-noise ratio of the Fe2p high resolution regions (Figure S6) might make unreliable the results of the curve-fitting procedure. Nevertheless, the main peak is centered at the same binding energy (B.E. = ~710 eV) in the two samples.

Table 3. Relative Peak Areas (RPA%) of N1s peaks *.

Peak Number	Binding Energies/eV	Functional Group	S_GA	S_GA_Cu	S_GA_Fe	S_GA_FeCu
(1)	398.3–398.5	Pyridinic N	25 ± 1	26 ± 2	29 ± 1	27 ± 2
(2)	399.2–399.6	N_x-Me or amine	13 ± 4	16 ± 4	13 ± 1	14 ± 1
(3)	400.9–401.0	Pyrrolic N	37 ± 2	35 ± 2	36 ± 1	31 ± 4
(4)	402.0–403.0	Quaternary N	12 ± 3	10 ± 2	10 ± 1	11 ± 3
(5)	403.3–403.6	Graphitic N	6 ± 1	5 ± 1	5 ± 1	7 ± 1
(6)	404.8–405.1	Shake-up π-π *	3 ± 2	4 ± 2	5 ± 1	6 ± 1
(7)	406.8–406.9	Shake-up π-π *	4 ± 1	4 ± 1	2 ± 1	4 ± 1

* The error is expressed as the larger value between the error associated with a single curve-fitting procedure (1%, at worst) and one standard deviation averaged out of three replicates.

Figure 5A–G report SEM images of the samples. In accordance with the porosity distribution results, all carbons are characterized by a porous structure (Figure 5A,B,D,F) due to the use of a hard templating agent lixiviated at the end of the synthetic process [38]. Interestingly, with respect to the other synthetic path [39], filamentous nanotubes are also evidenced in metal-doped samples (Figure 5C,E,G). In iron-doped carbons, these nanostructures are in a larger amount, are more well-organized, and longer than in S_GA_Cu. The high catalytic activity of iron to form ordered carbon nanostructures [40] might account for this behavior.

The results of STEM analysis shown in Figure 6A–D corroborate the picture emerging from the SEM observations and reveal that the filaments present in the sample S_GA_Fe are featured by a bamboo-like structure peculiar to N-doped carbon nanotubes (N-CNTs). A few nanoparticles (~20–30 nm in diameter) are entrapped within the nanotubes. Since XRPD does not evidence any Bragg peaks related to iron metal phases (Figure 3B,C), if crystalline, their concentration should be below the detection limit of our synchrotron radiation diffraction measurements.

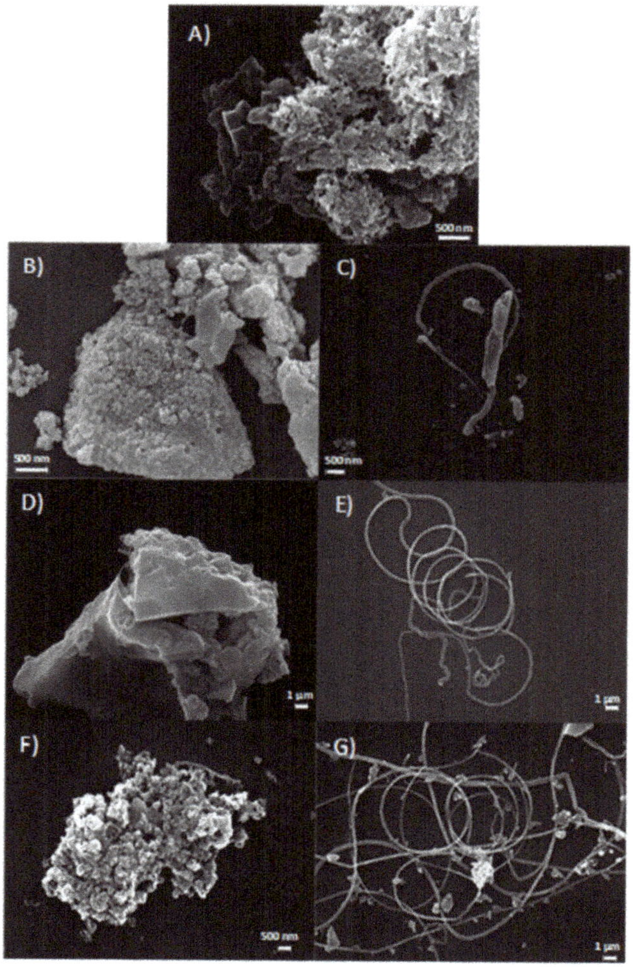

Figure 5. SEM images: (**A**) S_GA; (**B**,**C**): S_GA_Cu; (**D**,**E**): S_GA_Fe; (**F**,**G**): S_GA_FeCu.

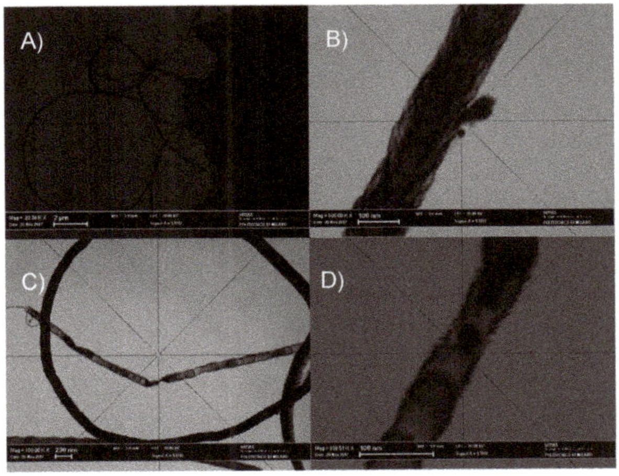

Figure 6. STEM images of S_GA_Fe. (**A**) Overview; (**B–D**) Details of (**A**).

Polarization curves in 0.1 mol dm^{-3} KOH are reported in Figure 7 and elaborated data are displayed in Table 4. A polarization curve for a commercial catalyst based on Pt (EC20, 20% Pt) and relative data are also shown for comparison.

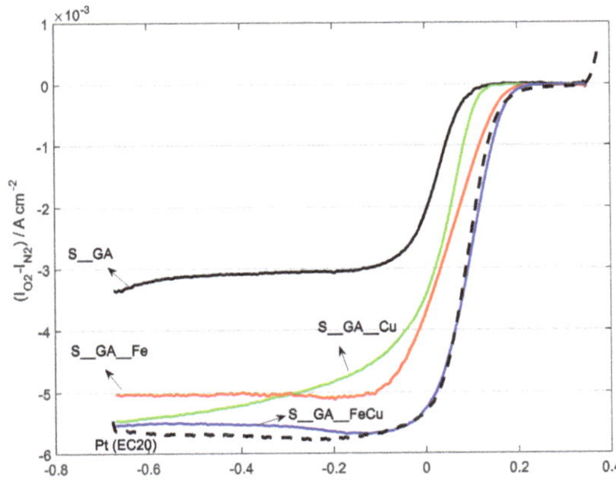

Figure 7. Oxygen reduction reaction (ORR) polarization curves recorded in oxygen-saturated 0.1 M KOH, v = 5 mV s^{-1}, ω = 1600 rpm, T = 25 °C.

Table 4. ORR Onset E_{onset} (at j = 1 mA cm^{-2}), Half Wave Potentials $E_{1/2}$, and Exchanged Electrons Number ne^{-}.

Sample Name	E_{onset} versus SHE/V at j = 1 mA cm^{-2}	$E_{1/2}$ versus SHE/V	n e^{-}
S_GA	0.044	0.033	2.72 ± 0.01
S_GA_Cu	0.082	0.068	3.82 ± 0.03
S_GA_Fe	0.117	0.066	3.98 ± 0.03
S_GA_FeCu	0.145	0.099	3.48 ± 0.02
Pt EC20	0.135	0.088	4.0 ± 0.1

Except for S_GA_Cu, all the samples show a well-defined limiting current. The best electrocatalytic performance pertains to S_GA_FeCu, with the most anodic onset potential (0.145 V versus SHE) and the highest $E_{1/2}$ (0.099 V versus SHE), also considering the Pt-based catalyst (0.135 V versus SHE and 0.088 V versus SHE, respectively). S_GA_Fe and S_GA_Cu are more cathodic (0.117 V and 0.082 V versus SHE, respectively) The exchanged electron number is about 3.5–4.0 for all metal-doped materials. S_GA, with the lowest onset potential (0.044 V versus SHE) and an exchanged electron number of ~2.7 (suggesting a predominant formation of peroxides) is the worst catalyst. Similar results on iron-doped samples have been obtained in a previous work by Liang et al. [41] on Fe–N-decorated hybrids of CNTs grown on hierarchically porous carbon. However, their proposed synthetic path is more complicated and time-consuming than the one used here. As a general behavior, iron-doped carbons (S_GA_Fe and S_GA_FeCu) exhibit better electrocatalytic properties with respect to S_GA_Cu. This could be justified by the presence both of iron species [15] and of a large amount of N-CNTs characterized by a high electrocatalytic activity towards ORRs [42–46]. This peculiarity of N-CNTs derives from the presence of substitutional nitrogen in their graphene planes, which, disturbing their uniform π-cloud [47] and increasing the localized density of states at the Fermi level [48], imparts a n-type dopant activity. As a consequence, their specific electronic properties improve and their electrocatalytic activity increases [42–49]. However, the presence of both iron and N-CNTs cannot account by itself for the improved electrocatalytic activity of S_GA_FeCu with respect to S_GA_Fe.

As reported in [20], the surface hydrophilicity of active carbons plays an important role in their electroactivity toward ORRs. Therefore, the hydrophilicity of the present samples was measured in order to assess its influence on their performance. Since the contact angle measurements on rough materials are very critical [50], following [20], the time necessary to spread completely a water drop onto the sample surface was used to quantify the hydrophilicity (the shorter this time, the more hydrophilic the material). Hydrophilicity was found to increase in the order S_GA < S_GA_Fe < S_GA_Cu < S_GA_FeCu (Table 5).

Table 5. Time Necessary to Spread a Water Drop on a Surface.

Sample Name	Time/s
S_GA	7.9
S_GA_Fe	5.8
S_GA_Cu	4.7
S_GA_FeCu	4.0

Since Fe(0) is more hydrophilic than Cu(0) [49] and N-CNTs favorably interact with hydrophilic molecules [51], the experimental hydrophilicity scale (Table 5) suggests that the presence of a metal phase (as Cu(0)) is more influential on hydrophilicity than the presence of N-CNTs. Moreover, since all the metal-doped samples exhibit similar amounts of surface oxygen and nitrogen, forming typical hydrophilic groups on the surface, the hydrophilicity of the active site, rather than the entire surface, should be responsible for the different hydrophilicities of carbons.

Coming back to the different electrocatalytic activity of S_GA_Fe and S_GA_FeCu carbons, our physico-chemical characterization shows that they have very similar structure and properties but there is a larger hydrophilicity in the S_GA_FeCu sample, which is probably related to the presence of Cu(0) phase. Thus, we propose that the improved performance of S_GA_FeCu has to be attributed to the different hydrophilicity of its active centers. Actually, an increase of the interactions between water and active centers could favor interactions with molecular oxygen and thus, its adsorption and, finally, its reduction. Although the influence of the active center's hydrophilicity on ORR activity of Pt-free materials should be deeply investigated, the results here presented pave the way to the comprehension of their properties.

4. Conclusions

In this work, catalysts based on nitrogen-modified carbons doped with different metals (Fe, Cu, and a mixture of them) were synthesized and characterized by BET, synchrotron radiation XRPD, micro-Raman, XPS, SEM, STEM, and hydrophilicity measurements. The N-modified carbon phases are characterized by porous structures and by the presence of N-CNTs. In the Fe-doped samples (S_GA_Fe and S_GA_FeCu), N-CNTs are more well-organized, in a larger amount, and longer than in the copper-doped ones. In Cu-doped (S_GA_Cu and S_GA_FeCu) samples, copper is present in the crystalline (ccp) phase.

In spite of the different nature of the metal-doping agent, catalysts exhibit few differences in terms of surface area, functional surface groups, and graphitization degree of the carbon phase. Electrocatalytic activity toward ORRs increases in the order S_GA_Cu < S_GA_Fe < S_GA_FeCu. The presence of a larger amount of N-CNTs can account for the better performance of iron-doped catalysts with respect to S_GA_Cu. Instead, also the hydrophilicity of the active sites, which decreases in the order S_GA_FeCu > S_GA_Cu > S_GA_Fe, must be considered in order to explain the higher electrocatalytic activity of S_GA_FeCu with respect to S_GA_Fe.

The presence of iron, N-CNTs, and an improved hydrophilicity synergically boost the electrocatalytic properties of ORR catalysts.

The preliminary results here presented highlight the influence of the hydrophilicity of active sites on ORRs and pave the way to the comprehension of the electrocatalytic properties of Pt-free materials.

Supplementary Materials: The following are available online at http://www.mdpi.com/2079-4991/8/9/643/s1, Figure S1: Porosity Distribution. Figure S2: Results of Raman spectra decomposition. Figure S3: XPS survey of S_GA_FeCu. Figure S4: XPS N1s region of S_GA_FeCu. Figure S5: Superimposition of XPS C1s spectra. Figure S6: XPS Fe2p spectra of S_GA_FeCu and S_GA_Fe. Table S1: Parameters inferred from Raman spectra fitting. Table S2: Lattice parameters, weight fraction (WF), average displacement parameters (Umean), and fit residuals for the refinements performed on crystalline phases.

Author Contributions: M.L. Funding Acquisition, Conceptualization, Electrochemical analysis, and BET measurements and analysis, C.C. Electrochemical Measurements, E.P. Hydrophilicity Measurements, M.C. and M.S. XRPD measurements and analysis, S.S. and S.P. Raman Measurements and analysis, N.D. and N.C. XPS measurements and analysis, A.F. STEM measurements. M.L., M.C., M.S., S.S., S.P. and N.D. contributed to the writing of the work.

Funding: This work was supported by MIUR under the Project SMARTNESS (PRIN 2015).

Acknowledgments: The authors gratefully acknowledge the European Synchrotron ESRF (Grenoble, France), beamline ID22, for providing beam time and Stefania Mazzini and Leonardo Scaglioni (UNIMI) for NMR determination of the sugar's composition.

Conflicts of Interest: The authors declare no conflict of interest.

References

1. Greeley, J.; Stephens, I.E.L.; Bondarenko, A.S.; Johansson, T.P.; Hansen, H.A.; Jaramillo, T.F.; Rossmeisl, J.; Chorkendorff, I.; Nørskov, J.K. Alloys of Platinum and Early Transition Metals as Oxygen Reduction Electrocatalysts. *Nat. Chem.* **2009**, *1*, 552–556. [CrossRef] [PubMed]
2. Wu, G.; Zelenay, P. Nanostructured Nonprecious Metal Catalysts for Oxygen Reduction Reaction. *Acc. Chem. Res.* **2013**, *46*, 1878–1889. [CrossRef] [PubMed]
3. Ghosh, T.; Vukmirovic, M.B.; Di Salvo, F.J.; Adzic, R.R. Intermetallics as Novel Supports for Pt Monolayer O_2 Reduction Electrocatalysts: Potential for Significantly Improving Properties. *J. Am. Chem. Soc.* **2010**, *132*, 906–907. [CrossRef] [PubMed]
4. Bashyam, R.; Zelenay, P. A Class of Non-Precious Metal Composite Catalysts for Fuel Cells. *Nature* **2006**, *443*, 63–66. [CrossRef] [PubMed]
5. Lefèvre, M.; Proietti, E.; Jaouen, F.; Dodelet, J.P. Iron-Based Catalysts with Improved Oxygen Reduction Activity in Polymer Electrolyte Fuel Cells. *Science* **2009**, *324*, 71–74. [CrossRef] [PubMed]

6. Jaouen, F.; Herranz, J.; Lefèvre, M.; Dodelet, J.-P.; Kramm, U.I.; Herrmann, I.; Bogdanoff, P.; Maruyama, J.; Nagaoka, T.; Garsuch, A.; et al. Cross-Laboratory Experimental Study of Non-Noble-Metal Electrocatalysts for the Oxygen Reduction Reaction. *ACS Appl. Mater. Interfaces* **2009**, *1*, 1623–1639. [CrossRef] [PubMed]
7. Wu, G.; More, K.L.; Johnston, C.M.; Zelenay, P. High-Performance Electrocatalysts for Oxygen Reduction Derived from Polyaniline, Iron, and Cobalt. *Science* **2011**, *332*, 443–447. [CrossRef] [PubMed]
8. Cheon, J.Y.; Kim, T.; Choi, Y.; Jeong, H.Y.; Kim, M.G.; Sa, Y.J.; Kim, J.; Lee, Z.; Yang, T.H.; Kwon, K.; et al. Ordered Mesoporous Porphyrinic Carbons with Very High Electrocatalytic Activity for the Oxygen Reduction Reaction. *Sci. Rep.* **2013**, *3*, 2715. [CrossRef] [PubMed]
9. Guan, B.Y.; Yu, L.; Lou, X.W. A Dual-Metal–Organic-Framework Derived Electrocatalyst for Oxygen Reduction. *Energy Environ. Sci.* **2016**, *9*, 3092–3096. [CrossRef]
10. Xia, B.Y.; Yan, Y.; Li, N.; Wu, H.B.; Lou, X.W.; Wang, X. A Metal–Organic Framework-Derived Bifunctional Oxygen Electrocatalyst. *Nat. Energy* **2016**, *1*, 15006. [CrossRef]
11. Liu, T.; Zhao, P.; Hua, X.; Luo, W.; Chen, S.; Cheng, G. An Fe–N–C Hybrid Electrocatalyst Derived from a Bimetal–Organic Framework for Efficient Oxygen Reduction. *J. Mater. Chem. A* **2016**, *4*, 11357–11364. [CrossRef]
12. Masa, J.; Xia, W.; Muhler, M.; Schuhmann, W. On the Role of Metals in Nitrogen-Doped Carbon Electrocatalysts for Oxygen Reduction. *Angew. Chem. Int. Ed.* **2015**, *54*, 10102–10120. [CrossRef] [PubMed]
13. Tylus, U.; Jia, Q.; Strickland, K.; Ramaswamy, N.; Serov, A.; Atanassov, P.; Mukerjee, S. Elucidating Oxygen Reduction Active Sites in Pyrolyzed Metal-Nitrogen Coordinated Non-Precious-Metal Electrocatalyst Systems. *J. Phys. Chem. C* **2014**, *118*, 8999–9008. [CrossRef] [PubMed]
14. Choi, C.H.; Choi, W.S.; Kasian, O.; Mechler, A.K.; Sougrati, M.T.; Brüller, S.; Strickland, K.; Jia, Q.; Mukerjee, S.; Mayrhofer, K.J.J.; et al. Unraveling the Nature of Sites Active toward Hydrogen Peroxide Reduction in Fe-N-C Catalysts. *Angew. Chem. Int. Ed.* **2017**, *56*, 8809–8812. [CrossRef] [PubMed]
15. Workman, M.J.; Dzara, M.; Ngo, C.; Pylypenko, S.; Serov, A.; McKinney, S.; Gordon, J.; Atanassov, P.; Artyushkova, K. Platinum Group Metal-Free Electrocatalysts: Effects of Synthesis on Structure and Performance in Proton-Exchange Membrane Fuel Cell Cathodes. *J. Power Sources* **2017**, *348*, 30–39. [CrossRef]
16. Longhi, M.; Marzorati, S.; Checchia, S.; Sacchi, B.; Santo, N.; Zaffino, C.; Scavini, M. Sugar-Based Catalysts for Oxygen Reduction Reaction. Effects of the Functionalization of the Nitrogen Precursors on the Electrocatalytic Activity. *Electrochim. Acta* **2016**, *222*, 781–792. [CrossRef]
17. Su, C.-Y.; Cheng, H.; Li, W.; Liu, Z.-Q.; Li, N.; Hou, Z.; Bai, F.-Q.; Zhang, H.-X.; Ma, T.-Y. Atomic Modulation of FeCo–Nitrogen–Carbon Bifunctional Oxygen Electrodes for Rechargeable and Flexible All-Solid-State Zinc–Air Battery. *Adv. Energy Mater.* **2017**, *7*, 1602420. [CrossRef]
18. Witkowska, A.; Giuli, G.; Renzi, M.; Marzorati, S.; Yiming, W.; Nobili, F.; Longhi, M. Fe Local Structure in Pt-Free Nitrogen-Modified Carbon Based Electrocatalysts: XAFS Study. *J. Phys. Conf. Ser.* **2015**, 012131/1–012131/4. [CrossRef]
19. Artyushkova, K.; Serov, A.; Rojas-Carbonell, S.; Atanassov, P. Chemistry of Multidinous Active Sites for Oxygen Reduction Reaction in Transition Metal-Nitrogen-Carbon Electrocatalysts. *J. Phys. Chem. C* **2015**, *119*, 25917–25928. [CrossRef]
20. Hao, G.-P.; Sahraie, N.R.; Zhang, Q.; Krause, S.; Oschatz, M.; Bachmatiuk, A.; Strasser, P.; Kaskel, S. Hydrophilic Non-Precious Metal Nitrogen-Doped Carbon Electrocatalysts for Enhanced Efficiency in Oxygen Reduction Reaction. *Chem. Commun.* **2015**, *51*, 17285–17288. [CrossRef] [PubMed]
21. Kumar, K.V.; Preuss, K.; Guo, Z.X.; Titirici, M.M. Understanding the Hydrophilicity and Water Adsorption Behavior of Nanoporous Nitrogen-Doped Carbons. *J. Phys. Chem. C* **2016**, *120*, 18167–18179. [CrossRef]
22. Bogdanov, K.; Fedorov, A.; Osipov, V.; Enoki, T.; Takai, K.; Hayashi, T.; Ermakov, V.; Moshkalev, S.; Baranov, A. Annealing-Induced Structural Changes of Carbon Onions: High-Resolution Transmission Electron, Microscopy and Raman Studies. *Carbon* **2014**, *73*, 78–86. [CrossRef]
23. Larson, A.C.; Von Dreele, R.B. *General Structural Analysis System (GSAS)*; Los Alamos National Laboratory, Report LAUR: Los Alamos, NM, USA, 2004.
24. Sing, K.S.W.; Everett, D.H.; Haul, R.A.W.; Moscou, L.; Pierotti, R.A.; Rouquerol, J.; Siemieniewska, T. Reporting Physisorption Data for Gas/Solid Systems with Special Reference to the Determination of Surface Area and Porosity. *Pure Appl. Chem.* **1985**, *57*, 603–619. [CrossRef]
25. Thommes, M. Physical Adsorption Characterization of Nanoporous Materials. *Chem. Ing. Tech.* **2010**, *82*, 1059–1073. [CrossRef]

26. Leonard, N.D.; Nallathambi, V.; Calabrese Burton, S. Carbon Supports for Non-Precious Metal Proton Exchange Membrane Fuel Cells. *ECS Trans.* **2011**, *41*, 1175–1181. [CrossRef]
27. Xy, J.B.; Zhao, T.S. Mesoporous Carbon with Uniquely Combined Electrochemical and Mass Transport Characteristics for Polymer Electrolyte Membrane Fuel Cells. *RSC Adv.* **2013**, *3*, 16–24.
28. Marzorati, S.; Longhi, M. Templating Induced Behavior of Platinum-free Carbons for Oxygen Reduction Reaction. *J. Electroanal. Chem.* **2016**, *775*, 350–355. [CrossRef]
29. Ferrari, A.C.; Robertson, J. Interpretation of Raman Spectra of Disordered and Amorphous Carbon. *Phys. Rev. B* **2001**, *61*, 14095–14107. [CrossRef]
30. Santangelo, S. Controlled Surface Functionalisation of Carbon Nanotubes by Nitric Acid Vapors Generated from Sub-Azeotropic Solution. *Surf. Interface Anal.* **2016**, *48*, 17–25. [CrossRef]
31. Cançado, L.G.; Takai, K.; Enoki, T.; Endo, M.; Kim, Y.A.; Mizusaki, H.; Jorio, A.; Coelho, L.N.; Magalhães-Paniago, R.; Pimenta, M.A. General Equation for the Determination of the Crystallite Size L_a of Nanographite by Raman Spectroscopy. *Appl. Phys. Lett.* **2006**, *8*, 163106–163108. [CrossRef]
32. Klement, W., Jr. *Technical Report 401844*; WM Keck Laboratory for Engineering Materials, California Institute of Technology: Pasadina, CA, USA, 1963. Available online: http://www.dtic.mil/dtic/tr/fulltext/u2/401844.pdf (accessed on 18 July 2018).
33. Chen, Q.; Jin, Z. The Fe-Cu System: A Thermodynamic Evaluation. *Metal. Mater. Trans. A* **1995**, *26*, 417–426. [CrossRef]
34. Peng, H.; Liu, F.; Liu, X.; Liao, S.; You, C.; Tian, X.; Nan, H.; Luo, F.; Song, H.; Fu, Z.; et al. Effect of Transition Metals on the Structure and Performance of the Doped Carbon Catalysts Derived from Polyaniline and Melamine for ORR Application. *ACS Catal.* **2014**, *4*, 3797–3805. [CrossRef]
35. Lahaye, J.; Nanse, G.; Bagreev, A.; Strelko, V. Porous Structure and Surface Chemistry of Nitrogen Containing Carbons from Polymers. *Carbon* **1999**, *37*, 585–590. [CrossRef]
36. Díaz, J.; Paolicelli, G.; Ferrer, S.; Comin, F. Separation of the sp^3 and sp^2 Components in the C1s Photoemission Spectra of Amorphous Carbon Films. *Phys. Rev. B* **1996**, *54*, 8064–8069. [CrossRef]
37. Biesinger, M.C.; Lau, L.W.M.; Gerson, A.R.; Smart, R.S.C. Resolving Surface Chemical States in XPS Analysis of First Row Transition Metals, Oxides and Hydroxides: Sc, Ti, V, Cu, and Zn. *Appl. Surf. Sci.* **2010**, *257*, 887–898. [CrossRef]
38. Galbiati, I.; Bianchi, C.L.; Longhi, M.; Carra, A.; Formaro, L. Iron and Copper Containing Oxygen Reduction Catalysts from Templated Glucose-Histidine. *Fuel Cells* **2010**, *10*, 251–258. [CrossRef]
39. Formaro, L.; Longhi, M.; Messina, P.; Galbiati, I. Catalysts Free from Noble Metals Suitable for the Electrochemical Reduction of Oxygen. U.S. Patent US 20140162869 A1, 12 June 2014.
40. Andrews, R.; Jacques, D.; Qian, D.; Rantell, T. Multiwall Carbon Nanotubes: Synthesis and Application. *Acc. Chem. Res.* **2002**, *35*, 1008–1017. [CrossRef] [PubMed]
41. Liang, J.; Zhou, R.F.; Chen, X.M.; Tang, Y.H.; Qiao, S.Z. Fe-N Decorated Hybrids of CNTs Grown on Hierarchically Porous Carbon for High-Performance Oxygen Reduction. *Adv. Mater.* **2014**, *26*, 6074–6079. [CrossRef] [PubMed]
42. Maldonado, S.; Stevenson, K.J. Influence of Nitrogen Doping on Oxygen Reduction Electrocatalysis at Carbon Nanofiber Electrodes. *J. Phys. Chem. B* **2005**, *109*, 4707–4716. [CrossRef] [PubMed]
43. Van Dommele, S.; de Jong, K.P.; Bitter, J.H. Nitrogen-Containing Carbon Nanotubes as Solid Base Catalysts. *Chem. Commun.* **2006**, 4859–4861. [CrossRef]
44. Shao, Y.; Sui, J.; Yin, G.; Gao, Y. Nitrogen-Doped Carbon Nanostructures and their Composites as Catalytic Materials for Proton Exchange Membrane Fuel Cell. *Appl. Catal. B* **2008**, *79*, 89–99. [CrossRef]
45. Sidik, R.A.; Anderson, A.B.; Subramanian, N.P.; Kumaraguru, S.P.; Popov, B.N. O_2 Reduction on Graphite and Nitrogen-Doped Graphite: Experiment and Theory. *J. Phys. Chem. B* **2006**, *110*, 1787–1793. [CrossRef] [PubMed]
46. Liu, R.; Wu, D.; Feng, X.; Müllen, K. Nitrogen-Doped Ordered Mesoporous Graphitic Arrays with High Electrocatalytic Activity for Oxygen Reduction. *Angew. Chem.* **2010**, *122*, 2619–2623. [CrossRef]
47. Chizari, K.; Janowska, I.; Houllé, M.; Florea, I.; Ersen, O.; Romero, T.; Bernhardt, P.; Ledoux, M.J.; Pham-Huu, C. Tuning of Nitrogen-Doped Carbon Nanotubes as Catalyst Support for Liquid-Phase Reaction. *Appl. Catal. A Gen.* **2010**, *380*, 72–80. [CrossRef]

48. Terrones, M.; Ajayan, P.M.; Banhart, F.; Blase, X.; Carroll, D.L.; Charlier, J.C.; Czerw, R.; Foley, B.; Grobert, N.; Kamalakaran, R.; et al. N-Doping and Coalescence of Carbon Nanotubes: Synthesis and Electronic Properties. *Appl. Phys. A Mater. Sci. Process.* **2002**, *74*, 355–361. [CrossRef]
49. Ismagilov, Z.R.; Shalagina, A.E.; Podyacheva, O.Y.; Ischenko, A.V.; Kibis, L.S.; Boronin, A.I.; Chesalov, Y.A.; Kochubey, D.I.; Romanenko, A.I.; Anikeeva, O.B.; et al. Structure and Electrical Conductivity of Nitrogen-Doped Carbon Nanofibers. *Carbon* **2009**, *47*, 1922–1929. [CrossRef]
50. Afanasev, B.N.; Akulova, Y.P. A Correlation Between the Hydrophilicity of a Metal and its Surface Tension. Calculation of the Bond Energy of Water Molecules Adsorbed on an Uncharged Metal Surface. *Prot. Met.* **2000**, *36*, 25–30. [CrossRef]
51. Lee, W.J.; Maiti, U.N.; Lee, J.M.; Lim, J.; Han, T.H.; Sang, O.K. Nitrogen-Doped Carbon Nanotubes and Graphene Composite Structures for Energy and Catalytic Applications. *Chem. Commun.* **2014**, *50*, 6818–6830. [CrossRef] [PubMed]

© 2018 by the authors. Licensee MDPI, Basel, Switzerland. This article is an open access article distributed under the terms and conditions of the Creative Commons Attribution (CC BY) license (http://creativecommons.org/licenses/by/4.0/).

nanomaterials

MDPI

Article

Nano-MnO_2 Decoration of TiO_2 Microparticles to Promote Gaseous Ethanol Visible Photoremoval

Marta Stucchi [1,*], Daria C. Boffito [2], Eleonora Pargoletti [1,3], Giuseppina Cerrato [4], Claudia L. Bianchi [1,3] and Giuseppe Cappelletti [1,3]

1 Chemistry Department, University of Milan, via Golgi 19, 20133 Milano, Italy; eleonora.pargoletti@unimi.it (E.P.); claudia.bianchi@unimi.it (C.L.B.); giuseppe.cappelletti@unimi.it (G.C.)

2 Chemical Engineering Department, Ècole Polytechnique Montréal, Pavillon Principal, Centre-Ville, Montréal, QC H3C 3A4, Canada; daria-camilla.boffito@polymtl.ca

3 National Consortium for Materials Science and Technology (INSTM), via Giusti 9, 50121 Florence, Italy

4 Chemistry Department & NIS Centre, University of Turin, via Pietro Giuria 7, 10125 Turin, Italy; giuseppina.cerrato@unito.it

* Correspondence: marta.stucchi@polymtl.ca; Tel.: +39-347-600-0699

Received: 11 August 2018; Accepted: 31 August 2018; Published: 3 September 2018

Abstract: TiO_2-based photocatalysis under visible light is an attractive way to abate air pollutants. Moreover, developing photocatalytic materials on a large-scale requires safe and low-cost precursors. Both high-performance TiO_2 nanopowders and visible-light active noble metals do not match these requirements. Here, we report the design of novel Mn-decorated micrometric TiO_2 particles. Pigmentary TiO_2 replaced unsafe nano-TiO_2 and firmly supported MnO_x particles. Mn replaced noble metals such as Au or Ag, opening the way for the development of lower cost catalysts. Varying Mn loading or pH during the impregnation affected the final activity, thus giving important information to optimize the synthesis. Photocatalytic activity screening occurred on the gas-phase degradation of ethanol as a reference molecule, both under ultraviolet (UV) (6 h) and Light Emitting Diode (LED) (24 h) irradiation. Mn-doped TiO_2 reached a maximum ethanol degradation of 35% under visible light after 24 h for the sample containing 20% of Mn. Also, we found that an acidic pH increased both ethanol degradation and mineralization to CO_2, while an alkaline pH drastically slowed down the reaction. A strict correlation between photocatalytic results and physico-chemical characterizations of the synthesized powders were drawn.

Keywords: micrometric TiO_2; Mn decoration; visible light photocatalysis; impregnation pH; surface hydroxyl groups

1. Introduction

Environmental photocatalysis based on TiO_2 and its applications for pollution abatement relies on some key aspects: (i) The availability of inexpensive and stable TiO_2 [1]; (ii) the photoelectric effect, for which photocatalysis depends upon the energy of the incident photons, but not on their intensity (thus, even a small amount of photons of the required energy induces photocatalysis) [2]; (iii) the effectiveness of photocatalysis in abating pollution as it accumulates, as opposed to treating large volumes [3,4].

In 2001, Wolkoff and Nielsen [5] estimated indoor concentrations for twelve Volatile Organic Compounds (VOCs). Concentrations of organics such as benzene, ethylbenzene, and *o*-xylene were less than 5 $\mu g\ m^{-3}$; toluene and xylene isomers, instead, exceeded 5 $\mu g\ m^{-3}$, with concentrations greater than 15 $\mu g\ m^{-3}$ in most cases. Recently, Höllbacher et al. [6] studied indoor air quality considering specific human activities, and they monitored VOCs, particulate matter (PM), carbonyl compounds and CO_2. The highest VOC concentration recorded was around 3500 $\mu g\ m^{-3}$; they detected a maximum formaldehyde concentration of 76 $\mu g\ m^{-3}$, and the PM concentration was 378 $\mu g\ m^{-3}$.

The concentration level of many pollutants changes the incidence of diseases among people who spend time inside buildings [7].

The conventional air-purification systems are usually provided with filters. However, in such filters, pollutants can be accumulated. Moreover, treatment of old filters could cause the risk of secondary pollution. Although TiO_2 is excellent to photocatalytically cleave organic compounds, photocatalysis appears as hardly applicable indoors, because visible light is not sufficient to activate TiO_2. Indeed, anatase TiO_2 absorbs ultraviolet light with wavelengths shorter than 380 nm [8]. Considering the sunlight, the Ultraviolet (UV) content barely covers 5% of the total irradiation spectrum, whereas the ultraviolet light in indoor illumination is significantly smaller because the fluorescent lamps mainly emit visible light [9]. Moreover, considering the growing concerns about energy consumption utilization, Light Emitting Diode lamps (LEDs) are increasingly replacing all other kinds of classical lighting system. LED emits in the range of 400–700 nm wavelengths only, without any UV component (IUPAC Gold Book, Light-emitting diode, LED). The number of papers regarding the modification of TiO_2 to improve its photocatalytic properties is countless. Noble metal nanoparticles are actually the only promising way for the enhancement of TiO_2 properties in the visible spectrum. For example, Pan and Xu [10] deposited noble-metal nanoparticles (Ag, Pt, and Pd) on a TiO_2 surface. The resulting M-TiO_2 (M = Ag, Pt, and Pd) nanocomposites were visible-light active for the selective oxidation of benzyl alcohol. Elahifard et al. [11] synthesized Ag/AgBr/TiO_2 covered apatite and proved their high photocatalytic activity as anti-bactericidal under visible light. Also, Au/TiO_2 catalysts degraded formaldehyde under visible light in a single pass continuous flow reactor, with conversions above 80% [12].

However, the high cost of metals such as Au, Ag, and Pt, limits the large-scale production of photoactive TiO_2-based building materials for pollution abatement in indoor environments. The use of some transition metal oxides could increase the photocatalytic activity. Among them, manganese oxides (MnO_x) recently emerged as quite active in the degradation of organic (VOCs) or inorganic (NO_x) pollutants, as reported in the recent literature [13]. Since MnO_2 is rich in non-stoichiometric oxygen vacancies, it could be exploited to increase the carrier density and electrical conductivity of semiconductors [14–16]. Mn can form different oxides, such as Mn_3O_4, Mn_2O_3, and MnO_2 that show different behavior in relation to the catalytic reaction in which they are involved [17–19].

Papers report several synthetic procedures to obtain Mn-TiO_2 nanoparticles from Ti-based precursors [20–22]. Binas et al. [20] reported a sol–gel method for the synthesis of Mn-doped TiO_2 photocatalysts with a grain size ranging between 20 and 30 nm, active for degradation of Methylene Blue under visible light. Xue et al. [21] synthesized mesoporous structured MnO_2/TiO_2 by a modified sol–gel method, showing excellent adsorption properties over the entire visible light range. Another recent paper reported the sol-gel synthesis of 20 wt.% Mn/TiO_2 nanocomposites at different pH. The nanocomposite synthesized at pH = 7, which was smaller in crystal size and had a larger pore volume, was the most active in methylene blue degradation. This study showed that pH influenced the amount of dopants on TiO_2, the amount of OH functional groups on its surfaces, crystal sizes, and pore volume [22].

On the basis of this preliminary research and with the aim of replacing nanopowders with safer micro-TiO_2, here we report for the first time Mn-decorated samples by impregnation of micro-pigmentary TiO_2 particles at different pH. We studied the photocatalytic activity both under UV and LED irradiation, selecting ethanol as a model VOC molecule. All the powders were finely characterized and a strict correlation between their physicochemical properties and photoactivity was drawn.

2. Materials and Methods

All chemicals were of reagent grade purity and were used without further purification; doubly distilled water passed through a Milli-Q apparatus that was used to prepare solutions and suspensions. Manganese (II) nitrate tetrahydrate ($Mn(NO_3)_2 \cdot 4H_2O$, 99%, Sigma-Aldrich, Sigma-Aldrich SRL, Milan, Italy) was the salt precursor; the impregnation procedure required acetone (CH_3COCH_3) (≥99.9%, Sigma-Aldrich, Sigma-Aldrich SRL, Milan, Italy), potassium hydroxide (KOH), and hydrochloric acid

(HCl) (reagent grade, 37%, Sigma-Aldrich, Sigma-Aldrich SRL, Milan, Italy) to set the acidic or alkaline pH. Micrometric TiO_2 (by Kronos, identified with the acronym 1077) consists of 100% anatase spherical primary particles of 130 nm, with a surface area of 11 m^2 g^{-1}, as reported by Bianchi et al. [23].

2.1. Synthesis of Mn-Impregnated Titania Samples

The Mn-modified samples were obtained through an impregnation method: (i) A fixed amount (1 g) of 1077 was suspended in an aqueous solution (to cover completely the 1077 powder layer at the bottom of the glass flask) with a few drops of acetone in order to increase the wettability of the titania particles; (ii) then, a $Mn(NO_3)_2 \cdot 4H_2O$ solution was added with a different Mn weight amount (in the range 5–30%). The natural pH of the initial suspension was 4. The dried Mn-doped TiO_2 particles were obtained by means of a Rotavapor for 26 h (at 40 °C for 24 h and 80 °C for the final 2 h), followed by a thermal treatment in an oven at 100 °C overnight. Every sample was calcined at 400 °C for 2 h in air.

Further, fixing the Mn amount to 20%, the pH of the suspension changed from the spontaneous value (around pH 4) to both acidic (pH 3) and alkaline (pH 9, 12) by addition of HCl and KOH, respectively. The solvent evaporation procedure and the heating treatment were the same as previously reported [24].

The samples were labeled as to Mn_x1077_y, where x is the Mn weight percentage and y is the pH value.

2.2. Physico-Chemical Characterizations

X-ray powder diffraction (XRPD) (Malvern Panalytical Srl, Milan, Italy) analysis was performed on a Philips PW 3710 Bragg–Brentano goniometer equipped with a scintillation counter and 1° divergence slit, 0.2 mm receiving slit, and 0.04° soller slit systems. We employed graphite-monochromated Cu Kα radiation (Cu Kα1 λ = 1.54056 Å, Kα2 λ = 1.54433 Å) at 40 kV × 40 mA nominal X-rays power. Diffraction patterns were collected between 20° and 80° with a step size of 0.1°.

A JEOL 3010-UHR Instrument (acceleration potential 300 kV, LaB6 filament) (JEOL Ltd., Milan, Italy) was utilized to acquire the High Resolution-Transmission Electron Microscopy (HR-TEM) images to evaluate the morphology of all the adopted catalysts.

The BET surface area (S_{BET}) (Beckan Coulter, Indianapolis, IN, USA) was determined from nitrogen adsorption–desorption isotherms at 77 K, by using Coulter SA 3100 apparatus.

Diffuse Reflectance spectra (DRS) (Shimadzu Europe, Duisburg, Germany) of the nanopowders were measured on a UV–Vis spectrophotometer Shimadzu UV-2600 equipped with an integrating sphere; $BaSO_4$ was used as a "total white" reference.

X-ray Photoelectron Spectroscopy (XPS) (Thermo Scientific, Massachusetts, USA) analysis was carried out by means of an M-probe apparatus (Surface Science Instrument), using a monochromatic Al Kα radiation source (1486.6 eV). The XPS binding energy scale was charge corrected using the standard calibration, fixing the C-1s peak at 284.6 eV.

2.3. Photocatalytic Tests

Ethanol (CH_3CH_2OH) was selected to evaluate the photocatalytic properties of the Mn-modified TiO_2 samples in the gaseous phase. The starting pollutant concentration was 400 ppm. The photodegradation reaction was performed in a four openings PIREX glass reactor of 5 L. A gas chromatograph (Agilent 3000 A Micro-GC, Santa Clara, CA, USA), directly connected to the reactor, monitored the internal concentration of the organic molecule, as well as of any by-products and CO_2, as reported in our previous work [24]. Two different irradiation sources, a UV lamp (HG 500 Jelosil SRL, 315–400 nm and irradiation intensity of 30 W m^{-2}, Milan, Italy) and an LED one (MW Mean Well, 400–700 nm, 350 mA, 16.8 W and irradiation intensity of 15,000 lx, to simulate the visible light, Taiwan, China) were used. According to the adopted source, photocatalytic tests lasted 6 h and 24 h, respectively. A hygrometer was placed inside the reactor to continuously monitor the relative humidity (RH around 40% for the whole reaction).

All the Mn-TiO_2 powders, suspended in 2-propanol, were deposited on a 100 cm^2 glass slab to obtain a homogeneous thin film. The covered glass support was placed at the bottom of the reactor. A cleaning procedure through the exploitation of an inert gas was performed to remove CO_2 traces. The desired amount of the organic pollutant was introduced through a microliter syringe.

Dark tests together with photolysis experiments (both under UV and visible light) were carried out to evaluate the molecules adsorption and their direct degradation, respectively. For both the irradiation sources, ethanol adsorption and photolysis were always negligible (<2%).

3. Results and Discussion

3.1. Ethanol Photodegradation Under UV and LED Sources

Firstly, Mn_x1077_4 (X refers to the Mn%, while number 4 refers to pH, with x in the range 5–30%) powders were tested on the photodegradation of gaseous ethanol molecules both under UV and visible light. Star Mn_x1077_4 ting from the ultraviolet-assisted experiments, as already stated in our previous work [23], the abatement of the pollutant and its complete conversion to CO Mn_x1077_4_2 within Mn_x1077_4n 6 h, see Table 1, were reached by bare TiO_2 (1077), notwithstanding the micrometric distribution of primary particles and aggregates (see Section 3.2). On the contrary, by impregnating the pristine TiO_2 with Mn, the photocatalytic performance drastically decreased, upon increasing dopant loading, see Table 1, 2nd and 3rd columns. As reported in the literature, the OH species distributed onto the TiO_2 surface mostly affect the photocatalytic activity [22,25]. Thus, the partial covering of the active sites, which can be activated by UV light, is much more adverse compared to the increment given by Mn in terms of electron capture.

Table 1. Gaseous ethanol photodegradation and mineralization degrees by both pure and Mn-doped 1077 samples, under ultraviolet (UV) (after 6 h) and Light Emitting Diode (LED) (after 24 h) irradiation sources.

Sample	UV (After 6 h)		LED (After 24 h)	
	% Degradation	% Mineralization	% Degradation	% Mineralization
1077	100	100	<2	<2
$Mn_5$1077_4	51	14	6	<2
Mn_{10}1077_4	50	13	12	<2
Mn_{20}1077_4	39	8	35	7
Mn_{30}1077_4	10	<2	10	<2

Further, the aim of the present work was the exploitation of the prepared photocatalysts under LED light. As expected, pure 1077 sample showed a scarce amount of activity under visible light (negligible ethanol molecules degradation and mineralization after prolonged irradiation; Table 1, 4th and 5th columns). Instead, by either comparing both the disappearance and mineralization of Mn-doped TiO_2 or considering the different Mn loadings, Mn_{20}1077_4 powder exhibited the best photocatalytic performance (Table 1, 4th and 5th columns). Moreover, EtOH degradation was lower for Mn loadings higher than 20% (Table 1, Mn_{30}1077_4 sample). Again, as widely reported, the decoration of the TiO_2 surface by Mn-based nanoparticles (NPs) increased the visible light absorption but partially covered the active surface sites [22]. Indeed, Mn NPs and active sites have a synergistic effect on the final photocatalytic activity. Thus, when both contributions are balanced, the maximum efficiency is achieved, as reported in the recent literature [26]. Metal and metal oxides species can positively act on the photoactivity of TiO_2, as demonstrated in the case of Cu by Chiang and Doong [27], because they are able to capture more photons from visible light, leading to the formation of more electron-hole couples, as we also showed in our previous work [26]. Therefore, in a similar way, Mn and MnO_2 species deposited on the micrometric anatase particles could act as electron traps [28], limiting the e^-/h^+ recombination. Moreover, whatever species modifies the bare TiO_2 support strongly affects the material properties, especially in relation to its amount. Indeed, the dopant content in each TiO_2-matrix sample is related to the efficiency of the electron-hole recombination process, as well as to the formation of the hydroxyl radicals under an irradiation source [29].

The photocatalytic oxidation pathway was demonstrated to be a two-step process, where acetaldehyde is the main intermediate molecule, and CO_2 and H_2O the final products [30–32]:

$$CH_3CH_2OH \rightarrow CH_3CHO \rightarrow 2CO_2$$

Figure 1 corroborates the previous mechanism showing how the progressive UV-assisted photodegradation of ethanol (in the case of Mn_{20}1077_4, as a representative sample) led to the formation of an acetaldehyde by-product and then to the complete conversion to CO_2. Instead, by means of LED irradiation a drastic decrease of the starting molecule degradation occurred accompanied by a slow acetaldehyde production and partial mineralization, see Table 1, 5th column.

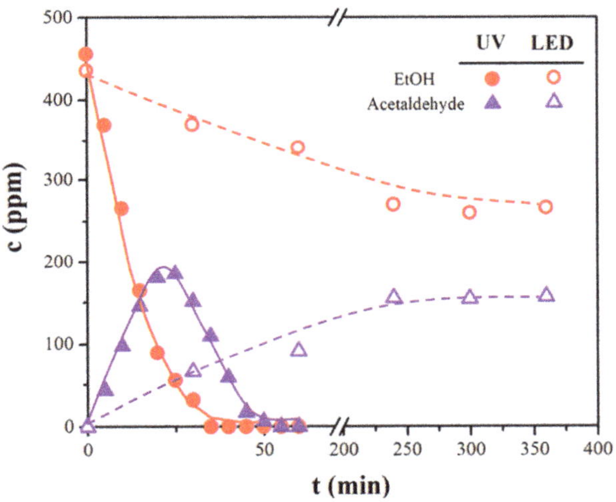

Figure 1. Ethanol disappearance and acetaldehyde formation in the case of Mn_{20}1077_4 under both UV and LED sources.

The second part of the research work was devoted to evaluating the role of the impregnation pH on the photocatalytic activity. Based on the promising results obtained with manganese-doped powders under LED light, the 20% Mn loading (Mn_{20}1077_y series) was selected. A different pH impregnation environment seemed to influence the final ethanol photodegradation, see Figure 2.

Particularly, an acidic pH (samples Mn_{20}1077_3 and Mn_{20}1077_4) led to an increase of both EtOH disappearance and the degree of mineralization (up to 35% and 10%, respectively for Mn_{20}1077_3; Figure 2 and Table 1 in inset). On the contrary, moving towards alkaline values, the overall photocatalytic performances seemed to be drastically inhibited (down to 7% and <3% for disappearance and mineralization by Mn_{20}1077_12, respectively). To explain this phenomenon, we believe that the adsorption of ethanol molecules is mostly dependent on the concentration of hydroxyls groups on the surface of the catalyst. Thus, in order to evaluate the role played by either Mn amount or the pH of impregnation on the structural, morphological, and optical properties, several physicochemical analyses were performed.

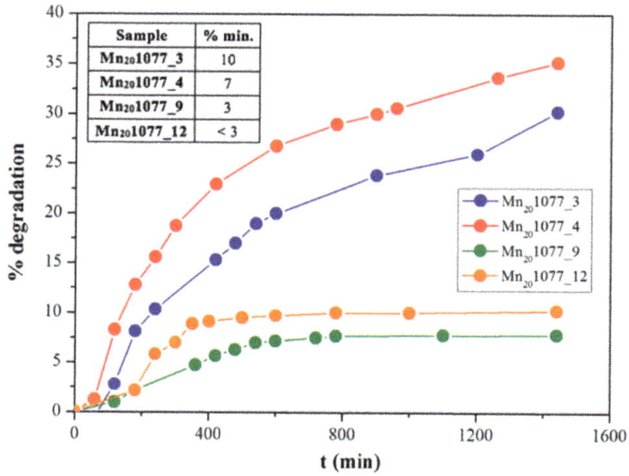

Figure 2. Ethanol degradation percentages after 24 h of Light Emitting Diode(LED) photocatalysis, by different pH Mn-impregnated TiO_2 samples. Table in inset: Relative mineralization degrees.

3.2. Mn-TiO_2 Physico-Chemical Properties

As a key parameter to explain the photoactivity of a catalyst, especially in the visible region, band gap evaluation by diffuse reflectance spectra (DRS) analysis was performed and is shown in Figure 3.

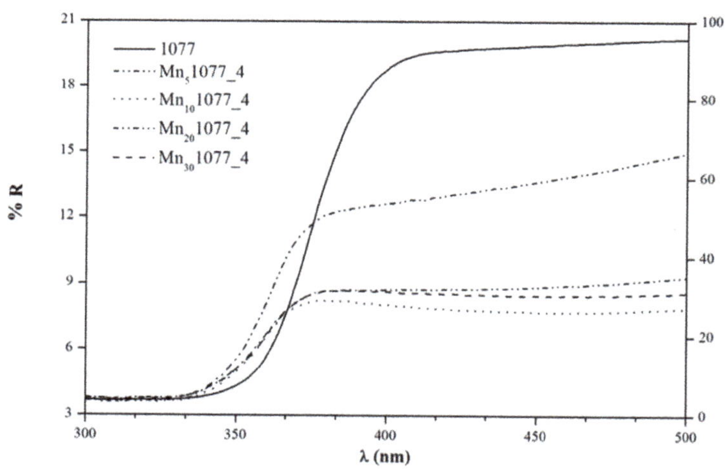

Figure 3. Diffuse reflectance spectra of impregnated TiO_2 powders with different Mn-dopant concentrations.

As expected, considering the ethanol degradation results under LED light, an increasing amount of manganese led to a decrease of the band gap values, see Table 2, 2nd column and Figure 3. This reduction should improve the photocatalytic performance of doped TiO_2 particles. On the contrary, varying pH did not seem to influence the optical properties, as shown in Table 2. Hence, the sole DRS measurements were not enough to explain the photocatalytic performances under LED light by varying either the Mn content or the impregnation pH, see Figure 2.

Table 2. Band gap values (by diffuse reflectance spectra (DRS)), specific surface area values (S_{BET}) and manganese-to-titanium (Mn/Ti) atomic ratios (by EDS and XPS) for all the adopted samples.

Sample	Band Gap (eV)	S_{BET} (m^2 g^{-1})	Mn/Ti	
			EDX	XPS
1077	3.28	10	–	–
$Mn_5$1077_4	3.05	15	0.04	0.06
Mn_{10}1077_4	2.96	14	0.09	0.12
Mn_{20}1077_4	2.91	16	0.18	0.22
Mn_{30}1077_4	2.83	11	0.30	0.37
Mn_{20}1077_3	3.19	10	0.18	0.21
Mn_{20}1077_9	3.08	9	0.06	0.29
Mn_{20}1077_12	2.94	7	0.05	0.50

From a structural point of view, X-ray diffraction patterns, shown in Figure 4, revealed that the Mn content did not affect the TiO_2 phase composition. Indeed, all patterns show the peak at around 25° ascribable to the anatase phase (ICDD Anatase file No. 21-1272). Interestingly, for Mn amounts higher than 10%, peaks at 2θ around 43° and 56° appeared, as shown in Figure 4, which could be assigned to pyrolusite MnO_2 species. Instead, pH did not significantly induce structural modifications.

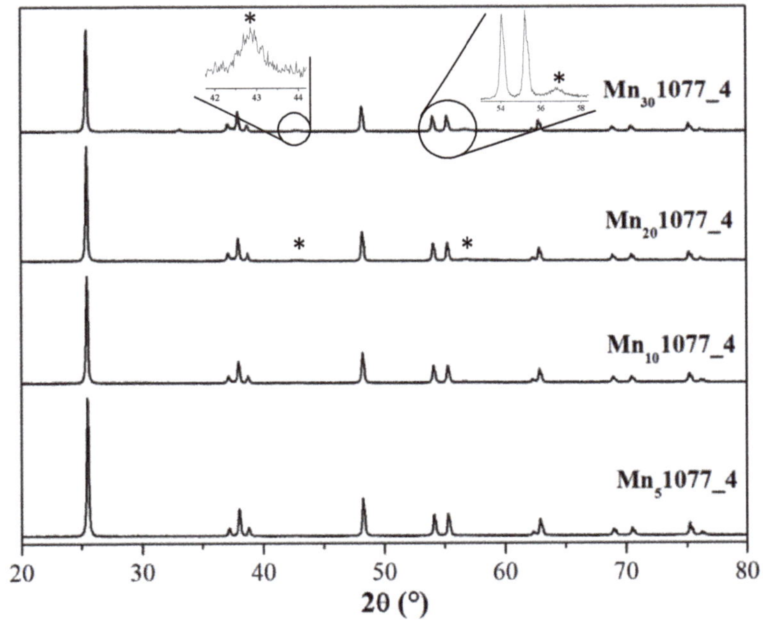

Figure 4. X-ray diffraction lines of Mn_x1077_4 samples (* = MnO_2 pyrolusite polymorph).

Then, HR-TEM analyses were performed to evaluate the morphological features of the photocatalysts, see Figure 5. All samples showed the peculiar characteristics of the bare TiO_2 material, which is made up of ordered particles with large average size, namely in the micrometric range (50–100 nm or more in diameter) and smoothed roundish contours: The (101) family of anatase crystal planes (0.352 nm; ICDD anatase file No. 21-1272) [33,34] was observed, see the electron diffraction pattern reported in Figure 5b, which is relevant for all the ordered TiO_2 microcrystals present in Figure 5. The TiO_2 morphology remains apparently unmodified for Mn amounts below 10%, see Figure 5a, even though in some regions extra morphological features start to appear, as indicated by the arrows. For higher concentrations (Mn_{20}1077_4, Figure 5b), the lattice fringes relative to the (101) family of crystal planes belonging to the pyrolusite MnO_2 polymorph (0.240 nm; ICDD 24-0735) could be observed (as also confirmed by the electron diffraction pattern reported on the left-hand side of the image), thus corroborating the previous XRPD results. Hence, we could hypothesize that a small presence of this secondary oxide may enhance the photodegradation performances, as previously discussed.

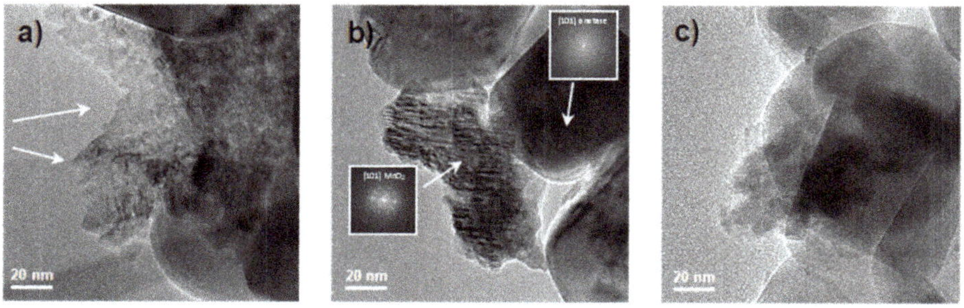

Figure 5. Transmission Electron Microscopy (TEM) images of (**a**) Mn_{10}1077_4; (**b**) Mn_{20}1077_4; and (**c**) Mn_{20}1077_12.

Moreover, in acidic experimental conditions (Mn_{20}1077_4 as a representative sample in Figure 5b), TiO_2 particles did not exhibit any morphological changes, whereas the Mn species were very evident as small envelopes of thin, closely packed particles (average dimensions 3–6 nm) with clear fringe patterns of pyrolusite MnO_2. In addition, with alkaline impregnation pH, see Figure 5c, a second family plane (111) of manganese dioxide polymorph was also observable.

Furthermore, BET analyses showed relatively low values of surface area for all the adopted samples, confirming the micrometric texture of the aggregates; a slight increase was appreciable for Mn-impregnated TiO_2 (from 10 to 16 $m^2\ g^{-1}$; Table 2, 3rd column). However, at alkaline pH, this parameter decreased down to 7 $m^2\ g^{-1}$ for the Mn_{20}1077_12 sample, see Table 2. These results could be explained by taking into account the actual Mn content obtained by both EDS and XPS analyses, see Table 2, 4th column. The bulk amount of manganese species was consistent with the one expected from the synthetic procedure for all the Mn concentrations. Regarding the impregnated samples at different pH, the interactions between the TiO_2 micrometric particles and the adopted $Mn(NO_2)_3$ salt precursor were deeply influenced by the pH value. Indeed, the final Mn bulk amount is similar to the starting concentration at an acidic pH, while it decreases at alkaline pH values, see Table 2, 4th column.

On the contrary, the surface Mn/Ti ratios were always higher than those obtained by the EDS analyses in accordance with the experimental method (impregnation) to decorate the micrometric titania powders. Then, notwithstanding the low amount of Mn at alkaline pH, all the guest metal was segregated to the surface forming a dense coverage of MnO_2 nanopowders. According to Li et al. [35], when manganese content increases, it becomes the recombination center for electron–hole pairs decreasing the separation efficiency of photo-generated charges, therefore invariably reducing/preventing the formation of hydroxyl radicals.

Hence, to corroborate the previous data and for a complete comprehension of the photocatalytic results, high-resolution O 1s spectra, shown in Figure 6, were fitted using four or five Gaussian/Lorentzian functions superimposed to a Shirley background. On the basis of the recent literature, the peaks could be assigned to; (i) oxygen bound to the metal ions (Ti [36,37] or Mn [38]) in the lattice (peak I at 528.5 eV); (ii) a high binding energy component (HBEC) developed with the increasing loss of oxygen or creation of oxygen vacancies (peak II at 530.0 eV) [39,40]; (iii) a low binding energy component (LBEC) due to adsorption of OH^- on the surface (peak III at 531.1 eV) [37,39]; and (iv) water adsorption (peaks IV and V at binding energy (B.E.) higher than 532.0 eV) [37,41]. It is worth noting that the sequence of the ratios between the oxygen of surface OH groups (peak III, pink area), and the sum of the previous one and the oxygen of the chemisorbed water molecules (peak IV and V, violet areas) at increasing pH was fully in accordance with the photocatalytic activities ($Mn_{20}1077\_9 < Mn_{20}1077\_12 < Mn_{20}1077\_3 < Mn_{20}1077\_4$, see Figure 2). Hydroxyl groups on the surface of a photocatalyst play an important role in the photocatalytic reaction since they can capture the photo-induced holes from the surface of the material, inhibiting electron–hole recombination as well as forming hydroxyl radicals with high oxidation potentials [42,43]. Thus, at alkaline pH, the high surface Mn loading and the paucity of OH groups inhibit the photocatalytic activity.

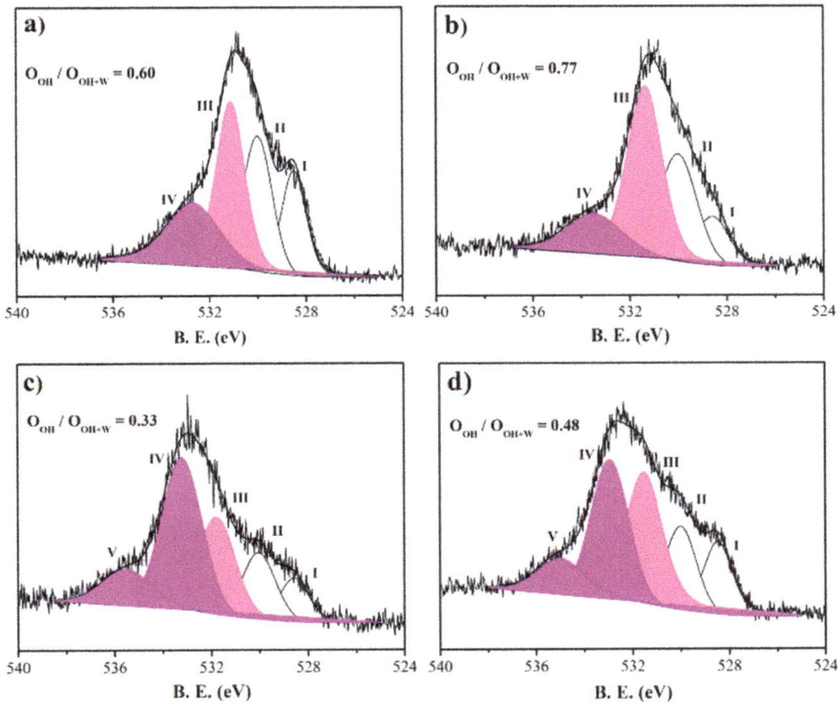

Figure 6. XPS high-resolution spectra of oxygen (O1s) relative to (**a**) $Mn_{20}1077\_3$; (**b**) $Mn_{20}1077\_4$; (**c**) $Mn_{20}1077\_9$; and (**d**) $Mn_{20}1077\_12$. Inset: Atomic ratio between the oxygen of surface OH groups (O_{OH}, pink area), and the sum of the previous one and the oxygen of the chemisorbed water molecules (O_W, violet areas).

4. Conclusions

Mn species activates anatase micro-TiO_2 for the visible light photodegradation of organic pollutants. Particularly, the final photocatalytic performance strongly depends on the Mn amount. Indeed, we surveyed Mn loadings between 5% and 30% under both UV and LED irradiation. Nevertheless, bare 1077 powder was the best performing photocatalyst, and the increasing Mn content led to a drastic decrease of the ethanol photoremoval under UV light. $Mn_{20}1077\_4$ showed the highest EtOH degradation under visible wavelengths. In fact, the presence of Mn species provoked a reduction of the band gap value, thus leading to higher efficiency under LED light.

A further phase of the present work was the investigation of the impregnation pH on the final photocatalysts performance. Only a fixed concentration of Mn (i.e., 20%) at an acidic pH (3 and 4) resulted in more active samples, whereas alkaline pH (up to 12) reduced the photocatalytic performance.

Indeed, we believe that the powders prepared at a higher pH of impregnation are less active because of the segregation of a high amount of Mn species and concomitantly a low amount of surface hydroxyl groups.

The novel results reported herein are a step forward in the photodegradation of the indoor gaseous pollutants by a micrometric Mn-decorated TiO_2 catalyst under visible light. Moreover, the positive effect obtained at an acidic pH could be further investigated, to assess whether lower values improve the final catalytic performance.

Author Contributions: Conceptualization, M.S.; Methodology, G.C. (Giuseppina Cerrato) and E.P.; Validation, E.P.; Investigation, M.S., G.C. (Giuseppina Cerrato) and E.P.; Writing-Original Draft Preparation, M.S., G.C. (Giuseppe Cappelletti) and G.C. (Giuseppina Cerrato); Writing-Review & Editing, M.S., C.L.B.; Supervision, G.C. (Giuseppe Cappelletti); Project Administration, C.L.B.; Funding Acquisition, D.C.B. and C.L.B.

Funding: This research was funded by the Coopération Québec-Italie 2017-2019, within the project QU17MO09 and it was undertaken, in part, thanks to funding from the Canada Research Chairs program.

Acknowledgments: The authors gratefully acknowledge the support of the Natural Sciences and Engineering Research Council of Canada (NSERC).

Conflicts of Interest: The authors declare no conflict of interest.

References

1. Wold, A. Photocatalytic properties of titanium dioxide (TiO_2). *Chem. Mater.* **1993**, *5*, 280–283. [CrossRef]
2. Fujishima, A.; Rao, T.N.; Tryk, D.A. Titanium dioxide photocatalysis. *J. Photochem. Photobiol. C Photochem. Rev.* **2000**, *1*, 1–21. [CrossRef]
3. Fujishima, A.; Zhang, X. Titanium dioxide photocatalysis: present situation and future approaches. *C. R. Chim.* **2006**, *9*, 750–760. [CrossRef]
4. Nakata, K.; Fujishima, A. TiO_2 photocatalysis: Design and applications. *J. Photochem. Photobiol. C Photochem. Rev.* **2012**, *13*, 169–189. [CrossRef]
5. Wolkoff, P.; Nielsen, G.D. Organic compounds in indoor air—Their relevance for perceived indoor air quality? *Atmos. Environ.* **2001**, *35*, 4407–4417. [CrossRef]
6. Höllbacher, E.; Ters, T.; Rieder-Gradinger, C.; Srebotnik, E. Emissions of indoor air pollutants from six user scenarios in a model room. *Atmos. Environ.* **2017**, *150*, 389–394. [CrossRef]
7. Śmiełowska, M.; Marć, M.; Zabiegała, B. Indoor air quality in public utility environments—A review. *Environ. Sci. Pollut. Res.* **2017**, *24*, 11166–11176. [CrossRef] [PubMed]
8. Sanjinés, R.; Tang, H.; Berger, H.; Gozzo, F.; Margaritondo, G.; Lévy, F. Electronic structure of anatase TiO_2 oxide. *J. Appl. Phys.* **1994**, *75*, 2945–2951. [CrossRef]
9. Nakata, K.; Ochiai, T.; Murakami, T.; Fujishima, A. Photoenergy conversion with TiO_2 photocatalysis: New materials and recent applications. *Electrochim. Acta* **2012**, *84*, 103–111. [CrossRef]
10. Pan, X.; Xu, Y.-J. Defect-Mediated Growth of Noble-Metal (Ag, Pt, and Pd) Nanoparticles on TiO_2 with Oxygen Vacancies for Photocatalytic Redox Reactions under Visible Light. *J. Phys. Chem. C* **2013**, *117*, 17996–18005. [CrossRef]
11. Elahifard, M.R.; Rahimnejad, S.; Haghighi, S.; Gholami, M.R. Apatite-Coated Ag/AgBr/TiO_2 Visible-Light Photocatalyst for Destruction of Bacteria. *J. Am. Chem. Soc.* **2007**, *129*, 9552–9553. [CrossRef] [PubMed]
12. Zhu, X.; Jin, C.; Li, X.S.; Liu, J.L.; Sun, Z.G.; Shi, C.; Li, X.; Zhu, A.M. Photocatalytic formaldehyde oxidation over plasmonic Au/TiO_2 under visible light: Moisture indispensability and light enhancement. *ACS Catal.* **2017**, *7*, 6514–6524. [CrossRef]
13. Bastos, S.S.T.; Órfão, J.J.M.; Freitas, M.M.A.; Pereira, M.F.R.; Figueiredo, J.L. Manganese oxide catalysts synthesized by exotemplating for the total oxidation of ethanol. *Appl. Catal. B Environ.* **2009**, *93*, 30–37. [CrossRef]
14. Wang, G.; Ling, Y.; Li, Y. Oxygen-deficient metal oxide nanostructures for photoelectrochemical water oxidation and other applications. *Nanoscale* **2012**, *4*, 6682–6691. [CrossRef] [PubMed]
15. Pargoletti, E.; Cappelletti, G.; Minguzzi, A.; Rondinini, S.; Leoni, M.; Marelli, M.; Vertova, A. High-performance of bare and Ti-doped α-MnO_2 nanoparticles in catalyzing the Oxygen Reduction Reaction. *J. Power. Sources* **2016**, *325*, 116–128. [CrossRef]

16. Orsini, S.; Pargoletti, E.; Vertova, A.; Minguzzi, A.; Locatelli, C.; Rondinini, S.; Cappelletti, G. Ad hoc tailored electrocatalytic MnO_2 nanorods for the oxygen reduction in aqueous and organic media. *J. Electroanal. Chem.* **2018**, *808*, 439–445. [CrossRef]
17. Liao, Y.; Zhang, X.; Peng, R.; Zhao, M.; Ye, D. Catalytic properties of manganese oxide polyhedra with hollow and solid morphologies in toluene removal. *Appl. Surf. Sci.* **2017**, *405*, 20–28. [CrossRef]
18. Kim, S.C.; Shim, W.G. Catalytic combustion of VOCs over a series of manganese oxide catalysts. *Appl. Catal. B Environ.* **2010**, *98*, 180–185. [CrossRef]
19. Pargoletti, E.; Pifferi, V.; Falciola, L.; Facchinetti, G.; Re Depaolini, A.; Davoli, E.; Marelli, M.; Cappelletti, G. A detailed investigation of MnO_2 nanorods to be grown onto activated carbon. High efficiency towards aqueous methyl orange adsorption/degradation. *Appl. Surf. Sci.* **2018**. [CrossRef]
20. Binas, V.D.; Sambani, K.; Maggos, T.; Katsanaki, A.; Kiriakidis, G. Synthesis and photocatalytic activity of Mn-doped TiO_2 nanostructured powders under UV and visible light. *Appl. Catal. B Environ.* **2012**, *113–114*, 79–86. [CrossRef]
21. Xue, M.; Huang, L.; Wang, J.-Q.; Wang, Y.; Gao, L.; Zhu, J.; Zou, Z.-G. The direct synthesis of mesoporous structured MnO_2/TiO_2 nanocomposite: a novel visible-light active photocatalyst with large pore size. *Nanotechnology* **2008**, *19*. [CrossRef] [PubMed]
22. Oseghe, E.O.; Ndungu, P.G.; Jonnalagadda, S.B. Synthesis of mesoporous Mn/TiO_2 nanocomposites and investigating the photocatalytic properties in aqueous systems. *Environ. Sci. Pollut. Res.* **2015**, *22*, 211–222. [CrossRef] [PubMed]
23. Bianchi, C.L.; Pirola, C.; Galli, F.; Stucchi, M.; Morandi, S. Nano and micro-TiO_2 for the photodegradation of ethanol: experimental data and kinetic modelling. *RSC Adv.* **2015**, *5*, 53419–53425. [CrossRef]
24. Stucchi, M.; Bianchi, C.L.; Pirola, C.; Vitali, S.; Cerrato, G.; Morandi, S.; Argirusis, C.; Sourkouni, G.; Sakkas, P.M.; Capucci, V. Surface decoration of commercial micro-sized TiO_2 by means of high energy ultrasound: A way to enhance its photocatalytic activity under visible light. *Appl. Catal. B Environ.* **2015**, *178*, 124–132. [CrossRef]
25. Turchi, C.S.; Ollis, D.F. Photocatalytic degradation of organic water contaminants: Mechanisms involving hydroxyl radical attack. *J. Catal.* **1990**, *122*, 178–192. [CrossRef]
26. Stucchi, M.; Bianchi, C.L.; Pirola, C.; Cerrato, G.; Morandi, S.; Argirusis, C.; Sourkouni, G.; Naldoni, A.; Capucci, V. Copper NPs decorated titania: A novel synthesis by high energy US with a study of the photocatalytic activity under visible light. *Ultrason. Sonochem.* **2016**, *31*, 295–301. [CrossRef] [PubMed]
27. Chiang, L.; Doong, R. Cu–TiO_2 nanorods with enhanced ultraviolet- and visible-light photoactivity for bisphenol A degradation. *J. Hazard. Mater.* **2014**, *277*, 84–92. [CrossRef] [PubMed]
28. Zhang, W.; Liu, Y.; Yu, B.; Zhang, J.; Liang, W. Effects of silver substrates on the visible light photocatalytic activities of copper-doped titanium dioxide thin films. *Mater. Sci. Semicond. Process.* **2015**, *30*, 527–534. [CrossRef]
29. Mazierski, P.; Nischk, M.; Gołkowska, M.; Lisowski, W.; Gazda, M.; Winiarski, M.J.; Klimczuk, T.; Zaleska-Medynska, A. Photocatalytic activity of nitrogen doped TiO_2 nanotubes prepared by anodic oxidation: The effect of applied voltage, anodization time and amount of nitrogen dopant. *Appl. Catal. B Environ.* **2016**, *196*, 77–88. [CrossRef]
30. Vorontsov, A. Selectivity of photocatalytic oxidation of gaseous ethanol over pure and modified TiO_2. *J. Catal.* **2004**, *221*, 102–109. [CrossRef]
31. Adjimi, S.; Roux, J.-C.; Sergent, N.; Delpech, F.; Thivel, P.-X.; Pera-Titus, M. Photocatalytic oxidation of ethanol using paper-based nano-TiO_2 immobilized on porous silica: A modelling study. *Chem. Eng. J.* **2014**, *251*, 381–391. [CrossRef]
32. Katsiev, K.; Harrison, G.; Alghamdi, H.; Alsalik, Y.; Wilson, A.; Thornton, G.; Idriss, H. Mechanism of Ethanol Photooxidation on Single-Crystal Anatase TiO_2 (101). *J. Phys. Chem. C* **2017**, *121*, 2940–2950. [CrossRef]
33. Bianchi, C.L.; Gatto, S.; Pirola, C.; Naldoni, A.; Di Michele, A.; Cerrato, G.; Crocellà, V.; Capucci, V. Photocatalytic degradation of acetone, acetaldehyde and toluene in gas-phase: Comparison between nano and micro-sized TiO_2. *Appl. Catal. B Environ.* **2014**, *146*, 123–130. [CrossRef]
34. Cappelletti, G.; Ricci, C.; Ardizzone, S.; Parola, C.; Anedda, A. Aged titania nanoparticles: The simultaneous control of local and long-range properties. *J. Phys. Chem. B* **2005**, *109*, 4448–4454. [CrossRef] [PubMed]
35. Li, H.; Wang, D.; Fan, H.; Jiang, T.; Li, X.; Xie, T. Synthesis of ordered multivalent Mn-TiO_2 nanospheres with tunable size: A high performance visible-light photocatalyst. *Nano Res.* **2011**, *4*, 460–469. [CrossRef]

36. Cappelletti, G.; Bianchi, C.L.; Ardizzone, S. XPS study of the surfactant film adsorbed onto growing titania nanoparticles. *Appl. Surf. Sci.* **2006**, *253*, 519–524. [CrossRef]
37. Cappelletti, G.; Ardizzone, S.; Bianchi, C.L.; Gialanella, S.; Naldoni, A.; Pirola, C.; Ragaini, V. Photodegradation of pollutants in air: Enhanced properties of nano-TiO_2 prepared by ultrasound. *Nanoscale Res. Lett.* **2009**, *4*, 97–105. [CrossRef] [PubMed]
38. Wang, H.; Chen, J.; Hu, S.; Zhang, X.; Fan, X.; Du, J.; Huang, Y.-G.; Li, Q.-Y. Direct growth of flower-like 3D MnO_2 ultrathin nanosheets on carbon paper as efficient cathode catalyst for rechargeable Li–O_2 batteries. *RSC Adv.* **2015**, *5*, 72495–72499. [CrossRef]
39. Ishfaq, M.; Rizwan Khan, M.; Bhopal, M.F.; Nasim, F.; Ali, A.; Bhatti, A.S.; Ahmed, I.; Bhardwaj, S.; Cepek, C. 1.5 MeV proton irradiation effects on electrical and structural properties of TiO_2 /n-Si interface. *J. Appl. Phys.* **2014**, *115*, 174506. [CrossRef]
40. Naeem, M.; Hasanain, S.K.; Kobayashi, M.; Ishida, Y.; Fujimori, A.; Buzby, S.; Shah, S.I. Effect of reducing atmosphere on the magnetism of $Zn_{1-x}Co_xO$ ($0 \leq x \leq 0.10$) nanoparticles. *Nanotechnology* **2006**, *17*, 2675–2680. [CrossRef] [PubMed]
41. Kang, S.; Pawar, R.C.; Pyo, Y.; Khare, V.; Lee, C.S. Size-controlled BiOCl–RGO composites having enhanced photodegradative properties. *J. Exp. Nanosci.* **2016**, *11*, 259–275. [CrossRef]
42. Hou, Y.D.; Wang, X.C.; Wu, L.; Chen, X.F.; Ding, Z.X.; Wang, X.X.; Fu, X.Z. N-Doped SiO_2/TiO_2 mesoporous nanoparticles with enhanced photocatalytic activity under visible-light irradiation. *Chemosphere* **2008**, *72*, 414–421. [CrossRef] [PubMed]
43. Xu, Y.; Lei, B.; Guo, L.; Zhou, W.; Liu, Y. Preparation, characterization and photocatalytic activity of manganese doped TiO_2 immobilized on silica gel. *J. Hazard. Mater.* **2008**, *160*, 78–82. [CrossRef] [PubMed]

© 2018 by the authors. Licensee MDPI, Basel, Switzerland. This article is an open access article distributed under the terms and conditions of the Creative Commons Attribution (CC BY) license (http://creativecommons.org/licenses/by/4.0/).

MDPI
St. Alban-Anlage 66
4052 Basel
Switzerland
Tel. +41 61 683 77 34
Fax +41 61 302 89 18
www.mdpi.com

Nanomaterials Editorial Office
E-mail: nanomaterials@mdpi.com
www.mdpi.com/journal/nanomaterials

www.ingramcontent.com/pod-product-compliance
Lightning Source LLC
Chambersburg PA
CBHW061455210926
51850CB00008B/1726
9783039288311